Energy, Environment, and Sustainability

Series Editor

Avinash Kumar Agarwal, Department of Mechanical Engineering, Indian Institute of Technology Kanpur, Kanpur, Uttar Pradesh, India

AIMS AND SCOPE

This books series publishes cutting edge monographs and professional books focused on all aspects of energy and environmental sustainability, especially as it relates to energy concerns. The Series is published in partnership with the International Society for Energy, Environment, and Sustainability. The books in these series are edited or authored by top researchers and professional across the globe. The series aims at publishing state-of-the-art research and development in areas including, but not limited to:

- Renewable Energy
- Alternative Fuels
- Engines and Locomotives
- Combustion and Propulsion
- Fossil Fuels
- Carbon Capture
- Control and Automation for Energy
- Environmental Pollution
- Waste Management
- Transportation Sustainability

Review Process

The proposal for each volume is reviewed by the main editor and/or the advisory board. The chapters in each volume are individually reviewed single blind by expert reviewers (at least four reviews per chapter) and the main editor.

Ethics Statement for this series can be found in the Springer standard guidelines here https://www.springer.com/us/authors-editors/journal-author/journal-author-helpdesk/before-you-start/before-you-start/1330#c14214

More information about this series at http://www.springer.com/series/15901

Swatantra P. Singh · Karthik Rathinam ·
Tarun Gupta · Avinash Kumar Agarwal
Editors

Nanomaterials and Nanocomposites for Environmental Remediation

Springer

and cleaner environment. The conference presented one high-voltage plenary talk by Mrs. Rashmi Urdhwareshe, Director, Automotive Research Association of India (ARAI), Pune.

The conference included 28 technical sessions on topics related to energy and environmental sustainability including 1 plenary talk, 25 keynote talks and 54 invited talks from prominent scientists, in addition to 70+ contributed talks and 80+ poster presentation by students and researchers. The technical sessions in the conference included fuels, engine technology and emissions, coal and biomass combustion/gasification, atomization and sprays, combustion and modelling, alternative energy resources, water and water and wastewater treatment, automobile and other environmental applications, environmental challenges and sustainability, nuclear energy and other environmental challenges, clean fuels and other environmental challenges, water pollution and control, biomass and biotechnology, waste to wealth, microbiology, biotechnological and other environmental applications, waste and wastewater management, cleaner technology and environment, sustainable materials and processes, energy, environment and sustainability, technologies and approaches for clean, sensors and materials for environmental, biological processes and environmental sustainability. One of the highlights of the conference was the Rapid-Fire Poster Sessions in (i) engine/fuels/emissions, (ii) environment and (iii) biotechnology, where 50+ students participated with great enthusiasm and won many prizes in a fiercely competitive environment. Three hundred plus participants and speakers attended this three days conference, where 12 ISEES books published by Springer, Singapore, under a special dedicated series 'Energy, Environment and Sustainability' were released. This was third time in a row that such significant and high-quality outcome has been achieved by any society in India. The conference concluded with a panel discussion on 'Balancing Energy Security, Environmental Impacts & Economic Considerations: Indian Perspective', where the panellists were Dr. Anjan Ray, CSIR-IIP Dehradun; Dr. R. R. Sonde, Thermax Ltd.; Prof. Avinash Kumar Agarwal, IIT Kanpur; Dr. R. Srikanth, National Institute of Advanced Studies, Bengaluru; and Dr. Rakesh Kumar, NEERI, Nagpur. The panel discussion was moderated by Prof. Ashok Pandey, Chairman, ISEES. This conference laid out the roadmap for technology development, opportunities and challenges in energy, environment and sustainability domain. All these topics are very relevant for the country and the world in present context. We acknowledge the support received from various funding agencies and organizations for the successful conduct of the Fourth International Conference on 'Sustainable Energy and Environmental Challenges' (IV-SEEC), where these books germinated. We would therefore like to acknowledge SERB, Government of India (special thanks to Dr. Sandeep Verma, Secretary); NEERI, Nagpur (special thanks to Dr. Rakesh Kumar, Director), CSIR, and our publishing partner Springer (special thanks to Swati Mehershi).

The editors would like to express their sincere gratitude to large number of authors from all over the world for submitting their high-quality work in a timely manner and revising it appropriately at short notice. We would like to express our special thanks to Dr. A. Sinha, Dr. A. Shriwastav, Dr. Simant K. Shrivastv, Dr. M. K. Tiwari, Dr. K. K. Singh, Dr. S. Abraham, Dr. P. Kalbar, Dr. M. Sahu, Dr. A. Chakraborty,

Dr. T. Nawaz, Dr. S. Singh and Dr. V. S. Vamsi Botlaguduru who reviewed various chapters of this monograph and provided their valuable suggestions to improve the manuscripts.

This book focussed on the recent development of the nanomaterials and nanocomposites for the pollution measurement and their control in water, air and soil. The book comprises 12 chapters and has been incorporated carbon-based, metal-based and metal-organic framework-based nanomaterials and nanocomposites for emerging contaminants (pharmaceuticals and personal care products) degradation, disinfection and other traditional pollutants degradation and removal. Overall, this book offers updated literature for researchers and academicians working in the field of environmental remediation by nanomaterials.

Chapters include recent results more focussed on current trends of environmental nanotechnology. In this book, readers will get the idea about the different metal-based and non-metal-based nanoparticles for the environmental remediation. We hope that the book would be of great interest to the professionals, postgraduate students involved in material science and engineering, chemical engineering and environmental nanotechnology research.

Mumbai, India — Swatantra P. Singh
Duisburg, Germany — Karthik Rathinam
Kanpur, India — Tarun Gupta
Kanpur, India — Avinash Kumar Agarwal

Contents

Editors and Contributors

About the Editors

Dr. Swatantra P. Singh is an environmental engineer with experience in membrane fabrication, environmental nanotechnology, fate, and transport of pollutants and emerging contaminants in the environment. Currently, he is an assistant professor in the Environmental Science and Engineering Department at Indian Institute of Technology Bombay, India. He has developed a key technology to fabricate the membranes for water purification and print graphene in-situ in a single step. He has four US patents (two granted and two provisional) on membrane and laser-based graphene fabrication techniques. He has authored 19 journal articles, one book, and three book chapters. He recently won the INAE Young Engineer Award (2020) and ISEES Young Scientists Award (2020).

Dr. Karthik Rathinam is currently leading adsorption technology group at the Chair for Process and Mechanical Engineering/Water Technology, University of Duisburg-Essen, Germany. His research interest is mainly focussed on water and wastewater treatment including precipitation of inorganic minerals, development of antiscaling and antifouling membranes, development of porous eco-friendly based adsorbents, thermal reactivation of activated carbon, emerging contaminants removal, etc. After his Ph.D. from the Gandhigram Rural Institute, India, he worked as a post-doctoral researcher in Zuckerberg Institute for Water Research, Israel and BASF-SE, Germany. He is serving as a reviewer for several Scopus indexed journals. He is a recipient of many national and international awards and has published 18 research articles in a highly reputed peer-reviewed journals and one book chapter.

Prof. Tarun Gupta is doctorate from Harvard University, USA, and M.Tech. from Indian Institute of Technology (IIT) Bombay, India. He has authored more than 125 ISI indexed journal publications, three books, ten book chapters, four patents, and has been reviewer of more than 36 journals. He has guided six Ph.D. and 34 M.Tech. theses. A submicron aerosol sampler designed, developed and evaluated by him at IIT Kanpur has been commercialized by Envirotech. He has developed a high volume fine PM sampler and transferred technology to BARC. He is currently P. K. Kelkar research fellow and selected member of INYAS and INAE Associate. He has recently won INAE Innovator and Entrepreneur Award (2018), VNMM award (2017), NASI-SCOPUS Award (2015), INSA Medal for Young Scientist (2011), INAE Young Engineer Award (2009) and IEI Young Engineer Award (2008). He is currently N. C. Nigam Chair Professor and serving as Associate Dean of Research and Development at IIT Kanpur.

Prof. Avinash Kumar Agarwal joined the Indian Institute of Technology (IIT) Kanpur, India in 2001 after working as a post-doctoral fellow at the Engine Research Center, University of Wisconsin at Madison, USA. His interests are IC engines, combustion, alternate and conventional fuels, lubricating oil tribology, optical diagnostics, laser ignition, HCCI, emissions and particulate control, and large bore engines. Professor Agarwal has published 290+ peer reviewed international journal and conference papers, 42 edited books, 78 books chapters and has 10000+ Scopus and 15300+ Google scholar citations. He is a Fellow of SAE (2012), Fellow of ASME (2013), Fellow of ISEES (2015), Fellow of INAE (2015), Fellow of NASI (2018), Fellow of Royal Society of Chemistry (2018), and a Fellow of American Association of Advancement in Science (2020). He is recipient of several prestigious awards such as Clarivate Analytics India Citation Award-2017 in Engineering and Technology, NASI-Reliance Industries Platinum Jubilee Award-2012; INAE Silver

Jubilee Young Engineer Award-2012; Dr. C. V. Raman Young Teachers Award: 2011; SAE Ralph R. Teetor Educational Award-2008; INSA Young Scientist Award-2007; UICT Young Scientist Award-2007; INAE Young Engineer Award-2005. Professor Agarwal received Prestigious Shanti Swarup Bhatnagar Award-2016 in Engineering Sciences. For his outstanding contributions, Prof. Agarwal is conferred upon Sir J. C. Bose National Fellowship (2019) by SERB.

Contributors

Shiju Abraham Department of Physics, St. Pius X College Rajapuram (Kannur University), Kasaragod, Kerala, India

Avinash Kumar Agarwal Department of Mechanical Engineering, IIT Kanpur, Kanpur, Uttar Pradesh, India

Imran Ali Department of Chemistry, College of Sciences, Taibah University, Al-Medina Al-Munawara, Saudi Arabia;
Department of Chemistry, Jamia Millia Islamia (Central University), New Delhi, India

Nandini Dixit Environmental Science and Engineering Department, IIT Bombay, Mumbai, India

Ibrahim M. A. ElSherbiny Chair for Mechanical Process Engineering and Water Technology, University of Duisburg-Essen, Duisburg, Germany

Tarun Gupta Department of Civil Engineering, IIT Kanpur, Kanpur, Uttar Pradesh, India

Ansaf V. Karim Environmental Science and Engineering Department, Indian Institute of Technology Bombay, Mumbai, India

Perumal Karthikeyan Department of Chemistry, The Gandhigram Rural Institute—Deemed to be University, Dindigul, Tamil Nadu, India

Sukanya Krishnan Environmental Science and Engineering Department, Indian Institute of Technology Bombay, Mumbai, India

Ashish Kumar Department of Environmental Science and Engineering (ESED), Indian Institute of Technology Bombay, Maharashtra, India

Kamlesh Kumar Department of Chemistry, Institute of Science, Banaras Hindu University (BHU), Varanasi, India

Pradip Kumar Integrated Approach for Design and Product Development, Council of Scientific and Industrial Research-Advanced Materials and Processes Research Institute, Bhopal, Madhya Pradesh, India

Tadimeti Divya Kusuma Department of Civil and Environmental Engineering, Indian Institute of Technology Tirupati, Tirupati, India

Shihabudheen M. Maliyekkal Department of Civil and Environmental Engineering, Indian Institute of Technology Tirupati, Tirupati, India

Chinmoy Mandal Environmental Science and Engineering Department, Indian Institute of Technology, Bombay, India

Sankaran Meenakshi Department of Chemistry, The Gandhigram Rural Institute—Deemed to be University, Dindigul, Tamil Nadu, India

Shivangi Mishra Nanomaterial Toxicology Group, CSIR-Indian Institute of Toxicology Research (CSIR-IITR), Lucknow, Uttar Pradesh, India

D. P. Mondal Lightweight Metallic Materials, Council of Scientific and Industrial Research-Advanced Materials and Processes Research Institute, Bhopal, Madhya Pradesh, India

Chettithodi Poovathumkuzhi Nabeena Department of Chemistry, Pocker Sahib Memorial Orphanage College, Malappuram, Kerala, India

M. S. V. Naga Jyothi Department of Civil and Environmental Engineering, Indian Institute of Technology Tirupati, Tirupati, India

Meenakshi M. Nara Chair for Mechanical Process Engineering and Water Technology, University of Duisburg-Essen, Duisburg, Germany

Stefan Panglisch Chair for Mechanical Process Engineering and Water Technology, University of Duisburg-Essen, Duisburg, Germany

Chebrolu Pulla Rao Department of Chemistry, Indian Institute of Technology Tirupati, Tirupati, India

Karthik Rathinam Chair for Mechanical Process Engineering and Water Technology, University of Duisburg-Essen, Duisburg, Germany

Manoranjan Sahu Environmental Science and Engineering Department, Indian Institute of Technology, Bombay, India

Amritanshu Shriwastav Environmental Science and Engineering Department, Indian Institute of Technology Bombay, Mumbai, India

Kamlesh Kumar Singh Department of Chemistry, Institute of Science, Banaras Hindu University, Varanasi, India

Shiv Singh Lightweight Metallic Materials, Council of Scientific and Industrial Research-Advanced Materials and Processes Research Institute, Bhopal, Madhya Pradesh, India

Swatantra P. Singh Department of Environmental Science and Engineering (ESED), Indian Institute of Technology Bombay, Mumbai, Maharashtra, India

Palliyalil Sirajudheen Department of Chemistry, The Gandhigram Rural Institute—Deemed to be University, Dindigul, Tamil Nadu, India;
Department of Chemistry, Pocker Sahib Memorial Orphanage College, Malappuram, Kerala, India

Simant Kumar Srivastav Department of Chemistry, Lalit Narayan Mithila University, Darbhanga, Bihar, India

Sivakumar Vigneshwaran Department of Chemistry, The Gandhigram Rural Institute—Deemed to be University, Dindigul, Tamil Nadu, India

Chapter 1
Nanomaterials and Nanocomposites for Environmental Remediation

Swatantra P. Singh, Karthik Rathinam, Tarun Gupta, and Avinash Kumar Agarwal

Abstract The environmentally unsustainable growth of the world resulted in a polluted ecosystem. In recent years, nanomaterials and nanocomposites have shown their potentials for environmental remediation. Carbon-based nanomaterials (CBNs), metal and metal oxide-based nanomaterials, and other nanocomposites have been used widely since the last decades. These nanomaterials and nanocomposites have shown their effectiveness through adsorption, reactive oxygen species generation through electro- and photocatalytic processes. This book focused on the recent development of the nanomaterials and nanocomposites for the pollution measurement and their control in water, air, and soil. The book comprises 12 chapters and has been incorporated carbon-based, metal-based, and metal–organic framework-based nanomaterials and nanocomposites for emerging contaminants (pharmaceuticals and personal care products) degradation, disinfection and other traditional pollutants degradation and removal. Overall, this book offers updated literature for researchers and academicians working in the field of environmental remediation by nanomaterials.

Keywords Nanomaterials · Nanocomposites · Environmental remediation · Pharmaceuticals and personal care products · Disinfection

S. P. Singh (✉)
Environmental Science and Engineering Department, IIT Bombay, Mumbai 400076, India
e-mail: swatantra@iitb.ac.in

K. Rathinam (✉)
Chair for Mechanical Process Engineering and Water Technology, University of Duisburg-Essen, Lotharstr.1, 47057 Duisburg, Germany
e-mail: karthik.rathinam@uni-due.de

T. Gupta (✉)
Department of Civil Engineering, IIT Kanpur, Kanpur 208016, UP, India
e-mail: tarun@iitk.ac.in

A. K. Agarwal
Department of Mechanical Engineering, IIT Kanpur, Kanpur 208016, UP, India

S. P. Singh et al. (eds.), *Nanomaterials and Nanocomposites for Environmental Remediation*, Energy, Environment, and Sustainability,
https://doi.org/10.1007/978-981-16-3256-3_1

1.1 Introduction

The importance of clean air, water, and the soil is the key to the better health of humans and the ecosystem. The health problems from the consumption of polluted air, water, and food are a significant cause of human misery and extinction of many other species from the Earth. The advancements in nanoscience and nanotechnology, in particular, the ability of detection, measurement, and degradation of pollutants have increased significantly over the years. Carbon-based nanomaterials (CBNs), such as carbon nanomaterials (CNMs), viz carbon nanofibers (CNFs), fullerenes, carbon nanotubes (CNTs), nanodiamonds/onions, graphene/graphene oxide, and their derivatives have strongly impacted the field of nanotechnology due to its physical, electronic, and chemical properties. The strong π–π interactions, with the electrostatic interaction of the functional groups and porous properties by the CNMs, make it a prominent adsorbent, and also the conductivity makes them a better candidate for the electrochemical sensor.

In Chap. 2, the authors have discussed the role of the CNMs for the quantification and detection of different pollutants. Among various carbon-based nanomaterials, graphene-based nanocomposite materials are gaining the attention of researchers globally as novel materials for biomedical, energy, electronic and environmental applications. Graphene has unique properties, especially its exceptionally high specific surface area, mechanical strength, electron mobility, and thermal conductivity, and recently scientists have demonstrated a new fabrication technique for the graphene called laser-induced graphene(LIG). Chapter 3 deals with the uses of titanium doped LIG for the environmental applications along with the advantages and disadvantages of different synthesis methods for graphene.

Disinfection is important for any water purification technologies, and many technologies such as UV, chlorination, and ozonation, have been implemented, but possess limitations in terms of operations, cost, and carcinogenic by-products formation. Electrochemical disinfection is an attractive technology to overcome many challenges. Graphene extraordinary electrical properties can be a solution to design next-generation electrodes and filters. Additionally, Metal-based nanoparticles have shown their potential for antimicrobial agents as they generate reactive oxygen species (ROS) and create oxidative stress in the microbes. Chapter 4 discusses the use of carbon and metal-based nanomaterials in disinfection processes and how LIG with the metal nanoparticles can provide next-generation electrodes and surfaces for the effective disinfection process.

The removal of multiple pollutants in water resources is a great challenge, at present, to protect the environment and human health. The emerging contaminants are not desirable even at low concentrations due to their toxicity; therefore, they must be either removed or degraded. The polymeric materials have gained their uses nowadays in water treatment applications for the removal of multiple pollutants. Chapter 5 in the book explains the graphitic carbon nitride (g-C3N4), a two-dimensional organic polymeric material for an effective adsorbent and photo-catalyst for both removal and degradation of multiple contaminants in the water. This chapter has covered the recent applications of g-C3N4-based materials in water and wastewater treatment processes, along with the possible future challenges for the g-C3N4.

The antibiotic-resistant bacteria is now a big challenge in current time due to the excessive use of antibiotics and the disposal of untreated wastewater. The antibiotics are designed to target the cell wall, protein synthesis, and DNA replication, but with the exposure, they develop the resistance for the antibiotics. The metal nanoparticles have shown great potential for antibiotic-resistant bacteria and inactivate them via oxidative stress, dissolved metal ions, or non-oxidative. Chapter 6 focuses on mechanisms for how bacteria develop antibiotic-resistant, allow with the antibiotic action on them. This chapter also describes the role of size, composition, shape, morphology, zeta potential, and environmental conditions of metal nanoparticles (silver, copper, gold, aluminium) and metal oxides nanoparticles (copper oxide, titanium dioxide nanoparticles, zinc oxide and magnesium oxide nanoparticles) on the antibacterial action.

Bi-based oxide nanostructures have been gaining their environmental applications such as catalytic oxidation of pollutants and their ability for the recovery from the system due to magnetic nature. The catalyst regeneration capacity makes it a cost-effective catalyst and could be one of the right candidates for sustainable pollutant remediation. Due to the narrow band gap, multiferroic behaviour, and better chemical stability, bismuth (Bi)-based oxides catalysts are the potential photocatalysts for environmental application. Chapter 7 describe the crystal structure, synthesis, characterizations, and photocatalytic degradation mechanism of $BiFeO_3$. The strategies for the improvement of photocatalytic performances of $BiFeO_3$ for organic pollutants are also discussed in this chapter.

Photocatalytic oxidation processes produce reactive oxygen species for the mineralization of organic pollutants. The visible light active and cost-effective photocatalysts development is still under research, and scientists are trying to improve the visible light active photocatalysts efficiency. Chapter 8 provides a comprehensive review on the basics of photocatalysis, its mechanism, and surface modification approaches for improving photocatalysts. This chapter also discussed the photoreduction for inorganic compounds removal, photocatalytic oxidation of organic compounds, and the factors affecting the photocatalysis process.

The metal-doped polymeric and bio-polymeric-based adsorbent and photocatalyst have sown their potential materials for the removal of toxic organic dyes from water. Chitosan and cellulose composites-based photocatalysts with metal can perform photocatalysis with adsorption process, and shown a synchronous effect in improving photocatalytic activity with doping. The modified biopolymers with metals have more active sites at the surface of the adsorbent, leads to enhance dye and semiconductor interaction along with suppressing electron–hole recombination rates during the photocatalytic process. Chapter 9 has discussed the application of various metal activated composites for photodegradation of dyes in the wastewater. It further emphasizes the existing obstruction and impending prediction for the deprivation of dyes via photocatalytic techniques.

Many 2D-based materials such as graphene, metal–organic frameworks (MOFs) were effectively used for capturing many toxic heavy metal ions from the water. The

MOFs are gaining their presence in the water and wastewater treatment. Chapter 10 has provided the latest methods of MOFs fabrication and their application for the removal of inorganic and emerging organic contaminants present in the wastewater.

Chapter 11 discussed the surface modification for the diverse environmental remedial applications like adsorption, photocatalytic degradation, oil–water separation and capture, self-cleaning, and anti-microbial/anti-bacterial or anti-foulants properties. This chapter provides a detailed outlook on the current status on environmental remedial strategies by using various modified surfaces.

The advanced oxidation process (AOPs) is a promising technique and has gained importance due to its ability to degrade and mineralize complex organic pollutants, including emerging contaminants (ECs). AOPs effectiveness depends on the in-situ generation of reactive chemical species (ROS) such as hydroxyl radicals. Chapter 12 discusses the working principles, and mechanisms of various AOPs, including ozonation, Fenton, photolysis, plasma, sonolysis, and photocatalysis, and their combinations. The role of different nanoparticles in these processes has been discussed. The application of AOPS for the degrading ECs are discussed along with the origin, fate, and human and ecological health impacts of ECs in water bodies. Overall, this book offers a state of the art literature for researchers and academicians working in the field of environmental remediation by nanomaterials and nanocomposites.

Chapter 2
Carbon Nanomaterials: A Prominent Emerging Materials Towards Environmental Pollution Study and Control

Shivangi Mishra, D. P. Mondal, Pradip Kumar, and Shiv Singh

Abstract Increasing global issues based on pollution are threat to not only the present scenario, but also the results are going to be more saddening in the coming future. A major portion of the solution can be achieved by focusing the aim on the recyclability of the natural resources. Whether it is a concern to water pollution leading to water scarcity, or air pollution leading to fresh and breathable air scarcity, the data are not significantly acceptable. In recent years, carbon nanomaterials (CNMs), viz; carbon nanofibers (CNFs), fullerenes, carbon nanotubes (CNTs), nanodiamonds/onions, graphene/graphene oxide, and their derivatives have been widely investigated as adsorbents, sensors, bio-chemical/electro-catalysts, fuel cells, and many more. The admirable properties of the aforementioned CNMs such as, high thermal and chemical stability, reusability, ease of modification, low density, high-electrical conductivity, sensitivity, and selectivity, are some positive aspects, which integrate CNMs as promising materials for detection and quantification of the pollutants. The strong π–π interactions and porous properties shown by the CNMs make it a prominent adsorbent, and also the high sensitivity and selectivity allow it to be a better option as an electrochemical sensor. In this book chapter, the role of the CNMs has been discussed thoroughly as a class of material to quantify and detect different pollutants, added with the pinch of its properties as an environmental beneficiary.

S. Mishra
Nanomaterial Toxicology Group, CSIR-Indian Institute of Toxicology Research (CSIR-IITR), Vishvigyan Bhawan, 31 Mahatma Gandhi Marg, Lucknow, Uttar Pradesh 226001, India

D. P. Mondal · S. Singh (✉)
Lightweight Metallic Materials, Council of Scientific and Industrial Research-Advanced Materials and Processes Research Institute, Hoshangabad Road, Bhopal, Madhya Pradesh 462026, India
e-mail: sshiv.singh@ampri.res.in

P. Kumar (✉)
Integrated Approach for Design and Product Development, Council of Scientific and Industrial Research-Advanced Materials and Processes Research Institute, Hoshangabad Road, Bhopal, Madhya Pradesh 462026, India
e-mail: pradip.kg@ampri.res.in

S. P. Singh et al. (eds.), *Nanomaterials and Nanocomposites for Environmental Remediation*, Energy, Environment, and Sustainability,
https://doi.org/10.1007/978-981-16-3256-3_2

2.1 Introduction

In the present era, the sustainability of clean and breathable surroundings is still one of the grand global challenges. The rate at which the world population is growing every day, it demands the need for development in urbanization and industrialization. This fact increases the concern for not only clean water and air supply but also of the availability of the resources for the remediation of the environment around the globe. The discharge of wastewater from the industries without pre-treatment is one of the main causes of water pollution, adding to it the discharge or dumping of the waste, and also leaching of different chemicals from the farms are some significant sources. The pollutants include heavy metals, fertilizers, pesticide residues, detergents, phenols, dyes, oil, and grease (Schweitzer and Noblet 2018; Srivastava et al. 2020a; Singh et al. 2018a; Kumar and Verma 2020). The presence of these pollutants even at trace levels is very threatening, as they have the property of accumulation in soil, water-bodies, and in human body, which may cause problems in future (Ren et al. 2018; Reddy 2017; Asfaram et al. 2018). By keeping in mind the present environmental conditions, there is an urgent need to develop highly efficient and sensitive, reliable, user-friendly methodologies for quantification and monitoring of these pollutants in different mediums.

The preparation and application of effective materials for wastewater treatment are an evolving area for research these days. Various carbon-based nanomaterials (CNMs) have achieved incredible attention in a variety of fields due to their exceptional properties (Kumar et al. 2008, 2012; Kumar and Bohidar 2010; Mishra et al. 2020; Pandey et al 2020a; Singh et al. 2020a; Singh et al. 2020b). The CNMs show distinct thermal and electrical conductivity, and also they differ in action from their bulky counterparts in mechanical and optical properties (Buzea et al. 2007; Kumar et al. 2015; Singh et al. 2020c). The former leader of the carbon allotropes was sp^2 hybridized graphite and sp^3 hybridized diamond. The novel CNMs includes a class of materials that are entirely sp^2 hybridized and range from zero-dimensional (0D) to three-dimensional (3D) CNMs, viz, fullerenes, graphenes and its derivatives, CNTs, CNFs, carbon dots, and many more (Novoselov et al. 2005; Yuan et al. 2014; Geim and Novoselov 2007; Geim 2009; Kumar et al. 2016, 2014; Pankaj et al. 2018; Pandey et al 2020b; Pophali et al. 2020; Ashfaq et al. 2013). Figure 2.1 shows the allotropes of carbon (Rauti et al. 2019). The structure, molecular composition, size, surface chemistry, and more importantly, the structure of the CNMs is controllable and can be tailored accordingly for the desired application (Li et al. 2019a).

Among several nanomaterials, CNMs have been proven as a next-generation material for a wide class of applications, such as adsorbents (Baig et al 2019; Bikshapathi et al. 2012; Singh et al. 2014a), drug delivery (Deb and Vimala 2018; Singh et al. 2019a), catalytic reactions (Li et al. 2015; Jeng et al. 2011), biosensors (Singh et al. 2019b; Xie et al. 2019; Bairagi et al. 2019), antibacterial agents (Singh et al. 2013, 2014b), wastewater treatments (Kumar et al. 2020) and fuel cells (Xie et al. 2019; Gupta et al. 2017; Singh et al. 2018b, 2016; Modi et al. 2016; Singh and Verma 2015a, 2015b). With the increasing interest in nanotechnology, CNMs have emerged

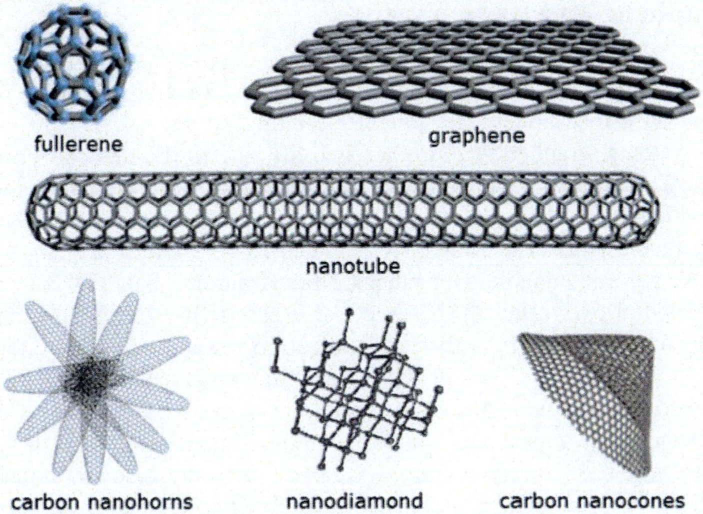

Fig. 2.1 Allotropes of carbon. Reproduced with permission from Rauti et al. (2019). Copyright 2019 Elsevier

out as an efficient material for their applications, mainly in wastewater remediation (Farghali et al. 2013; Dil et al. 2016; Taher et al. 2018). The properties, as mentioned above of the CNMs, are admirable and thus are provoking their use in every possible field (Singh and Verma 2015b).

Carbon-based nanomaterials and nanotechnology provide an excellent perspective for wastewater treatment with high efficacy and lead to the availability of more clean water sources. As compared to the conventional technologies, CNMs offer far better and cost-effective methodologies for the wastewater treatment at small scale, because they hold high-surface area, and are highly reactive. CNMs are proven to be an efficient adsorbent because of their porous structure and their tendency toward surface modification (Goh et al. 2016). These appealing characteristics and the assurance offered by this class of nanomaterials have to lead to the fabrication of various CNMs for wastewater treatments.

Many researchers have reported CNMs for their fascinating utilities in wastewater treatment and other water related applications. Nevertheless, the number of book chapters on the application of CNMs for wastewater remediations is few in numbers. This chapter provides a comprehensive overview of the recent advances in nanotechnologies and nanostructured carbon materials for the quantification and treatment of pollutants in water resources.

2.2 Graphene and Its Derivatives

Graphene has a flat two-dimensional (2D) structure, in which carbon atoms are packed layerwise in a honeycomb manner (Geim and Novoselov 2007). Among other nanosystems, graphene has higher stability due to the delocalization of the π-electrons. The thermal conductivity in graphene is high with remarkable electron mobility, which is 3000 WmK^{-1} and 200,000 $cm^2\ V^{-1}\ s^{-1}$, respectively (Kumar 2019). The theoretical surface area is calculated to be 2630 $m^2\ g^{-1}$ for graphene, which is also recommendable over other CNMs (Baig and Saleh 2018). Graphene is also considered over other CNMs because it can be used as a base material to derive other forms of CNMs. Various modification is also possible by doping and creating defects; hence this can tune up the electronic and reactive properties of the graphene and its derivatives (Stoller et al. 2008; Cho et al. 2017, 2016). Coming to the 2-D graphene nanomaterials, which have a high-aspect-ratio due to the sheet-like structure with a single-layered or few-atom-thick layers, also this structure adds superior electronic and surface properties to graphene-based nanomaterials (Zhang et al. 2019; Chen et al. 2018; Su et al. 2018).

With more upgradation in the technologies, 3-D graphene nanomaterials also provide a wider perspective for the wastewater treatments. These 3-D graphenes are found to be used as adsorbents. These 3-D graphenes nanomaterials are more prominent over the 2-D graphenes, as these, along with the retention of the intrinsic properties of graphene, also posses comparatively higher porosity, high-mechanical strength and electronic properties, larger surface area and lower density (Qiu et al. 2018; Salzano de Luna et al. 2019; Wang et al. 2019). Rolling over such tremendous features, the graphene-based CNMs are persistent fabrication materials, for wastewater remediation. This book chapter provides the view over the use of graphene-based CNMs in wastewater remediation as an adsorbent (Mandeep et al. 2020), sensors (Hajati 2020).

2.2.1 Adsorbents

As discussed earlier, the properties of the 2-D and 3-D graphenes make them favorable for the use in wastewater remediation, as adsorbents. The high porosity and higher surface area, low-production cost allows these materials for the removal of toxic pollutants such as dyes, pesticides, halogenated compounds, heavy metals, etc. from the water sources (Vicente-Martínez et al. 2020). The property of graphene allows it to get fabricated into more usable for by doping or attaching of metals or oxides or peroxides. The introduction of oxygenated functionalities into the graphene structure for the making of graphene oxides (GO) provides a hydrophilic nature to the fabricated material. The oxidized graphene-based CNMs can easily disperse in the aqueous media basically in the organic solvents. Figure 2.2 shows the schematic adsorption mechanism of graphene oxide on methylene blue (Jun et al. 2018). Further,

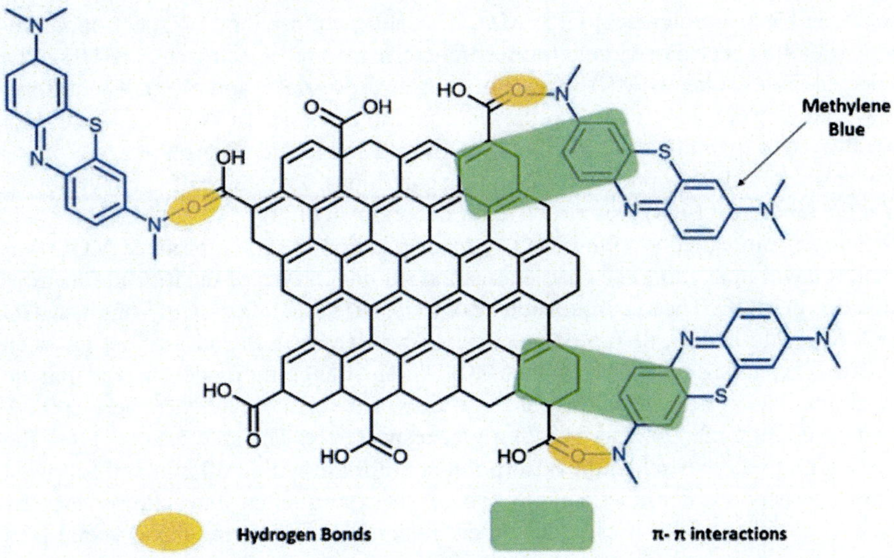

Fig. 2.2 Illustration of the adsorption mechanism of graphene oxide on methylene blue. Reproduced with permission from Jun et al. (2018). Copyright 2018 Elsevier

the recovery of the honeycomb structure, oxidized graphenes are reduced to removal forces, hydrogen bonding, πer the electrical conductivity. These reduced forms of the oxidized graphenes are known as reduced graphene oxides or r-GO. Graphene shows six types of interaction with the adsorbates, i.e., electrostatic interactions, van der Waals forces, π- bonding interactions, Lewis-acid-base interactions, hydrogen bonding, and hydrophobic effects. These kinds of interactions allow feasible adsorption of organic pollutants on the graphene-based CNMs. Also, the presence of defects and functional groups on the surface or within the structure of graphene-based CNMs acts as a site for the growth of nanoparticles via nucleation (Mandeep et al. 2020).

In this section of the chapter, we will discuss the use of graphene and its derivatives in the field of wastewater remediation as adsorbents. Various literature provides a variety of data on how graphene-based adsorbents are used for the quantification and detection of organic pollutants such as nitro compounds, polycyclic aromatic hydrocarbons, pesticides, antibiotics, phenols, heavy metals, and many more (Jun et al. 2018; Gusain et al. 2020).

2.2.1.1 Heavy Metals

In recent research done on the removal of toxic M(II) (Pb^{+2}, Cr^{+3}, Cu^{+2}, Zn^{+2}, and Ni^{+2}), magnetic graphene oxides (MGO) were used as adsorbents. MGO was synthesized by the fabrication of iron particles (Fe^{3+}) on the surface of GO by the co-precipitation method (Su et al. 2017). The prepared MGO was characterized using scanning electron microscopy (SEM), X-ray diffraction, vibration spectrophotometry, UV-VIS spectrometry, and fourier-transform infrared spectroscopy (FTIR). The selected targeted heavy M(II) was (Pb^{+2}, Cr^{+3}, Cu^{+2}, Zn^{+2}, and Ni^{+2}). The adsorption studies were done under optimized conditions, i.e., pH (3–9), adsorbent dose (0.002–0.016 g), and contact time (10–65 min). Kinetic and adsorption experiments were also done under optimized adsorption conditions. The study effectively showed that the fabricated MGO was effective for the removal of Pb^{+2}, Cr^{+3}, Cu^{+2}, Zn^{+2}, and Ni^{+2} ions from aqueous solutions. The magnetization of GO to produce MGO was achieved within 20 min. The characterization studies confirmed the fruitful functionalization of MGO. The maximum uptake of Cr^{+3}, Ni^{+2}, Zn^{+2}, Pb^{+2}, Cu^{+2}ions by MGO took place at pH of 6, 8, 7, 5, respectively. An increase in the adsorption was seen with the increasing dose from 0002 to 0.016 g. The Langmuir model showed that the maximum adsorption capacities for Pb^{+2}, Cr^{+3}, Cu^{+2}, Zn^{+2}, and Ni^{+2} were 200.00, 24.330, 62.893, 63.694, and 51.020 mg/g, respectively. The study showed that the adsorption process reached the adsorption equilibrium within 30 min and followed a pseudo-second-order kinetic model. Thus, this experimental study showed that the MGO was successfully used as adsorbent material for removing heavy metal ions for wastewater remediation (Ain et al. 2020).

One more study based on the MGO as adsorbents was done. In this study, three types of graphene-based CNMs were fabricated, which were, MGO, magnetic chemically reduced graphene (MCRG), and magnetic annealing-reduced graphene (MARG). All three adsorbents were investigated to isolate adsorption capacities and different molecular mechanisms. Three subjects were finalized for the identification, quantification, and removal, namely tetracycline (TC), arsenate (Huang et al. 2019), and cadmium (Huang et al. 2019). The study revealed that the adsorption of the three contaminants was pH-dependent, and also MGO was found more suitable for studying the adsorption behaviors in three different binary systems (Fig. 2.3). The maximum adsorption capabilities of MGO were reported as 14 mg/g for As(V), 252 mg/g for TC, and 234 mg/g for Cd (II). The presence of various O-containing functional groups, high dispersibility, and the presence of thin nanosheets provide superiority to MGO over the other two graphenes. The presence of n-π electron-donor-acceptor (EDA) effect was more prominent over H-bonding and π-π interactions and was the main reason for the adsorption of TC on the MGO. It was found that, in the presence of TC, the adsorption of As(V) was suppressed, but on the contrary, As(V) merely showed any variation in the TC adsorption. In the co-adsorption system, the adsorption was increased by 65 and 30% for Cd(II) and As(V), respectively. These novel CNMs elucidated various interaction mechanisms and opened new paths for the fabrication of novel adsorbents for practical applications (Huang et al. 2019).

In another report, graphene marked as G-ASP2 was synthesized in the laboratory aiming to investigate the ability of G-ASP2 and its activated from AG-ASP2 as adsorbents for the removal of heavy metal ions, namely Ni (II), Pb(II), and Fe(II) ions from aqueous media. The characterization study for the prepared graphene-based CNMs (G-ASP2) was done by SEM, transmission electron microscopy (TEM), Raman spectroscopy, and thermogravimetric analysis (TGA). For the acid activation of the prepared G-ASP2, concentrated H_2SO_4 and HNO3 were used. The adsorption

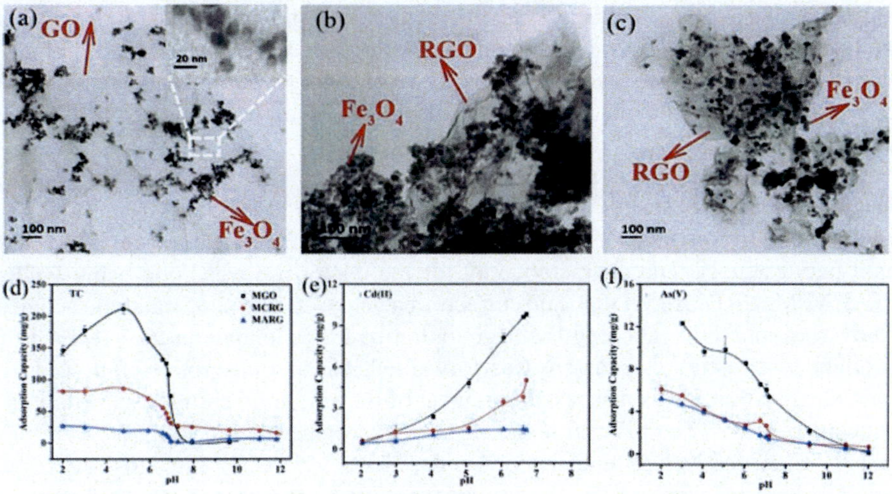

Fig. 2.3 TEM images of **a** magnetic graphene oxide (MGO), **b** magnetic chemically reduced graphene (MCRG), and **c** magnetic annealing reduced graphene (MARG) samples. (**d–f**) pH-dependent adsorptions of tetracycline (TC), Cd(II), and As(V), onto MGO, MCRG and MARG, respectively. The adsorption capacity was calculated for a contact time of 24 h. Reproduced with permission from Huang et al. (2019). Copyright 2019 Elsevier

is strongly governed by the surface area and functionality, and also on the pore distribution and structure of the adsorbents. It is found that activated graphene showed a much higher surface area, but AG-ASP2 lost the affinity toward the heavy metal ions due to the removal of the functional groups present on the graphene surface because of the aid activation. This lead to the conclusion that the acid activation of G-ASP2 failed to meet the expectations for increasing the adsorption capacity of graphene. For all analyzed systems, a total of 30 min was sufficient to reach the equilibrium. The pseudo-second-order model described the sorption kinetics of Fe(II), Pb(II), and Ni(II) on both the sorbents. Single-metal systems were found more efficient than the multi-metal systems for the removal of M(II), which thus, indicated the presence of an antagonistic effect. In the multi-metal solutions, the activated graphene as sorbent showed an increase in the removal effect for the Pb(II) ions, and this was evident for the synergistic effect. The result obtained from this study was helpful for understanding the role of graphene and activated graphene in the removal of heavy metal ions from wastewater (Atkovska et al. 2020).

2.2.1.2 Organic Pollutants

Organic pollutants cover a wide class of pollutants, which include PAHs, antibiotics, dyes, phenolic compounds, and pesticides, etc. PAHs are semi-volatile and carcinogenic organic compounds, which are categorized by the presence of fused aromatic rings. Many kinds of literature have been found, which overlays that graphene and its derivatives can be used successfully for the removal of the aforementioned organic pollutants from the polluted water. This section reveals the potential of graphene-based nanomaterials as an adsorbent for different classes of organic pollutants.

A novel biocomposite polyacrylic acid-grafted-carboxylic graphene/titanium nanotube (PAA-g-CGR/TNT) was synthesized for the adsorption of enrofloxacin, an antibiotic from aqueous solution. This novel fabricated biocomposite also acted as a photocatalyst. The characterization of the composite was done using FTIR, SEM, TEM, XRD, XPS, DRS, and surface area analyser. The photocatalytic activity study was done by evaluating the degradation of the pollutants in the presence of sunlight. A two-step mechanism was forwarded, the first adsorption, followed by photodegradation for the successful removal of the antibiotic enrofloxacin from the aqueous solution. The pH and temperature for the maximum adsorption were 5.0 and 30 °C, respectively. Pseudo-second-order kinetics were best fitted to the adsorption mechanism, whereas for the photodegradation process, the kinetic behavior followed first-order kinetics with pH value 5.0. The bandgap of the biocomposite (PAA-g-CGR/TNT) was 2.60 eV, which falls in the visible range. The adsorbent – cum-photocatalyst was tested for up to five cycles to check its reuse and regeneration (Anirudhan et al. 2017).

A study was performed for the adsorption of some selected organophosphorus pesticides (OPPs), namely parathion, chlorpyrifos, and malathion from aqueous medium. For this study, a novel composite graphene oxide-based silica-coated magnetic nanoparticles (Fe_3O_4@SiO_2@GO) functionalized with 2-phenylethylamine (PEA) was fabricated. The characterization of the fabricated (Fe_3O_4@SiO_2@GO) composite was carried by using XRD, Zeta potential, TEM, FTIR, vibrating sample magnetometer (VSM), and Nitrogen adsorption/desorption at 77 K. The maximum adsorption was achieved at 15 min, with an adsorbent dose of 15 mg using 1 μg/mL. The synthesized material can be used on a variety of samples, as no significant pH difference was found. Sorption isotherm data were found to be suitable to fit with Freundlich, Dubinin-Radushkevich, Redlich, Langmuir, Sips, and Peterson models when non-linear methods were studied. All the three pesticides were found to fit best with the Sips model, then other sorption isothermic models. The experimental data obtained using the non-linear methods showed that pseudo-first and pseudo-second-order kinetic model was more acceptable, and also the pseudo-second-order kinetic model was the best fit with all the three pesticides' data. The fabricated material was tested for its reusability, and it was found that after 10 cycles, there was a little lost in the recovery of the three pesticides. The adsorbent was tested for real samples, and the recovery obtained was remarkable. Thus, Fe_3O_4@SiO_2@GO composite can be used efficiently as an adsorbent for the quantification and other analysis of the pesticides (Wanjeri et al. 2018).

Apart from the use of graphene as an adsorbent for pesticides and other organic pollutants, it was also proven to be an efficient alternative for dyes, as compared to other available alternatives. In a report, graphene oxide was a proficient adsorbent for the removal of two dyes, namely, basic red 12 (BR12), and methyl orange (MO) from aqueous solution. The study was conducted under optimized conditions, and the impact of various parameters including, contact time, initial dye concentrations, temperature, and pH were well observed. Adsorption process took 100 min, under optimized conditions, and maximum removal of the dye occurred at pH 3. It was found that the adsorption capacity for BR 12 dye gradually increased due to a hike in the interactions between the adsorbent surface and dye particles. On the contrary to this, due to the negative charge over methyl orange and negative charge carried by the graphene oxide surface, there was a repulsion, which lead to a decrease in the adsorption of methyl orange. The effect of temperature was endothermic. The linear regression and least value of the chi-square statistic were $R^2 = 0.999$ and $X^2 = 2.28$, respectively, which were remarkably high. The observed statistic and linear values showed that type (I) Langmuir isotherm was best fitted to the adsorption equilibrium (Robati et al. 2016). Also, it was found that the uptake of methyl blue from aqueous phase gradually increased with an increase in the oxidation degrees of graphene oxide (Yan et al. 2014).

2.2.2 Sensors

The excellent electronic properties of graphene welcome its application as electrochemical biosensors for environmental remediations. It was reported that graphene field-effect transistor (GEFT) biosensor was fabricated for the detection of various water pollutants. The variation of environmental conditions causes some nonspecific disturbances, which hinders the sensitivity of such sensors. The fabricated differential GEFT sensor was designed in such a manner that it can selectively detect the water pollutants and that too in the presence of nonspecific disturbances. This differential sensor was used for the selective detection of 17β-estradiol in both buffer and tap water. Consistent detection results were obtained in accordance with the variations (Li et al. 2019b).

Due to the dumping and discharge of industrial wastes into water sources, various carcinogenic metals and compounds can be found in the water system. In a report, a graphene monolayer based sensor was designed for the detection of carcinogenic heavy metals including, Cr, Hg, As, Cd, and Pb in the aqueous phase. The transport and electronic properties of the graphene monolayer were studied in both the aqueous phase and vacuum. The analysis of the graphene made clear that graphene, in its pure form, has no sensitivity for water molecules, but it can interact with the carcinogenic heavy metal Cr. The interaction between the Cr metal and monolayer was calculated through Bader analysis, for the significant charge transfer. The substantial variations in the density of states (DOS) profile, current–voltage, and band-structure characteristics obtained for the graphene monolayer in the presence of Cr, confirmed that graphene monolayer is the comparatively better material for the detection of Cr, even at a small applied voltage of 400 mV in water. This report proved that the graphene monolayer could be used efficiently for the detection of Cr with a response time of 82% (Srivastava et al. 2020b).

2.3 Carbon Nanotubes

Carbon nanotubes CNTs are defined as graphite sheets rolled up into cylindrical shapes. There are two types of carbon nanotubes that are classified on the basis of carbon layers present. First one is single-walled carbon nanotubes (SWCNTs) is made up of single layer of graphene whose diameter varies between 0.4 and 2 nm, and the second one is multi-walled carbon nanotubes (MWCNTs) is made up of two or more graphene sheet cylinders, varying between 1 and 3 nm of diameter. The advantageous characteristics of CNTs include high-tensile strength, special electronic structures, ultra-lightweight, high chemical, and thermal stability (Polizu et al. 2006; He et al.2013; Che et al. 2000). CNTs are one of the most exploited CNMs for several applications. In this portion, an overview of how CNTs are used for wastewater remediation is discussed.

2.3.1 *Adsorbents*

As discussed earlier on how CNMs are useful in the environmental remediation, CNTs as a class of CNMs cover a major area of attribution toward the wastewater treatment as adsorbents. In one of the reports, chemically modified carboxylated multi-walled carbon nanotubes (MWCNTs-COOH) were used for the removal of Cd(II) from wastewater. 3-aminopyrazole was used to fabricate MWCNTs-f. Three nano adsorbents, namely; MWCNTs, MWCNTs-f, MWCNTs-COOH, were separately used for the adsorption of Cd (II) ions (Alimohammady et al. 2017).

In another report, CNTs were utilized for the removal of diquat dibromide (DQ) herbicide from the water. Oxidized multi-walled carbon nanotube (OMWCNTs) was used as adsorbent carrying non-magnetic properties, and on the other hand, OMWCNTs-Fe_3O_4 and OMWCNT-κ-carrageenan-Fe_3O_4 nanocomposites were carrying non-magnetic properties. For each set of adsorption system, the effects of the initial concentration of DQ, temperature, and contact time on the adsorption were studied. Experimental data showed that OMWCNTs had a faster rate of adsorption and also showed higher maximum adsorption capacity, over the magnetic adsorbents. It was found that the initial sorption rate gradually reported a fall from 0.674 to 0.612 mg/g/min for OMWCNT-κ-carrageenan-Fe_3O_4, from 1.21 to 0.823 mg/g/min for OMWCNT-DQ system, and from 29.1 to 4.28 mg/g/min for the non-magnetic OMWCNTs-DQ system. At 25° C, the maximum adsorption capacity of OMWCNTs was approximately 5.4 and 2.8 fold higher than the magnetic OMWCNT-κ-carrageenan-Fe_3O_4 and OMWCNTs, respectively. It was found that isothermic data and the adsorption kinetics achieved for all the adsorption sets were best fit to Langmuir models, and pseudo-second-order, respectively. Furthermore, the endothermic process of adsorption was observed for DQ herbicide adsorption on the CNT based nanocomposite. Thus, this research indicated that CNTs with magnetic and non-magnetic characteristics could be a promising tool for DQ adsorption from water (Duman et al. 2019).

For extraction and removal of cobalt (II) from aqueous solution, a novel adsorbent was fabricated using MWCNTs and γ-alumina. These new nanoscale adsorbents were characterized using TEM, SEM, and FTIR analysis. A second-order polynomial model was suggested via the response surface methodology (RSM). The results obtained for the two adsorbents were ρ-value < 0.0001 and R^2 of 0.9992 for the γ-alumina adsorbents and for MWCNTs ρ-value < 0.0001 and R^2 of 0.9980. For the removal of Co(II), the maximum removal efficiency was 90%, when MWCNTs were used as adsorbent under the optimized conditions (pH = 10, MWCNTs dosage = 1.57 mg/L, contact time = 38.6 min, and initial Co(II) concentration = 56.57 mg/L). Furthermore, 93% of Co(II) removal was achieved when γ-alumina was used as adsorbent under optimized conditions (pH = 10, γ-alumina dosage = 1.63 g/L, contact time = 35.5 min, and initial Co(II) concentration = 52.15 mg/L). Pseudo second-order kinetics as per the McKay and Ho's kinetic model approach was best fitted to the experimental results obtained for both the adsorbents. Thus, such CNMs adsorbents are approachable for future remediation methods (Dehghani et al. 2020).

2.3.2 Sensors

The tunable electronic properties of CNTs welcome their use in environmental remediation as sensors. The multifunctional aspects offered from CNTs, including tremendous chemical and physical properties with the remarkable surface to volume ratios, make it more desirable. An electrochemical sensor for quantification of pollutants including, catechol (CC), hydroquinone (HQ), and resorcinol (RC) from water and other sources. In this report, the glassy carbon electrode (GC) was used for the very first time as a receptor for di-hydroxybenzene isomers. For fabricating the sensor, the GC electrode was modified with successive layers of poly-hydroquinone (PHQ), MWCNTs, and benzo-12-crown-4 carbon electrode, and thus, GC/CNT/PHQ/CE was obtained. The sensor was based on the host–guest theory. The fabricated layered sensor was able to determine the isomers simultaneously successfully, RC, HQ, and CC in the concentration ranges varying between 0.05–100 μM, 0.03–100 μM, and 0.01–100 μM with low detection limits of 0.427, 0.156, and 0.118 nM, respectively. The result obtained showed that the sensor demonstrated a high-electro-catalytic activity. This made the fabricated sensor a better option for faster and efficient determination of pollutants in water bodies (Atta et al. 2020).

In another report, for the assessment of toxicity of water pollutants, a highly sensitive and miniature cell-based electrochemical sensor was fabricated. In this biosensor, for biological recognition, human hepatoma (HepG2) cells were utilized to determine the changes in the electrochemical signals and to check the cell viability. The building of the graphene oxide quantum dots/carboxylated CNTs hybrid biosensor was achieved through green means. The sensor was efficiently used to examine the toxicity of six main water pollutants, namely Hg, Pd, 2,4,6-tricholorphenol, Cd,

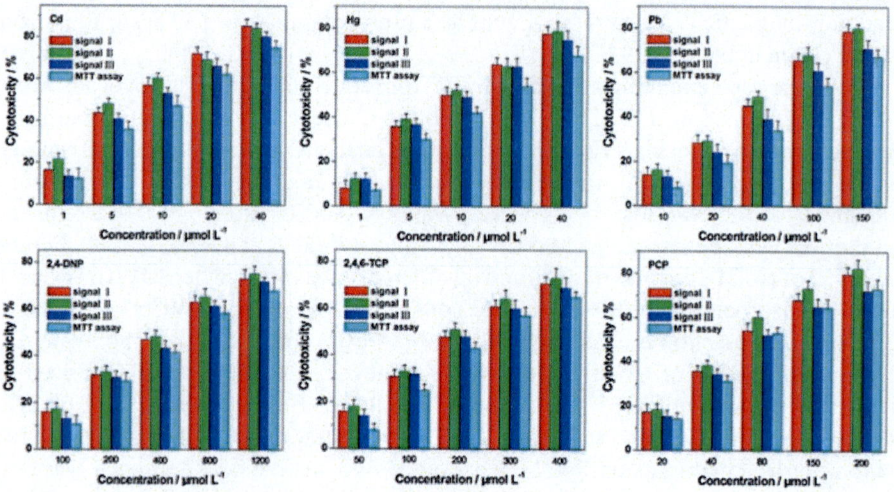

Fig. 2.4 Electrochemical biosensor and MTT assay detected cytotoxicity of Cd, Hg, Pb, 2,4-DNP, 2,4,6-TCP, and PCP on HepG2 cells. Reproduced with permission from Zhu et al. (2017) Copyright 2017 Elsevier

pentachlorophenol, and 4-dinitrophenol (Fig. 2.4). The efficient sensitivity of the electrochemical sensor was evaluated based on the 24 h IC_{50} values, which were much lower as compared to the conventional MTT assays. The sensor was found to be very sensitive for the detection of phenols and heavy metals (Zhu et al. 2017).

For the successful determination of clomazone, pesticides in water samples were fabricated using MWCNTs supported with Pt via, differential pulse voltammetry (DPSV). The determination of clomazone was done in 0.1 M phosphate buffer solution in the concentration range of 0.61–20.56 ng cm^{-3} at pH 7. The limit of quantification (LOQ) and limit of detection (LOD) values, 0.61 and 0.38, respectively, which were in the same range with HPLC/DAD used mainly for comparative studies. HRTEM data revealed that Pt nanoparticles were evenly decorated on the MWCNT, and also Pt NPs were the main source of the required changes in the structural and electronic properties. As per the Raman spectroscopy analysis, there were structural defects present in the fabricated sensor material, which lead to a better sensing response (Ranđelović et al. 2019).

2.4 Carbon Nanofibers

Carbon nanofibers (CNFs) are a known class of CNMs. CNFs are carbon filaments whose diameter lies within the nanoscale range. They have several quotable properties, such as high surface to volume ratios, high-graphitic character, hydrophobic, easy to functionalized outstanding chemical and mechanical stability, and brilliant electronic properties. CNFs are prepared through two methods, electrospinning, and chemical vapor deposition (CVD), respectively (Singh et al. 2020a, 2014a, 2016; Modi et al. 2016; Singh and Verma 2015a,2015b; Miao et al. 2010; Jong and Geus 2000). The aforementioned properties of the CNFs have made them very popular in the field of research on wastewater treatment.

2.4.1 Adsorbents

In a study performed for the removal of Cu(II) ions from wastewater, magnetic tubular carbon nanofibers (MTCNFs) were used as adsorbent. Hypercrosslinked α, ά-dichloro-p-xylene, was used for synthesizing this novel green adsorbent. The characterization of the adsorbent revealed that it had a surface area of 280.94 m^2/g, inner and outer surface areas were 25 and 110 nm, respectively. Also, for the fabricated MTCNFs, the saturation magnetization was 10.65 emu/g. The batch method was used to study the adsorption properties of the adsorbent. The highest removal efficiency of Cu(II) ions was observed to be $99.9 \pm 0.1\%$ in comparatively short contact time = 10 min, and at smaller adsorbent dosing of 0.5 g/L at pH = 6. The kinetic studies revealed that pseudo-first-order kinetics was found more suitable than the pseudo-second-order kinetic model. Isothermal studies revealed that Langmuir isotherms are best suited than the Freundlich isothermic model, evident with the highest adsorption value of Cu(II) to be 375.93 mg/g. the adsorbent showed endothermic thermodynamics, and also it was gathered that the adsorption was spontaneous (Ahmad et al. 2020).

In another report, an activated electrospun carbon nanofiber (A-ECNFs) was prepared through the electrospinning technique for the adsorption of methylene blue (MB) and congo red (CR) dyes. The adsorption pattern was investigated for pristine ECNFs and the new A-ECNFs for both the dyes in the aqueous solution. The A-ECNFs was having a higher surface area of 1615 m^2/g with micro/mesoporous structures. The characterization of the A-ECNFs and pristine ECNFs was done using various characterization techniques, including SEM, TEM, FTIR, raman spectroscopy, XRD, and PSA. The adsorption performance was allied with the surface properties including, surface area, surface charge and porosity, and also with the pH and temperature of the medium and nature of the dyes. Redlich-Peterson isotherm model was found to best fit out of the six non-linear isotherm models applied to study the adsorption behavior. Elovich kinetic model suited with the kinetic results of A-ECNFs adsorbent for CR and MB, whereas the pseudo-second-order kinetic model was for pristine ECNFs. The adsorbents were found to be reusable, thus are a significant adsorbent (Thamer et al. 2019).

The role of CNFs has been found evident from a number of works of literature. In recent work, polyimide (PI)-based carbon nanofibers (CNFs) were fabricated via electrospinning method of polyamic acids as an adsorbent for the removal of methylene blue (MB), 2,4-dicholrophenol (2,4-DCP), and tetracycline (TC) dyes from water sources. The fabricated PI-CNFs was having a high-specific surface

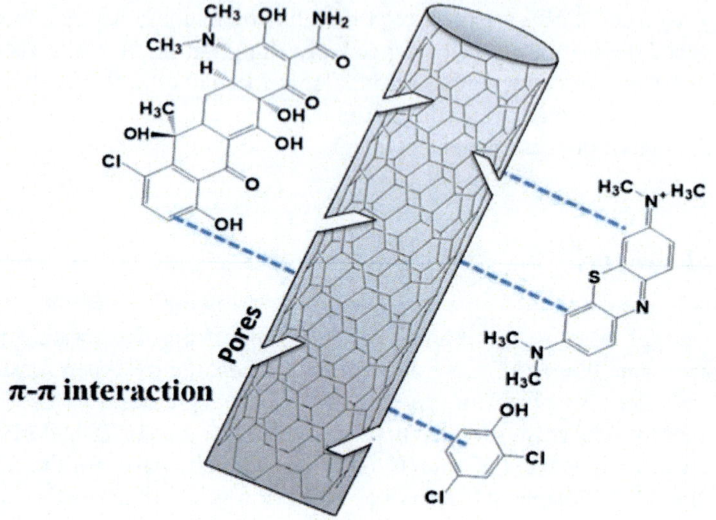

Fig. 2.5 Schemtatic of adsorption mechanism of 2,4-DCP, TC, and MB on CNFs. Reproduced with permrission from Zhang et al. (2018). Copyright 2018 Elsevier

area of 715.89 m^2/g, and showed maximum adsorption of 272.48, 4–483.09, and 146.63 mg/g for MB, 2,4-DCP and Tc, respectively. It was found that adsorption efficiency of MB was highly dependent on pH 11, whereas the other two dyes, 2,4-DCP and TC showed an efficient performance with the fabricated CNFs, within a wide pH range of 3–7 and 4–7, respectively. Langmuir isotherm and pseudo-second-order kinetic model were found to best fit with the isotherm and kinetic experiments, respectively. Also, thermodynamics of the experiment were spontaneous and endothermic. Thus, these CNFs can be employed for future aspects of environmental remediation (Zhang et al. 2018). Figure 2.5 shows the adsorption mechanism 2,4-DCP, TC, and MB on CNFs.

2.4.2 Sensors

CNFs offers a great electronic platform because of the easy electron flow and π-π interactions, H-bonding, etc. This section of the book chapter demonstrates the electronic part of the CNFs as sensors for wastewater treatment. In a report, a cerium oxide catalyzed 1-D CNFs based sensor was prepared for the wastewater remediation. The Ce-CNFs were fabricated via chemical vapor deposition (CVD) method, and the sensor was found to be selective and sensitive for the determination of heavy metal ions Pb(II) and Cu(II). Acetylene was used as the carbon source, and cerium oxide was used as a catalyst. For binding of the composite, polymethyl vinyl ether-alt-maleic anhydride (PMMVEA) was used. Three electrode assembly was used for the electrochemical analysis, out of which the glassy carbon electrode modified with the Ce-CNFs was used as the working electrode. This assembly was used for simultaneous and selective detection of Cu(II) and Pb(II) using the differential pulse voltammetry (DPV) and cyclic voltammetry (CV). The LOD was calculated to be 0.3 ppb and 0.6 ppb, respectively, and showed a linear behavior. The fabricated sensor can thus be used for real water samples (Singh et al. 2019b).

For the detection of polyphenols in water samples, polyaniline-encapsulated carbon/copper composite nanofibers (PANI/CuCNFs) were prepared by an efficient and simple method. The PANI/CuCNFs were then pasted over the laccase (Lac) based biosensor for detecting the polyphenol pollutants. The morphological characteristics were studied through a number of techniques, including SEM, TEM, XRD, and FTIR. The electrochemical behavior of the prepared biosensor was studied through chronoamperometry, cyclic voltammetry (CV), electrochemical impedance spectroscopy (EIS). The analysis revealed that the Lac-based biosensor was high sensitivity (41.65 μAmM^{-1}), a more extensive linear range from 500 nM to 110 μM, a low-detection limit (0.24 μM). Hydroquinone detection was successfully approached through the synthesized biosensor leading to efficient detection of pollutants in river water (Fu et al. 2018).

In another work, a nanocomposite of ionic liquid 1-ethyl-3methylimidazolium bis (trifluoromethylsulfonyl) imide (EMIM][NTf_2], bismuth particles and carbon nanofibers modified carbon paste (BiPs-CNFs/[EMIM][NTf_2]/CPE) based sensor for the detection Pb^{+2} and Cd^{+2} M(II) in water samples. The sensor possed a high-surface area and decent electrical conductivity, and due to these characteristics, the fabricated sensor exhibited tremendous electroanalytical performances toward the determination of heavy metal ions. The LOD values for the Pb^{+2} and Cd^{+2} M(II) under optimal conditions were 0.12 and 0.25 $\mu g\ L^{-1}$, respectively. It was found that even in the presence of interfering ions, the sensor showed appreciable reproducibility and repeatability (Oularbi et al. 2020). Few more sensor studies were perfomed on CVD grown CNFs for determination of biomolecules (Shrivastava et al 2020).

2.5 Conclusion and Future Perspective

The problem of water pollution and wastewater treatment has become one of the major concerns of everybody and every section, including researchers, industrialists, policymakers, and the public, both at national and international levels. The proper treatment of wastewater before disposal from the industries has become a top priority all around the globe. The demand for cleaner and safer water resources is rising every day and is the biggest challenge the whole world is currently facing. The removal of pollutants that is toxic and sometimes can cause hazardous health issues has become one of the greatest demand for the researchers and industrialists. Adsorption is a well-established methodology for the removal of pollutants from wastewater. Also, electrochemical sensing is a boon technology in the field of wastewater treatment. For the measurement of the adsorption of the pollutants, various methodologies have been employed for such as kinetics and thermodynamic behavior of the adsorbent along with isotherm models. Also, for calculating the capacity of the sensor composite, various electrochemical methodologies are there, including cyclic voltammetry, differential pulse voltammetry, and other.

In recent times, a number of composites based on CNMs has been employed for the removal of the hazardous organic and inorganic pollutants. CNMs, including graphene and its derivative, CNTs and CNMs, fullerenes, nanodiamonds, have drawn extensive attention for the elimination of pollutants from the wastewater, due to their extraordinary surface and electronic properties. This book chapter has summarized the advancements in nanotechnology and carbon nanomaterials in the field of wastewater treatments. CNMs due to their adjustable physiochemical properties have been exploited smartly for trapping and eliminating a wide class of pollutants such as pesticides, PAHs, pharmaceutical wastes, heavy metals, and dyes. Modification in the CNMs that existed in the past had led to easier and feasible material availability, and modification to the current existing CNMs can turn them into a tycoon in the coming future. CNMs have proven their potency from the laboratory to the field, and thus, it can be said definitely that they will lead the race of wastewater treatment composites.

References

Ahmad M, Wang J, Xu J, Zhang Q, Zhang B (2020) Magnetic tubular carbon nanofibers as efficient Cu(II) ion adsorbent from wastewater. J Cleaner Prod 252:119825

Ain Q-U, Farooq MU, Jalees MI (2020) Application of magnetic graphene oxide for water purification: heavy metals removal and disinfection. J Water Proc Eng 33:101044

Alimohammady M, Jahangiri M, Kiani F, Tahermansouri H (2017) A new modified MWCNTs with 3-aminopyrazole as a nanoadsorbent for Cd(II) removal from aqueous solutions. J Environ Chem Eng 5:3405–3417

Anirudhan TS, Shainy F, Christa J (2017) Synthesis and characterization of polyacrylic acid- grafted-carboxylic graphene/titanium nanotube composite for the effective removal of enrofloxacin from aqueous solutions: adsorption and photocatalytic degradation studies. J Hazard Mater 324:117–130

Asfaram A, Ghaedi M, Dashtian K, Ghezelbash GR (2018) Preparation and characterization of Mn0.4Zn0.6Fe2O4 nanoparticles supported on dead cells of yarrowia lipolytica as a novel and efficient adsorbent/biosorbent composite for the removal of Azo food dyes: central composite design optimization study. ACS Sustain Chem Eng 6:4549–4563

Ashfaq M, Singh S, Sharma A, Verma N (2013) Cytotoxic evaluation of the hierarchical web of carbon micronanofibers. Ind Eng Chem Res 52:4672–4682

Atkovska K, Paunovik P, Dimitrov A, Lisichkov K, Alghuthaymi M, Grozdanov A (2020) 9–Graphene and activated graphene as adsorbents for removal of heavy metals from water resources. In: Abd-Elsalam KA (ed) Carbon nanomaterials for agri-food and environmental applications, Elsevier, pp 177–91

Atta NF, Galal A, El-Gohary ARM (2020) New insight for simultaneous determination of hazardous di-hydroxybenzene isomers at crown ether modified polymer/carbon nanotubes composite sensor. J Hazard Mater 388:122038

Baig N, Saleh TA (2018) Electrodes modified with 3D graphene composites: a review on methods for preparation, properties and sensing applications. Microchim Acta 185:283

Baig N, Ihsanullah, Sajid M, Saleh TA (2019) Graphene-based adsorbents for the removal of toxic organic pollutants: a review. J Environ Manag 244:370–82

Bairagi PK, Goyal A, Verma N (2019) Methyl nicotinate biomarker of tuberculosis voltammetrically detected on cobalt nanoparticle-dispersed reduced graphene oxide-based carbon film in blood. Sens Actuators B 297:126754

Bikshapathi M, Singh S, Bhaduri B, Mathur GN, Sharma A, Verma N (2012) Fe-nanoparticles dispersed carbon micro and nanofibers: surfactant-mediated preparation and application to the removal of gaseous VOCs. Colloids Surf Physicochem Eng Aspects 399:46–55

Buzea C, Pacheco II, Robbie K (2007) Nanomaterials and nanoparticles: sources and toxicity. Biointerphases. 2:MR17-MR71

Chen Y, Fan Z, Zhang Z, Niu W, Li C, Yang N et al (2018) Two-dimensional metal nanomaterials: synthesis, properties, and applications. Chem Rev 118:6409–6455

Che J, Çagin T, Goddard WA (2000) Thermal conductivity of carbon nanotubes. Nanotechnology 11:65–69

Cho KY, Seo HY, Yeom YS, Kumar P, Lee AS, Baek K-Y et al (2016) Stable 2D-structured supports incorporating ionic block copolymer-wrapped carbon nanotubes with graphene oxide toward compact decoration of metal nanoparticles and high-performance nano-catalysis. Carbon 105:340–352

Cho KY, Yeom YS, Seo HY, Kumar P, Lee AS, Baek K-Y et al (2017) Molybdenum-doped PdPt@Pt core-shell octahedra supported by ionic block copolymer-functionalized graphene as a highly active and durable oxygen reduction electrocatalyst. ACS Appl Mater Interfaces 9:1524–1535

Deb A, Vimala R (2018) Camptothecin loaded graphene oxide nanoparticle functionalized with polyethylene glycol and folic acid for anticancer drug delivery. J Drug Delivery Sci Technol 43:333–342

Dehghani MH, Yetilmezsoy K, Salari M, Heidarinejad Z, Yousefi M, Sillanpää M (2020) Adsorptive removal of cobalt (II) from aqueous solutions using multi-walled carbon nanotubes and γ-alumina as novel adsorbents: modelling and optimization based on response surface methodology and artificial neural network. J Mol Liq 299:112154

De Jong KP, Geus JW (2000) Carbon nanofibers: catalytic synthesis and applications. Catal Rev 42:481–510

Dil EA, Ghaedi M, Ghaedi AM, Asfaram A, Goudarzi A, Hajati S et al (2016) Modeling of quaternary dyes adsorption onto ZnO–NR–AC artificial neural network: analysis by derivative spectrophotometry. J Ind Eng Chem 34:186–197

Duman O, Özcan C, Gürkan Polat T, Tunç S (2019) Carbon nanotube-based magnetic and non-magnetic adsorbents for the high-efficiency removal of diquat dibromide herbicide from water: OMWCNT, OMWCNT-Fe3O4 and OMWCNT-κ-carrageenan-Fe3O4 nanocomposites. Environ Pollut 244:723–732

Farghali AA, Bahgat M, Enaiet Allah A, Khedr MH (2013) Adsorption of Pb(II) ions from aqueous solutions using copper oxide nanostructures. Beni-Suef Univ J Basic Appl Sci 2:61–71

Fu Y, An Q, Ni R, Zhang Y, Li Y, Ke H (2018) Preparation of polyaniline-encapsulated carbon/copper composite nanofibers for detection of polyphenol pollutant. Colloids Surf, A 559:289–296

Geim AK (2009) Graphene: status and prospects. Science 324:1530

Geim AK, Novoselov KS (2007) The rise of graphene. Nat Mater 6:183–191

Goh K, Karahan HE, Wei L, Bae T-H, Fane AG, Wang R et al (2016) Carbon nanomaterials for advancing separation membranes: a strategic perspective. Carbon 109:694–710

Gupta S, Yadav A, Singh S, Verma N (2017) Synthesis of silicon carbide-derived carbon as an electrode of a microbial fuel cell and an adsorbent of aqueous Cr(VI). Ind Eng Chem Res 56:1233–1244

Gusain R, Kumar N, Ray SS (2020) Recent advances in carbon nanomaterial-based adsorbents for water purification. Coord Chem Rev 405:213111

Hajati Y (2020) Tunable broadband multiresonance graphene terahertz sensor. Opt Mater 101:109725

Harikishore Kumar Reddy D (2017) Water pollution control technologies. In: Abraham MA (ed) Encyclopedia of sustainable technologies. Oxford, Elsevier, pp 3–22
He H, Pham-Huy LA, Dramou P, Xiao D, Zuo P, Pham-Huy C (2013) Carbon nanotubes: applications in pharmacy and medicine. BioMed Res Int 2013:578290
Huang D, Wu J, Wang L, Liu X, Meng J, Tang X et al (2019) Novel insight into adsorption and co-adsorption of heavy metal ions and an organic pollutant by magnetic graphene nanomaterials in water. Chem Eng J 358:1399–1409
Jeng K-T, Hsu N-Y, Chien C-C (2011) Synthesis and evaluation of carbon nanotube-supported RuSe catalyst for direct methanol fuel cell cathode. Int J Hydrogen Energy 36:3997–4006
Jun LY, Mubarak NM, Yee MJ, Yon LS, Bing CH, Khalid M et al (2018) An overview of functionalised carbon nanomaterial for organic pollutant removal. J Ind Eng Chem 67:175–186
Kumar P (2019) Ultrathin 2D Nanomaterials for electromagnetic interference shielding. Adv Mater Interfaces 6:1901454
Kumar P, Bohidar HB (2010) Aqueous dispersion stability of multi-carbon nanoparticles in anionic, cationic, neutral, bile salt and pulmonary surfactant solutions. Colloids Surf, A 361:13–24
Kumar A, Verma N (2020) Cu-Fe bimetal-carbon nanofiberous catalytic beads for enhanced oxidation of dichlorvos pesticide and simultaneous reduction of Cr(VI) in wet air. Catal Today 348:194–202
Kumar P, Karmakar S, Bohidar HB (2008) Anomalous self-aggregation of carbon nanoparticles in polar, nonpolar, and binary solvents. J Phys Chem C 112:15113–15121
Kumar P, Meena R, Paulraj R, Chanchal A, Verma AK, Bohidar HB (2012) Fluorescence behavior of non-functionalized carbon nanoparticles and their in vitro applications in imaging and cytotoxic analysis of cancer cells. Colloids Surf, B 91:34–40
Kumar P, Maiti UN, Lee KE, Kim SO (2014) Rheological properties of graphene oxide liquid crystal. Carbon 80:453–461
Kumar P, Shahzad F, Yu S, Hong SM, Kim Y-H, Koo CM (2015) Large-area reduced graphene oxide thin film with excellent thermal conductivity and electromagnetic interference shielding effectiveness. Carbon 94:494–500
Kumar P, Yu S, Shahzad F, Hong SM, Kim Y-H, Koo CM (2016) Ultrahigh electrically and thermally conductive self-aligned graphene/polymer composites using large-area reduced graphene oxides. Carbon 101:120–128
Kumar A, Omar RA, Verma N (2020) Efficient electro-oxidation of diclofenac persistent organic pollutant in wastewater using carbon film-supported Cu-rGO electrode. Chemosphere 248:126030
Li Z-F, Xin L, Yang F, Liu Y, Liu Y, Zhang H et al (2015) Hierarchical polybenzimidazole-grafted graphene hybrids as supports for Pt nanoparticle catalysts with excellent PEMFC performance. Nano Energy 16:281–292
Li Z, Wang L, Li Y, Feng Y, Feng W (2019a) Carbon-based functional nanomaterials: preparation, properties and applications. Compos Sci Technol 179:10–40
Li Y, Zhu Y, Wang C, He M, Lin Q (2019b) Selective detection of water pollutants using a differential aptamer-based graphene biosensor. Biosens Bioelectron 126:59–67
Mandeep, Gulati A, Kakkar R (2020) Graphene-based adsorbents for water remediation by removal of organic pollutants: Theoretical and experimental insights. Chem Eng Res Des 153:21–36
Miao J, Miyauchi M, Simmons TJ, Dordick JS, Linhardt RJ (2010) Electrospinning of nanomaterials and applications in electronic components and devices. J Nanosci Nanotechnol 10:5507–5519
Mishra S, Khare P, Singh S (2020) Catalytic chemical vapor deposition grown carbon nanofiber for bio-electro-chemical and energy applications. In: Mukhopadhyay A, Sen S, Basu DN, Mondal S (eds) Dynamics and control of energy systems. Springer Singapore, Singapore, pp 497–526
Modi A, Singh S, Verma N (2016) In situ nitrogen-doping of nickel nanoparticle-dispersed carbon nanofiber-based electrodes: its positive effects on the performance of a microbial fuel cell. Electrochim Acta 190:620–627
Novoselov KS, Geim AK, Morozov SV, Jiang D, Katsnelson MI, Grigorieva IV et al (2005) Two-dimensional gas of massless dirac fermions in graphene. Nature 438:197–200

Oularbi L, Turmine M, Salih FE, El Rhazi M (2020) Ionic liquid/carbon nanofibers/bismuth particles novel hybrid nanocomposite for voltammetric sensing of heavy metals. J Environ Chem Eng 8:103774

Pandey H, Khare P, Singh S, Singh SP (2020) Carbon nanomaterials integrated molecularly imprinted polymers for biological sample analysis: a critical review. Mater Chem Phys 239:121966

Pankaj A, Tewari K, Singh S, Singh SP (2018) Waste candle soot derived nitrogen doped carbon dots based fluorescent sensor probe: an efficient and inexpensive route to determine Hg(II) and Fe(III) from water. J Environ Chem Eng 6:5561–5569

Polizu S, Savadogo O, Poulin P, Yahia LH (2006) Applications of carbon nanotubes-based biomaterials in biomedical nanotechnology. J Nanosci Nanotechnol 6:1883–1904

Pophali A, Singh S, Verma N (2020) PAH simultaneous hydrogen generation and COD reduction in a photoanode-based microbial electrolysis cell. Int J Hydrogen Energy. https://doi.org/10.1016/j.ijhydene.2020.01.053

Qiu B, Xing M, Zhang J (2018) Recent advances in three-dimensional graphene based materials for catalysis applications. Chem Soc Rev 47:2165–2216

Ranđelović MS, Momčilović MZ, Milićević JS, Đurović-Pejčev RD, Mofarah SS, Sorrel CC (2019) Voltammetric sensor based on Pt nanoparticles suported MWCNT for determination of pesticide clomazone in water samples. J Taiwan Inst Chem Eng 105:115–123

Rauti R, Musto M, Bosi S, Prato M, Ballerini L (2019) Properties and behavior of carbon nanomaterials when interfacing neuronal cells: How far have we come? Carbon 143:430–446

Ren X, Zeng G, Tang L, Wang J, Wan J, Liu Y et al (2018) Sorption, transport and biodegradation—an insight into bioavailability of persistent organic pollutants in soil. Sci Total Environ 610–611:1154–1163

Robati D, Mirza B, Rajabi M, Moradi O, Tyagi I, Agarwal S et al (2016) Removal of hazardous dyes-BR 12 and methyl orange using graphene oxide as an adsorbent from aqueous phase. Chem Eng J 284:687–697

Salzano de Luna M, Wang Y, Zhai T, Verdolotti L, Buonocore GG, Lavorgna M et al (2019) Nanocomposite polymeric materials with 3D graphene-based architectures: from design strategies to tailored properties and potential applications. Prog Polymer Sci 89:213–249

Schweitzer L, Noblet J (2018) Chapter 3.6—water contamination and pollution. In: Török B, Dransfield T (eds) Green chemistry, Elsevier, pp 261–90

Shrivastava S, Bairagi PK, Verma N (2020) Spermine biomarker of cancerous cells voltammetrically detected on a poly (β-cyclodextrin)–electropolymerized carbon film dispersed with Cu–CNFs. Sens Actuators b: Chem 313:128055

Singh S, Verma N (2015a) Fabrication of Ni nanoparticles-dispersed carbon micro-nanofibers as the electrodes of a microbial fuel cell for bio-energy production. Int J Hydrogen Energy 40:1145–1153

Singh S, Verma N (2015b) Graphitic carbon micronanofibers asymmetrically dispersed with alumina-nickel nanoparticles: a novel electrode for mediatorless microbial fuel cells. Int J Hydrogen Energy 40:5928–5938

Singh S, Ashfaq M, Singh RK, Joshi HC, Srivastava A, Sharma A et al (2013) Preparation of surfactant-mediated silver and copper nanoparticles dispersed in hierarchical carbon micro-nanofibers for antibacterial applications. N Biotechnol 30:656–665

Singh S, Singh A, Bais VSS, Prakash B, Verma N (2014a) Multi-scale carbon micro/nanofibers-based adsorbents for protein immobilization. Mater Sci Eng, C 38:46–54

Singh S, Joshi HC, Srivastava A, Sharma A, Verma N (2014b) An efficient antibacterial multi-scale web of carbon fibers with asymmetrically dispersed Ag–Cu bimetal nanoparticles. Colloids Surf Physicochem Eng Aspects 443:311–319

Singh S, Modi A, Verma N (2016) Enhanced power generation using a novel polymer-coated nanoparticles dispersed-carbon micro-nanofibers-based air-cathode in a membrane-less single chamber microbial fuel cell. Int J Hydrogen Energy 41:1237–1247

Singh S, Bairagi PK, Verma N (2018b) Candle soot-derived carbon nanoparticles: an inexpensive and efficient electrode for microbial fuel cells. Electrochim Acta 264:119–127

Singh S, Srivastava A, Singh SP (2018a) Inexpensive, effective novel activated carbon fibers for sample cleanup: application to multipesticide residue analysis in food commodities using a QuEChERS method. Anal Bioanal Chem 410:2241–2251

Singh S, Singh D, Singh SP, Pandey AK (2019a) Candle soot derived carbon nanoparticles: Assessment of physico-chemical properties, cytotoxicity and genotoxicity. Chemosphere 214:130–135

Singh S, Pankaj A, Mishra S, Tewari K, Pratap Singh S (2019b) Cerium oxide-catalyzed chemical vapor deposition grown carbon nanofibers for electrochemical detection of Pb(II) and Cu(II). J Environ Chem Eng 7:103250

Singh S, Noori MT, Verma N (2020a) Efficient bio-electroreduction of CO2 to formate on a iron phthalocyanine-dispersed CDC in microbial electrolysis system. Electrochim Acta 338:135887

Singh S, Tewari K, Pant D (2020b) An overview of carbon and nanoparticles application in bioelectrochemical system for energy production and resource recovery. Microb Electrochem Technol 207

Singh S, Yin J, Lun F, Mondal DP, Kim D, Lee S-H (2020bc) Significance of modification of slurry infiltration process for the precursor impregnation and pyrolysis process of SiCf/SiC composites. J Eur Ceram Soc 40:2245–2251

Srivastava A, Singh M, Singh S, Singh SP (2020a) Recent advances in micro-extraction based analytical approaches for pesticides analysis in environmental samples. In: Gupta T, Singh SP, Rajput P, Agarwal AK (eds) Measurement, analysis and remediation of environmental pollutants. Springer Singapore, Singapore, pp 281–318

Srivastava M, Srivastava A, Pandey SK (2020b) Suitability of graphene monolayer as sensor for carcinogenic heavy metals in water: a DFT investigation. Appl Surf Sci 517:146021

Stoller MD, Park S, Zhu Y, An J, Ruoff RS (2008) Graphene-based ultracapacitors. Nano Lett 8:3498–3502

Su H, Ye Z, Hmidi N (2017) High-performance iron oxide–graphene oxide nanocomposite adsorbents for arsenic removal. Colloids Surf, A 522:161–172

Su T, Shao Q, Qin Z, Guo Z, Wu Z (2018) Role of interfaces in two-dimensional photocatalyst for water splitting. ACS Catal 8:2253–2276

Taher FA, Kamal FH, Badawy NA, Shrshr AE (2018) Hierarchical magnetic/chitosan/graphene oxide 3D nanostructure as highly effective adsorbent. Mater Res Bull 97:361–368

Thamer BM, Aldalbahi A, Moydeen AM, Al-Enizi AM, El-Hamshary H, Singh M et al (2019) Alkali-activated electrospun carbon nanofibers as an efficient bifunctional adsorbent for cationic and anionic dyes. Colloid Surf a: Physicochem Eng Aspects 582:123835

Vicente-Martínez Y, Caravaca M, Soto-Meca A, De Francisco-Ortiz O, Gimeno F (2020) Graphene oxide and graphene oxide functionalized with silver nanoparticles as adsorbents of phosphates in waters. A comparative study. Sci Total Environ 709:136111

Wang Z, Lin F, Huang L, Chang Z, Yang B, Liu S et al (2019) Cyclodextrin functionalized 3D-graphene for the removal of Cr(VI) with the easy and rapid separation strategy. Environ Pollut 254:112854

Wanjeri VWO, Sheppard CJ, Prinsloo ARE, Ngila JC, Ndungu PG (2018) Isotherm and kinetic investigations on the adsorption of organophosphorus pesticides on graphene oxide based silica coated magnetic nanoparticles functionalized with 2-phenylethylamine. J Environ Chem Eng 6:1333–1346

Xie F, Yang M, Jiang M, Huang X-J, Liu W-Q, Xie P-H (2019) Carbon-based nanomaterials—a promising electrochemical sensor toward persistent toxic substance. TrAC Trends Anal Chem 119:115624

Yan H, Tao X, Yang Z, Li K, Yang H, Li A et al (2014) Effects of the oxidation degree of graphene oxide on the adsorption of methylene blue. J Hazard Mater 268:191–198

Yuan C, Wu HB, Xie Y, Lou XW (2014) Mixed transition-metal oxides: design, synthesis, and energy-related applications. Angew Chem Int Ed 53:1488–1504

Zhang Y, Ou H, Liu H, Ke Y, Zhang W, Liao G et al (2018) Polyimide-based carbon nanofibers: a versatile adsorbent for highly efficient removals of chlorophenols, dyes and antibiotics. Colloids Surf, A 537:92–101

Zhang S, Song X, Liu S, Sun F, Liu G, Tan Z (2019) Template-assisted synthesized MoS2/polyaniline hollow microsphere electrode for high performance supercapacitors. Electrochim Acta 312:1–10

Zhu X, Wu G, Lu N, Yuan X, Li B (2017) A miniaturized electrochemical toxicity biosensor based on graphene oxide quantum dots/carboxylated carbon nanotubes for assessment of priority pollutants. J Hazard Mater 324:272–280

Chapter 3
Titanium Oxide Composites with Graphene and Laser-Induced Graphene for the Environmental Applications

Ashish Kumar, Simant Kumar Srivastav, Kamlesh Kumar Singh, and Swatantra P. Singh

Abstract Carbon-based nanomaterials have strongly impacted the field of nanotechnology due to its physical, electronic, and chemical properties. Amongst various carbon-based nanomaterials, graphene-based nanocomposite materials are gaining the attention of researchers globally as novel materials for biomedical, energy, electronic, and environmental applications. Graphene has unique physico-chemical properties, especially its exceptionally high-specific surface area, mechanical strength, electron mobility, and thermal conductivity. However, conventional methods such as micromechanical exfoliation, Hummers method, CVD etc. for graphene synthesis are tedious and multi-step processes, particularly when we are fabricating surfaces such as electrodes. Laser-induced graphene (LIG) is recently developed single-step facile method incorporates direct laser print graphene on any carbonaceous material by using 10.6 μm CO_2 infrared laser. However, graphene itself contains low-catalytic properties which generate demand for heteroatom doping, for instance, titanium oxide TiO_x (i.e. TiO_2 as a photocatalyst and Magnéli phases Ti_nO_{n-1} as an electrocatalyst) are considered as a noble catalyst for environmental application whose catalysis reaction results in some environmental friendly by-products such as CO_2 and H_2O in most of the cases. The incorporation of TiO_x with graphene enhances the catalysis reaction as this composite cause the working of TiO_2 more efficiently in solar light and provides a free pathway for electron movement enhancing the electrocatalytic property of Magnéli phase-graphene composite. The current chapter includes the basic introduction of graphene as carbon-based nanomaterials, advantages, and disadvantages of its synthesis by conventional methods and the latest method by

A. Kumar · S. P. Singh (✉)
Department of Environmental Science and Engineering (ESED), Indian Institute of Technology Bombay, Maharashtra 400076, India
e-mail: swatantra@iitb.ac.in

S. K. Srivastav
University Department of Chemistry, L. N. Mithila University Darbhanga, Bihar 846006, India

K. K. Singh
Department of Chemistry, Institute of Science, Banaras Hindu University, Varanasi 221005, India

S. P. Singh et al. (eds.), *Nanomaterials and Nanocomposites for Environmental Remediation*, Energy, Environment, and Sustainability,
https://doi.org/10.1007/978-981-16-3256-3_3

laser. Additionally, this chapter provides an insight of TiO_x doping in graphene and its effect on electrochemical and photochemical catalytic performances.

Keywords Carbon-based nanomaterials · Graphene · Laser-induced graphene · Infrared laser · Photothermal conversion · Magnéli phases · Organic dye degradation · Antimicrobial · Electrocatalyst

3.1 Introduction

Growing world's population, intensified development in industrial and agricultural activities; environmental issues such as contamination of air, soil and aquatic ecosystem and climate change are nowadays becoming a prime concern all over the globe. These issues have acquired scientific attention to study human impact on environment and develops the need of new mitigation measures and technologies. Nanotechnology has triggered augmented interest due to its unique physical and chemical properties (high surface to volume ratio, higher reactivity, etc.) for multiple environmental applications, such as energy production, water treatment and contaminant sensing (Perreault et al. 2015a, 2015b). Also, nanomaterials in various fashion can be used to promote environmental sustainability by using natural energy source such as solar light for TiO_2 photocatalysis and resulting green by-products (Zhang et al. 2010). Synthesis, design, and time-wise modification of nanomaterials lead to the enhancement of their application as well as widening their range to a different sector of the environment-related domain (Perreault et al. 2015a, 2015b). These wide varieties of application can be as semiconductor catalyst (CT) for the degradation of aromatic pollutants such as phenol or catechol (Dong et al. 2014); photocatalysts for the degradation of many hazardous pollutants including organic dyes from wastewater (Leary and Westwood 2011); metal based nanoparticles for degradation or organic contaminants (Singh et al. 2013; Singh and Bose 2015, 2016, 2017); absorbent such as graphene for many harmful contaminants from the water (Al-Khateeb et al. 2014); as an antibacterial and antifouling surfaces (Singh et al. 2018a, 2018b; Gupta et al. 2019); can act as a substrate for any other catalyst (Dong et al. 2014); having some novel semiconductor property, it can be made into a sensor device to sense harmful pollutants out of our edibles and drinkable items (Perreault et al. 2015a, 2015b; Thamaraiselvan et al. 2020) etc.

Graphene, a carbon-based nanomaterial (CBNs) having two-dimensional layer (single atom thickness) of sp^2 hybridised carbon atom arranged in a hexagonal crystalline structure (Zhao et al. 2012) as shown in Fig. 3.1. Graphene has got a lot of attention due to its excellent physico-chemical properties, for instance, the extremely high-surface area, high electron and thermal mobility, and exceptionally high-mechanical strength (Zhao et al. 2012). These extraordinary properties of graphene triggered a huge interest in energy storage and environmental pollution remediation application (Perreault et al. 2015a, 2015b). Graphene and graphene-based nanocomposites are being used to develop new sorbents, electrocatalytic or photocatalytic materials for

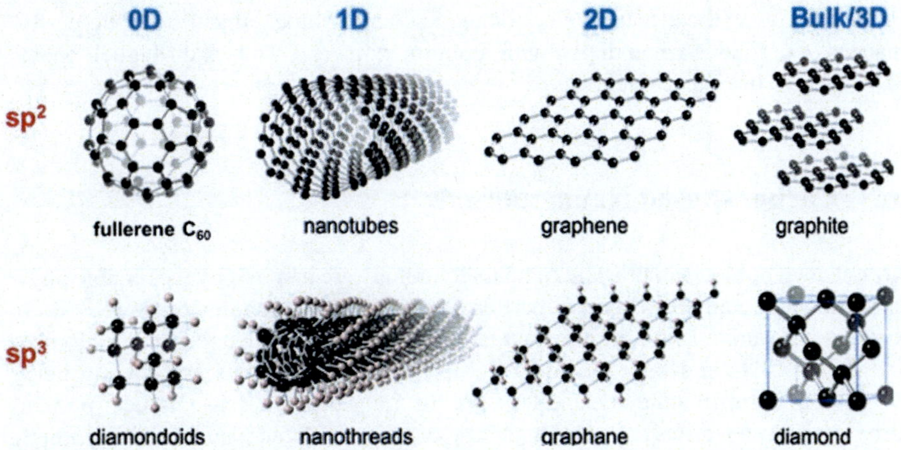

Fig. 3.1 Hybridization state of carbon-based nanomaterials (Mauter and Elimelech 2008). Copyright © 2008, American Chemical Society

the environment decontamination such as bacterial inactivation, organic dye degradation, prevention from biofouling, etc. (Yin et al. 2017; Wu et al. 2013). Graphene could be a building blocks for the next generation membranes for water treatment and as electrode materials for contaminants surveillance and removal (Perreault et al. 2015a, 2015b). Nonetheless, for practical applications of graphene and graphene-based materials, there should be a trade-off between quantity and quality of graphene. However, conventional methods of graphene synthesis such as chemical exfoliation of graphite by Hummers, Staudemaier and Brodie methods, chemical vapour deposition (CVD), epitaxial growth, etc. are very tedious, costly (Singh et al. 2017) and produces graphene with a high density of defects which results in a considerable reduction of exceptional properties of graphene and also hinder its scalable production.

Recently, a one-step facile and scalable approach to form and pattern 3D porous graphene on a carbonaceous surface is developed by using a commercial CO_2 infrared laser scriber (Lin et al. 2014). The graphene by this one-step process is called laser-induced graphene (LIG), which proved to be an advantage over conventionally synthesised graphene surfaces (Lin et al. 2014). The properties and application of LIG and LIG based nanocomposites and its comparison with conventionally available graphene, i.e. graphene oxide (GO) and reduced graphene oxide (rGO) are discussed in this chapter. Also, despite having such noble properties, graphene alone itself does not pose good catalytic properties which can be used for contaminants degradation (Huang et al. 2013). Hence, graphene most of the times incorporated with a different heteroatom, which can greatly enhance its electrochemical and photochemical properties. The focus of this chapter is to provide basic information of graphene, advantages, and disadvantages of its synthesis by conventional methods and the latest method by laser, the effect on electrochemical and photochemical catalytic performances of graphene when doped with titanium oxide (i.e. TiO_2 and Magnéli phases Ti_nO_{n-1}) also been discussed.

3.2 Carbon-Based Nanomaterials

Recent developments in the field of nanotechnology are inspired by the tunable physical, chemical, and electronic properties of carbon-based nanomaterials (CBNs) to develop innovative solution to persistent environmental challenges. In the biomedical field, CBNs and their functionally derive nanocomposites are already being explored to optimize "the fate and transport of drugs through dense tissues" particularly targeted cancerous cells and make use of functionalized nanotubes as synthetic transmembrane pore (Mauter and Elimelech 2008). Analogously, some environmental application including engineered removal of hazardous pollutants and novel water filtration membranes structure can be employed (Smith and Rodrigues 2015; Chowdhury et al. 2018). Carbon-based nanotechnologies have leading application in the fields of sorption, depth filtration, environmental sensing devices, high-flux membranes separation, pathogens control, renewable energy production, etc. (Brady-Estévez et al. 2008; Weber et al. 1991; McGinnis et al. 2007).

Carbon's structural conformation and its hybridisation state strongly influence the physical, chemical, and electronic properties of carbon-based nanomaterials. Ground-state's orbital configuration of carbon's electrons is $1s^2$, $2s^2$, $2p^2$. The fine energy gap between the 2 s and 2p electron layers facilitates the endorsement of one's orbital electron to its higher energy orbital p, which is empty in the ground state. According to the bonding relationships with neighbouring atoms of a carbon atom, carbon shows sp, sp^2, or sp^3 hybridization. Also, the energy obtained from the covalent bond with neighbouring atoms fulfils the requirement for the higher energy state of this electronic configuration (Shenderova et al. 2002). This compensated energy is almost equal for sp^2 and sp^3 hybridization states after taking into account the off-plane binding due to the π-bonds between unhybridized p orbitals (Mauter and Elimelech 2008). Figure 3.1 shows different hybridizations of CBNs.

3.2.1 Unique Properties of Carbon-Based Nanomaterials

Properties of carbonaceous nanomaterials (both single-molecule and bulk properties) are broadly classified into three classes which are usually cited in environmental application as size, shape and surface area; molecular interaction and sorption properties; and electronic, optical and thermal properties.

3.2.1.1 Size, Shape and Surface Area

For carbonaceous nanomaterials, molecular manipulation has a greater influence over the structure and conformity, including size, length, chirality, and the number of layers in the material. Variation in synthesis technique of these CBNs, temperature, catalyst, pressure, electron field and process gases can alter the structure, purity, and physical orientation of nanomaterials for specific applications (Bandow et al. 1998). For example, the diameter of single-walled nanotubes (SWNT) is strongly linked to synthesis technique, with high-pressure carbon monoxide (HiPCO) synthesis capable of producing nanotubes between 0.7 and 1 nm. Alternatively, laser ablation and other graphitic based methods yield slightly larger nanotubes nearly of the range of 1–2 nm (Sun et al. 2002). The size, shape, and surface area of carbon nanomaterials are highly reliant on aggregation and solvent chemistry. Additionally, impurities such as vapour, biomolecules, and metals that adsorb over the surface of nanomaterials may basically alter the aggregation behaviour, electric and thermal characteristics, mechanical strength, and physico-chemical properties of the nanomaterials (Chowdhury et al. 2018).

3.2.1.2 Molecular Interaction and Sorption Properties

Explanation of molecular interaction and sorption properties for CBNs is a combined effort of many theories and experiments (Mauter and Elimelech 2008). CBNs are normally consistent with traditional physico-chemical models and theories such as electrostatics (Keblinski et al. 2002), hydrophobicity (Furmaniak et al. 2006), adsorption (Walther et al. 2004) and solubility parameters (Ham et al. 2005). It was found by a study that both hydrophobicity and capillarity of CBNs contribute to the adsorption behaviour as well as the orientation of sorbates in microporous carbon. For environmental purposes, adsorption capacity plays an important role in contamination removal and hydrogen storage (Mauter and Elimelech 2008).

3.2.1.3 Electronic, Optical and Thermal Properties

The bonding configuration of different CBNs like fullerenes and nanotubes offers unique electronic properties to it. These properties can contribute to the field of the environmental sensing device, power generation, photo-oxidation of organics, electrocatalysis etc. (Franklin 2015; Kamat et al. 2004). Materials like fullerenes have strong absorption band for both UV and visible ranges, which are attributed to the electronic transition from HOMO to LUMO orbital (i.e. from highest occupied molecular orbital to lowest unoccupied molecular orbital) (Mauter and Elimelech 2008). The radiation emission produced by these substances are capable of produces a variety of reactive oxygen species-like nascent oxygen, superoxide radical anions and hydroxyl radicals in aqueous media, which tends to oxidize many environmental contaminants (Lee et al. 2007).

Unique hybridization properties of carbon and sensitivity of carbon structure to any change in synthesis condition allow for any tailored manipulation to the degree that is not yet been shown with the inorganic nanostructures. In conclusion, environmental application of CBNs is both proactive and retroactive (Mauter and Elimelech 2008), meaning that it can works ranges from preventive measures for environmental contaminants degradation, optimizing energy efficiency, etc. to the remedial measures like pollutants transformation, wastewater recycling, etc.).

3.3 Graphene

Graphene is a two-dimensional (single-atom thickness) carbon-based nanomaterials having sp^2 hybridized carbon atoms arranged in a hexagonal manner. It is found to be a building block of graphite, where multiple graphene sheets are attached by π-staking (out-of-plane bond) to hold the graphite lamellar strongly in place having an interlayer spacing of 3.34Å (Park et al. 2011) between the graphene sheets. Graphene has some unique properties such as exceptionally high-specific surface area (theoretically 2630 m^2g^{-1}) (Rao et al. 2004); high-electrical conductivity (2×10^5 cm^2V^{-1} s^{-1}) (Zhu et al. 2010); high-thermal conductivity (~5000 Wm^{-1} K^{-1}) (Zhao et al. 2012); and very high-mechanical strength (Young's modulus ~ 1.1 TPa) (Perreault et al. 2015a, 2015b). Graphene was initially demonstrated in a study done by Geim and Novoselov (2007) where, micromechanical exfoliation of graphite was performed by just performing a sequential cleavage of graphite to graphene using adhesive as can be clearly shown in Fig. 3.2.

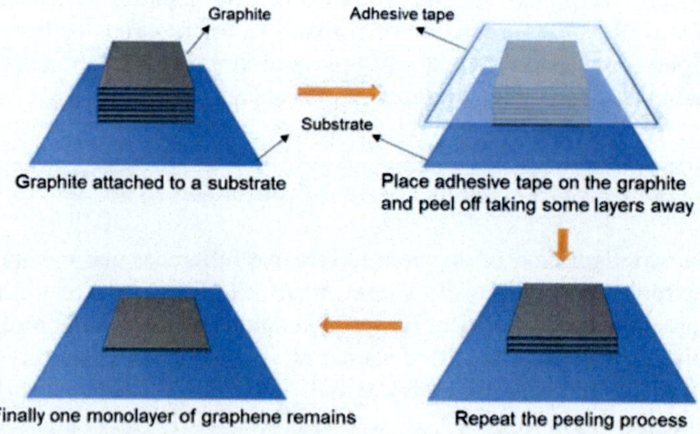

Fig. 3.2 Micromechanical exfoliation of graphite to graphene by adhesive tape. Adopted and modified from Dasari et al. (2017)

3.3.1 *Conventional Synthesis Methods of Graphene*

Micromechanical exfoliation is a labour-intensive process and cannot be used for scale-up production of graphene (Zhao et al. 2012). However, to get the various application of graphene to remain intact, there is a need to synthesise graphene-based nanomaterials with controlled size, morphology, and thickness. So, the development of effective synthesis approach is of prime importance. Figure 3.3 shows two-conventional strategies generally followed to synthesize graphene, i.e. top-down approaches, and bottom-up approaches (Perreault et al. 2015a, 2015b; Zhao et al. 2012).

3.3.1.1 Top-Down Approach

Top-down approaches involve chemical exfoliation followed by ultrasonication of graphitic material to graphene by using some organic solvents. Graphitic materials mostly used for this purpose are 0D fullerenes, 3D graphite, and 1D nanotubes which act as a carbon source for graphene synthesis (Zhao et al. 2012). The most common top-down approaches used is by the mechanical exfoliation by adhesive tape as already discussed above; and oxidative or chemical exfoliation of graphite which can be achieved by famous three methods named as Hummers method, Staudemaier method and Brodie's methods which differs only in term of chemicals used for oxidation of graphite block as shown in Table 3.1 (Zhao et al. 2012; Li et al. 2015; Zhu et al. 2010).

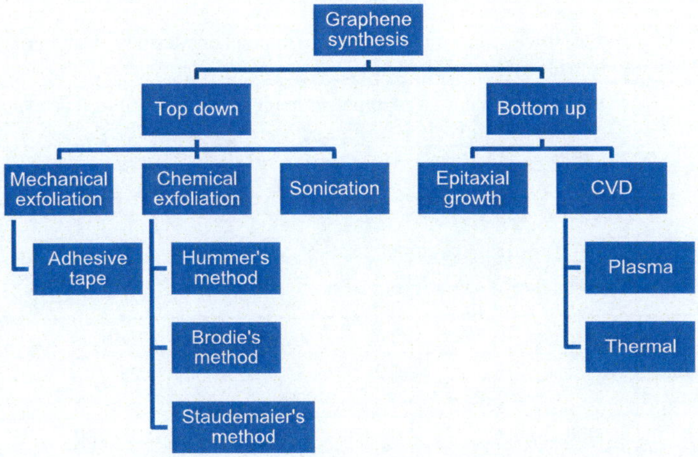

Fig. 3.3 Conventional strategies generally followed to synthesize graphene, i.e. top-down approaches, and bottom-up approaches. Adopted and modified from Zhao et al. (2012)

Table 3.1 Different methods for chemical exfoliation. Aadopted from Zhao et al. 2012; Li et al. 2015; Zhu et al. 2010

Methods for chemical exfoliation	Chemicals used for oxidation
Hummer's method	Conc. H_2SO_4, $KMnO_4$ and $NaNO_3$
Brodie's method	Conc. H_2SO_4, $KClO_3$ and Conc. nitric acid
Staudenmaier's method	Conc. H_2SO_4, $KMnO_4$ and fuming nitric acid
Modified Hummer's method	Conc. H_2SO_4, $KMnO_4$ and H_3PO_4

Chemical exfoliation involves two steps; First to reduce interlayer van der Waals forces in graphite, which can be achieved through oxidation of graphite by strong oxidising agents such as $KMnO_4$ and $NaNO_3$ in H_2SO_4 used in Hummer's method (Li et al. 2015). This will increase the interlayer distance to nearly 9.5Å (Bhuyan et al. 2016). This step of oxidising graphitic material yields in graphite oxide, which is then, followed by sonication resulting in the exfoliation of single or multilayer graphene called graphene oxide (GO). GO have numerous oxygen functional group in its structure, making it more hydrophilic (Lee et al. 2013). However, this oxygen functional group is treated as defects in the structure of GO (Lee et al. 2013). These defects can be removed by reduction of GO through chemical reducing agents, photo-reduction, thermal annealing, or hydrogen plasma treatment yield another modified graphene which is known to be as reduced graphene oxide (rGO). Table 3.2 shows properties comparison of pristine graphene GO and rGO.

Table 3.2 Comparison of structure and major properties of graphene-based nanomaterials (adopted and 2015a, 2015b modified from Perreault et al.)

Properties	Pristine graphene	Graphene oxide	Reduced graphene oxide
	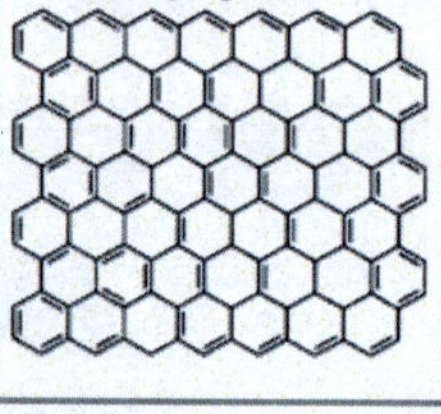	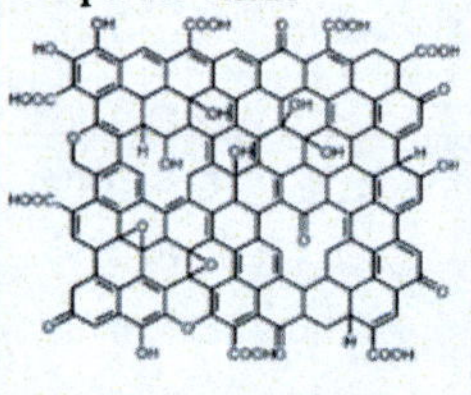	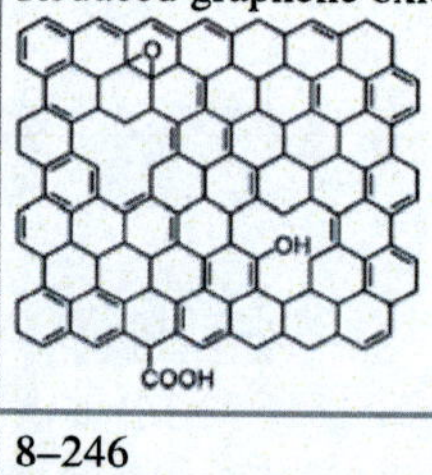
C:O ratio	No oxygen	2–4	8–246
Young's modulus (TPa)	1.1	0.2	0.25
Electron mobility ($cm^2v^{-1}\ s^{-1}$)	10,000–50,000	insulator	0.05–200
Production cost	High	Low	Low
Defects present	Negligible	Exceptionally large in number	Comparatively less

Many other methods are also available which comes under top-down approach such as ultrasonication of graphitic material to GO; exfoliation of graphitic oxide by laser excitation to get rGO; unzipping of carbon nanotubes (CNTs) by using an acid reaction to synthesise carbon nanoribbons, plasma treatment and other catalytic approaches, etc. (Perreault et al. 2015a, 2015b). However, methods involved in top-down approach have various disadvantages for instance involvement of hazardous and toxic chemicals (i.e. potassium permanganate, sulfuric acid and $NaNO_3$) in the synthesis process; the as-prepared graphene structure contains a large number of defects; which limits the use of graphene in many fields (Zhao et al. 2012).

3.3.1.2 Bottom-Up Approaches

Unlike the top-down approach, which starts from graphitic material to produce graphene, the bottom-up approach starts from a small organic molecular level to form graphene. In the bottom-up approach, deposition of hydrocarbons to form graphene or graphitic materials can be a catalyst by metal surfaces through epitaxial growth or chemical vapour deposition:

Chemical Vapour Deposition (CVD)

In CVD methods, the precursor hydrocarbon gases such as methane, ethylene, acetylene, and benzene pass over a heated transition metal substrate in a close chamber. Metal surface involved can be Cu, Pd, Ru, Ir, etc. (Zhao et al. 2012). The diffused hydrocarbon gas adsorbs on the metal surface, after which metal helps separation of carbon and hydrogen from the methane, followed by the desorption of by-product from the surface, as a result of which carbon starts building up on the metal substrate as a hexagonal graphene layer (Ago 2015). Lastly, the etching of metal occurs leaving a high-quality graphene sheet. Figure 3.4 shows CVD with methane as a precursor, Cu as a metal surface at the furnace temp of 900–1080 °C (Gu et al. 2014).

Epitaxial Growth

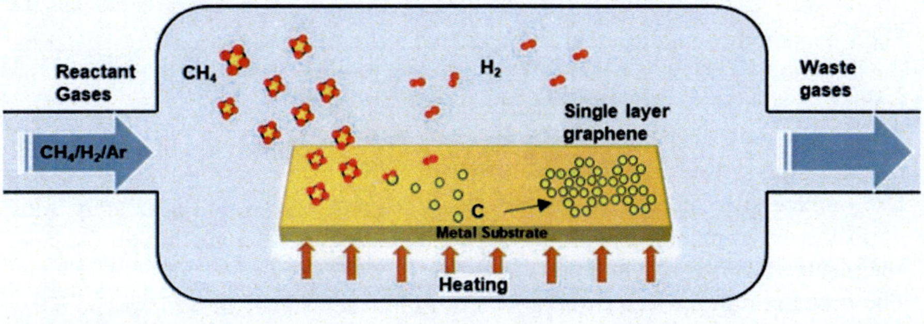

Fig. 3.4 Production of graphene by CVD on the metal surface with methane as a precursor. Adopted and modified from Ago (2015)

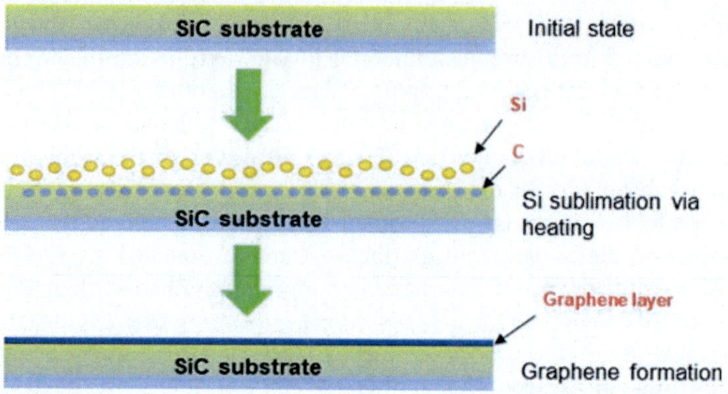

Fig. 3.5 Epitaxial deposition of SiC wafer occur on SiC substrate. Adopted and modified from Soler and Manuel (2014)

This method employs deposition of a nano-crystal film on a nano-crystal substrate, and this film is known as the epitaxial layer. Epitaxy comes from the Greek word, which is a combination of two words, *Epi* means "above", and *taxis* means deposition in an ordered manner (Bhuyan et al. 2016). In this method, epitaxial deposition of SiC wafer occurs on SiC substrate, followed by the desorption of Si atoms from the surface resulting in the formation of graphene on SiC surface as shown in Fig. 3.5 (Soler and Manuel 2014).

Bottom-up approach has an advantage over the top-down approach as high-quality graphene can be produced by this approach at an industrial level.

3.3.2 Laser-Induced Graphene

Despite having the tremendous application of graphene, their conventional methods of synthesis have many limitations as listed below:

- High-temperature conditions are required, which damages the structure by altering the morphology of the substance. Due to this defect are produced in the structure leading to changes in unique properties of graphene.
- Top-down approaches are multi-stepped chemical synthesis process which requires high skilled supervision. (Lin et al. 2014)
- Use of excessive and hazardous chemicals are required, which ultimately adds up in the environment generating extra cost expenses for remediation of these chemicals. (Rathinam et al. 2017)
- The processes are proven to be good for graphene at a small-scale preparation but lack proficiency in large-scale production due to change in morphology and lesser bulk availability.

- Production of 3D porous graphene by CVD can be achieved, but high temperature, continuous dying and etching can hinder its scalable production. (Han et al. 2018)

Due to all these limitations in conventional methods for graphene synthesis, laser-induced graphene (LIG) provides a facile one-step approach to directly print graphene on any carbonaceous substrate by using easily available CO_2 infrared laser without involving any reagent in the LIG formation. The formation of the graphene layer by using laser can be well-understood by knowing the basic functioning of the laser and the burning processes.

3.3.2.1 Laser

The term "laser" itself is an acronym for light amplification by stimulated emission of radiation. A laser is a device that emits intense beams of monochromatic and coherent light which is highly directional in nature. This beam is most often having a very pure wavelength and frequency when compared with the other sources of light (Allmen and Blatter 1995). A laser beam typically has negligible divergence, which makes it able to travel over a great distance maintaining high intensity (Julsgaard et al. 2007). Its beam can be focussed on a very tiny spot having brightness that can even exceed the brightness of the sun (Strickland et al. 2001). With these properties, the laser is used for an overly broad range of application, for instance, it is being used for drilling holes in metal and can be used for medical purpose-like laser treatment of eye retina (Allmen and Blatter 1995). Apart from these, the laser has a lot of medical, industrial, scientific, and military applications. Figure 3.6 shows a schematic of basic laser assembly having fully and partly reflecting mirror arranged at the two ends of chamber serving as a laser medium that can be gas, liquid, solid, etc.,

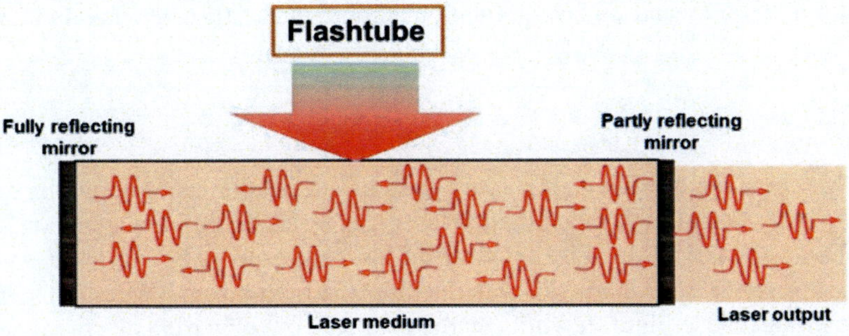

Fig. 3.6 Schematic diagram of basic laser. Adopted and modified from Drever et al. (1983)

Laser Burning

Laser mechanism depends on various factors like the wavelength of the laser used for the purpose, laser output power, spot size of the beam and the properties of burning material (Julsgaard et al. 2007). The mechanism behind laser burning is basically to displace atoms of burning materials permanently from its former position (Dubey and Yadava 2008). The laser shoots photons which when hit the atom of burning material, their electrons get excited. If these electrons get excited enough, they will have enough energy to dissociate from the parent atoms (Sellin et al. 2001). This makes individual atoms disassociate with whatever atom they were bonded to. Some atoms, like the one directly hit by the laser beam, goes to vapour and float away. All this atom displaced wanted to go in lower energy state, they likely go outside the range of the spot of laser resulting burning of the material. Wavelength has a very important role in laser burning as the burning ability is directly related to the energy of the laser beam which is inversely proportional to the wavelength of light as can see by equation $E = \frac{hc}{\lambda}$ (Julsgaard et al. 2007).

If one does a series of laser burning operation on a material creating a line of overlapping holes, it can be thought of as laser cutting. Laser cutting/burning is basically done with the help of a jet of air, oxygen, or dry nitrogen. Purpose of this jet is different for metal and non-metal surfaces, like for metal surface, laser beam heats the surface sufficient enough so as to cause the oxygen from the jet passes over it start burning and combustion process causing cutting/burning the surface can be started (Sellin et al. 2001). On the other hand, for non-metal surfaces, the role of the jet is only to blow away the debris from the cutting zone to improve the quality of the cut (Sellin et al. 2001). Carbon dioxide laser is most used for the purpose of laser cutting/burning applications. Solid woods, MDF and plywood, acrylic, cardboard, paper/cardstock, foam (Polymer types), cork, organic fabric/polyester, rubber, thin metals like stainless steel, etc. are some materials which are compatible with CO_2 laser cutting (Dubey and Yadava 2008; Snakenborg et al. 2004; Davim et al. 2008).

CO_2 Gas Laser

Lasers are broadly classified into four types based on the lasing medium during its lasing mechanism such as solid-state laser, gas laser, liquid laser, semiconductor laser. However, fabrication of LIG is done by gas laser, especially CO_2 laser in which an electric current is discharged inside the laser medium through a gas (CO_2 in our case) to produce laser light. It works on the principle of converting electrical energy into light energy. In gas lasers, the laser medium is a gaseous state which is made up of a mixture of gases. This mixture is packed up into a glass tube which further acts as an active medium or laser medium. Apart from a CO_2 laser, gas lasers are of different types notably Helium (He)–Neon (Ne) lasers, carbon monoxide lasers (CO lasers), argon ion lasers, excimer lasers, hydrogen lasers, nitrogen lasers, etc. The type of gas used to construct the laser medium can determine the laser wavelength or efficiency (Aldrich 2019).

3.3.2.2 Formation of Laser-Induced Graphene (LIG)

Initially, LIG was formed by using commercially available polyimide (PI) sheet as a precursor in which laser scribing was done by using 10.6 μm CO_2 infrared laser under ambient condition (Lin et al. 2014). LIG formation can be well-understood by a study done by Vashisth et al. (2020), where the evolution of LIG from five different polymers substrates (polybenzimidazole (PBM), polyether ether ketone (PEEK), poly (ether imide) (PEI), polycarbonate (PC), and polyimide (Kapton) were examined using ReaxFF (reactive force field) reactive molecular dynamics simulation. Their investigation reveals that the LIG formation entails the intermediate formation of an amorphous structure during the initial 0.2 ns of the simulations that eventually converted into an ordered graphitic structure with 5-,6-,7- membered rings. Additionally, the study also reveals that CO and H_2 molecules constitutes a key portion of gases evolved throughout LIG formation with CO formed quickly during the initial conversion into amorphous structure, whereas H_2 evolves gradually throughout the process of LIG evolving into a more ordered structure.

However, recent studies show LIG formation on different carbon precursor such as cloth, paper, wood, cork coconut shells, and even on edible items like bread and potato (Chyan et al. 2018). Figure 3.7 shows LIG on different carbon precursors such as coconut shell and bread. LIG formation was reported to be better for lignocelluloses carbon precursor by just controlling the laser setting and atmosphere of the system (Chyan et al. 2018). For LIG formation on the wood require inert atmosphere due to lower lignin content of the wood. On the other hand LIG formation on other carbon precursor having higher lignin content such as coconut shell, bread, paper, cloth and potato, multiple lasing using multiple passes or defocus methods is only require (Chyan et al. 2018). The mechanism suggested behind this conversion to LIG involves the initial conversion of carbon precursor to amorphous carbon, followed by final conversion to graphene because of subsequent lasing. LIG made in ambient

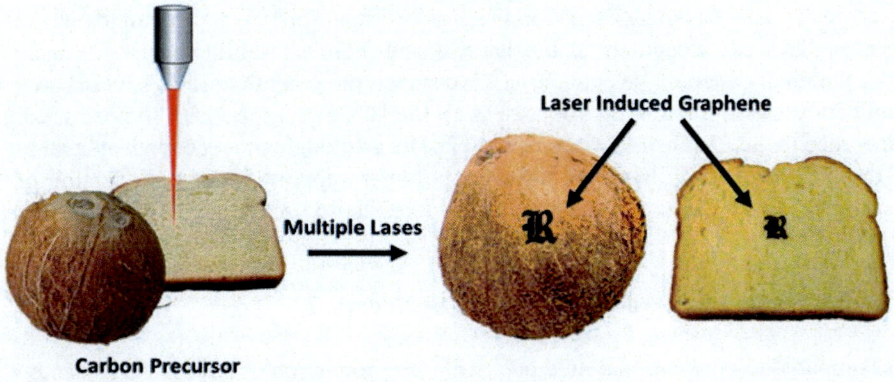

Fig. 3.7 LIG formation on carbon precursor say coconut shell and bread due to multiple lasing (Chyan et al. 2018). Copyright © 2018, American Chemical Society

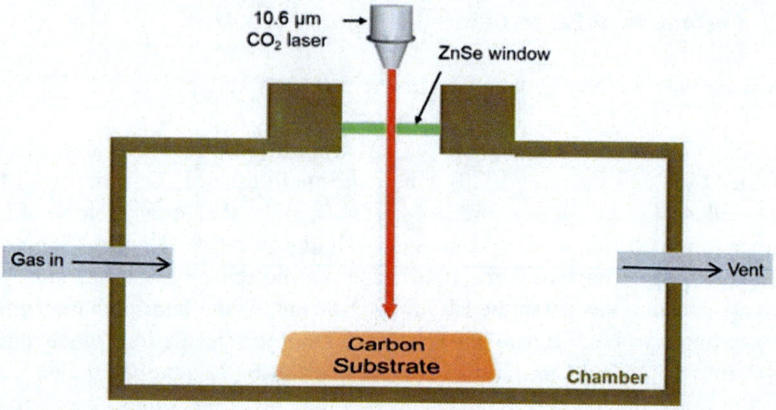

Fig. 3.8 Controlled atmosphere chamber for LIG fabrication. Adopted and modified form Li et al. (2017)

atmospheric condition always produces hydrophilic surfaces. Although surface characteristics can be changed from hydrophilic to hydrophobic by controlling the gas atmosphere in the chamber, as shown in Fig. 3.8. For example, if Ar or H_2 as gas atmosphere, LIG surface formed will be superhydrophobic (contact angle >150ºC) and superhydrophilic (contact angle >0 ºC) surface can be achieved by using O_2 gas in a chamber (Li et al. 2017).

Additionally, some recent studies show the formation of LIG based composites or metal ions doped LIG for enhancement of properties of individual LIG and respective doping elements. For example in one study (Singh et al. 2018a, 2018b), sulphur doped porous LIG was made by directly made on polysulfone, poly(ether sulfone) and polyphenylsulfone; In another study done by Han et al. (2018), metal-doped LIG was formed by direct laser scribing was done on oven-dried cedar wood soaked in a polymeric solution containing metal nitrate precursor.

Formerly, LIG has some limitations such as LIG made on PI was difficult to recycle because PI has excellent thermal, mechanical, and chemical stability hence raising an environmental concern (Ye et al. 2018). Moreover, the potential use of LIG in some application was limited by its robustness on the substrate as the LIG showed weak adherence to the PI substrate (Luong et al. 2019). Although, the recent advancement in the LIG technology by using different polymer support, better optimisation of lasing conditions, etc. has successfully conquered these limitations.

3.3.2.3 Large-Scale Production of LIG Surfaces

As compared to conventional methods, scale-up production of LIG surfaces can be achieved easily maintaining the quality of graphene. Most commonly, there are two methods for scale-up production of LIG surfaces, i.e. roll-to-roll production, and 3D printing (Ye et al. 2018). Roll-to-roll production includes LIG formation chamber

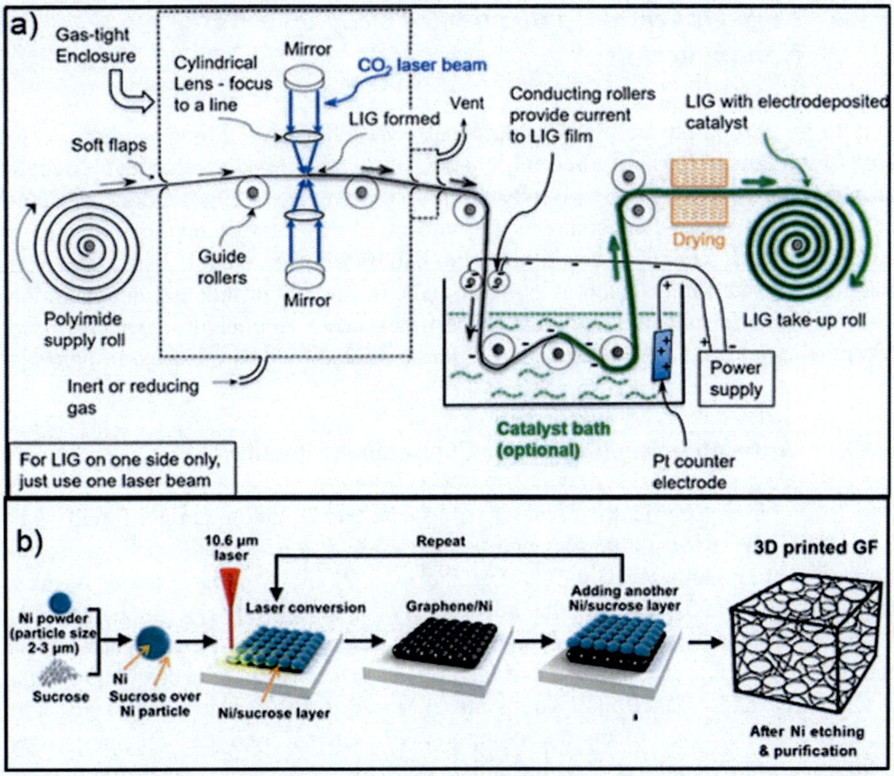

Fig. 3.9 Schematic diagram of both Roll-to-roll production of LIG (**a**), and of in situ synthesis of 3D graphene foam (GF) using a simulated 3D printing process (**b**). Reproduced from Ye et al. (2018). Copyright © 2018, American Chemical Society

where a static CO_2 laser is focused to a line where continuous feeding of Polymer sheets (for example PI sheet as shown in Fig. 9a) is done by the help of supply roll which results in continuous formation of 3D LIG film by two-sided laser irradiation. Whereas 3D printing of LIG briefly involves a 10.6 μm CO_2 laser which is used to heat the mixture of Ni and sucrose as shown in Fig. 9b, in which Ni acts catalyst that converts sucrose into graphene. The 3D structure can be formed by successively adding Ni/sucrose layer on already formed LIG and its subsequent conversion to Ni/graphene by laser irradiation. At last, after etching out Ni, 3D printed LIG is formed (Ye et al. 2018).

3.3.3 Environmental Application of Graphene-Based Nanomaterials

Due to its unique physico-chemical properties such as exceptionally high-surface area, high electrical and thermal conductivity, and high-mechanical strength, graphene-based nanomaterials (GBNs) found to be acquired a huge amount of interest of many researchers. Especially in the field of energy storage application such as supercapacitors, solar cells and lithium ion batteries (Devrim et al. 2018), and environmental related fields such as pollutant adsorption, environmental decontamination, antibacterial and antibiofouling properties, and environmental sensor electrodes (Perreaul et al. 2015). All these environmental related fields are discussed below.

3.3.3.1 Adsorption Application for Contaminant Removal

Due to its large specific surface area, graphene-based nanomaterials found to be a good adsorbent for pollutants such as heavy metal ions, organic pollutants, and gaseous contaminants socially CO_2 (Perreaul et al. 2015). Amongst various forms of graphene, GO usually used for the adsorption of heavy metal ions (Such as Cd(II), Pb(II), U(VI), Co(II)) because of having abundant oxygen functional groups such as epoxide, hydroxyl, carboxyl, and carbonyl on its surface (Ruparelia et al. 2008; Zhao et al. 2011). These functional groups are expected to form strong complexes with metal ions. In the case of organic compounds like dyes, hydrocarbons, pesticides, antibiotics, etc., the interaction mechanism depends on the structural properties, for instance, dipole–dipole moment, molecular conformation, functional group availability (Ramesha et al. 2011). A hence different form of graphene, i.e. whether it is pristine graphene, GO or rGO can have different adsorption capacity even for the same molecule. Also doping of graphene with the different functional group may have different mechanism and efficiency of adsorption. In a study (Chandra et al. 2010), Magnetite-rGO (M-RGO) was used for the removal of arsenic (III), and they reported adsorption capacity of 12–13 mg/g for 0.2 g/lit of M-RGO in water.

3.3.3.2 Pollutants Degradation for Graphene-Based Nanomaterials

Incorporation of photocatalysts such as TiO_2, ZnO, CdS etc. with graphene can be done to achieve photocatalytic degradation of many toxic compounds from the water. Zero band gap of graphene results in noble electronic properties of graphene, which helps photocatalysis to happen in a predominant manner. (Gao et al. 2012) synthesised GO-TiO_2 hybrid to get the photo-degradation of methyl orange and the results show that better photo-degradation of methyl orange as compared to suspended TiO_2 was due to electron adsorption and transferring from TiO_2 to GO which restricted the recombination of the electrons in TiO_2 up to a large extent.

3.3.3.3 Graphene for Desalination

Graphene can be applied as a barrier for gaseous and liquid contaminants because of its impermeable nature. Although having only one atom thickness, graphene is an impermeable material in its pristine form. The delocalised electron clouds of π orbitals block the space in the hexagonal rings of graphene, resulting in the blockage of the passage of the smallest molecular species (Gu et al. 2014). These unique properties lead to the design of ultrathin water separation membrane based on graphene. Figure 3.10 shows two strategies mainly used in graphene-based membranes in membrane filtration process, i.e. nanoporous graphene membranes and membranes made of stacking GO sheets. In nanoporous graphene, rejection of ions involves mechanism-like steric effects, electrostatic repulsion, and hydrodynamic interactions between species and pores (Sint et al. 2008). Whereas in stacked GO membranes, narrow hydrophobic channels between GO layers formed by unoxidized regions on GO provides pathway for water flow (Nair et al. 2012). Stacked GO membranes allows the rejection of both salts and organic molecules maintaining fast water transport (Joshi et al. 2014).

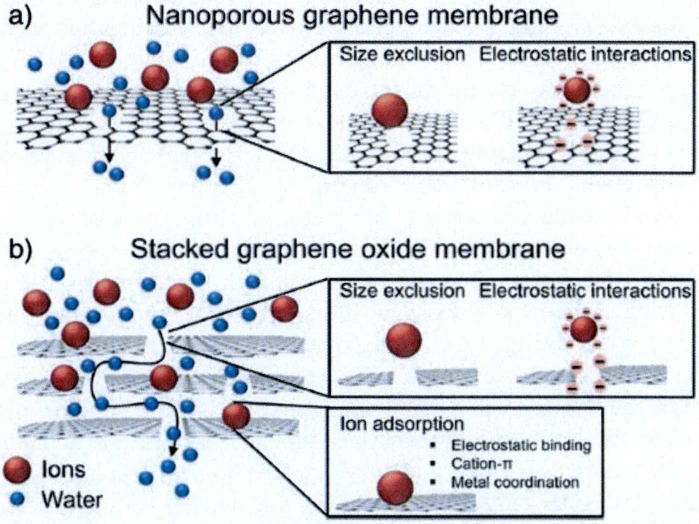

Fig. 3.10 Schematic representing types of graphene-based membranes. **a** Nanoporous graphene membranes. **b** Membranes made of stacking GO sheets (Perreault et al. 2015a, 2015b). Copyright © 2015, Royal Society of Chemistry

3.3.3.4 Graphene-Based Sensors for Environmental Monitoring

Graphene has an exceptionally large surface area and unique electronic properties, and efforts are continuously made to develop graphene-based sensors to detect environmental pollutants, such as heavy metals, toxic gases, and biomolecules. Pristine graphene, owing to extremely high-charge concentration, high charge carrier, the extremely high optical transmission factor and the nearly zero band gap energy (Perreault et al. 2015a, 2015b) makes it attractive nanomaterials for the construction of electrodes for biological and chemical sensors. Also, very less dependability of electron mobility of graphene on temperature, making graphene-related materials, even more, promising for environmental detection applications (Bhuyan et al. 2016).

3.4 Titanium Dioxide (TiO_2): A Photocatalyst

Recently, advanced oxidation processes (AOP) were developed in search of low cost and efficient method to fight the problem of recalcitrant organic compound present as a pollutant. By using AOPs, degradation of many contaminants like an antibiotic, dyes, organic pesticides, polycyclic aromatic hydrocarbon and so forth can be achieved (Yaparatne et al. 2018; Chong et al. 2010; Singh et al. 2013). Amongst these AOPs, heterogeneous photocatalysis employing semiconductor catalysts (TiO_2, ZnO, Fe_2O_3, CdS, GaP and ZnS) has demonstrated its efficiency in degrading a wide range of refractory organic micro-pollutants, water pathogens and disinfection by-product into readily biodegradable compounds, and finally mineralized them to harmless carbon dioxide and water (Chong et al. 2010). Amongst all these semiconductor catalysts, titanium dioxide (TiO_2) has taken the greatest interest in R&D of photocatalysis technology because of the number of advantages, for instance, its nontoxic nature; chemically resistant; its stability as it remains intact even after the repeated catalytic cycle; it has strong oxidizing power; high-catalytic activity; stability against photo corrosion; economical etc. (Singh et al. 2013). However, apart from the advantages, there are some limitations also in using the TiO_2 in powder form or in suspended form like post-treatment separation, less recycle rate of photocatalyst etc. So, new technologies are emerging in which the photocatalyst immobilisation or doping of TiO_2 into some suitable substrate is being done to get better performance (Oblak et al. 2018; Leary and Westwood 2011).

3.4.1 Photocatalysis Mechanism of TiO_2

The mechanism of photocatalysis is like when a photocatalyst is illuminated by light of equal or greater energy than its bandgap (like 3.2 eV for anatase TiO_2, 3.0 eV for rutile TiO_2), produce electron–hole pairs, holes in the valence band (h_{vb}^+) similarly electron in the conduction bands (e_{cb}^-) as shown in Fig. 3.11. Holes in the valance

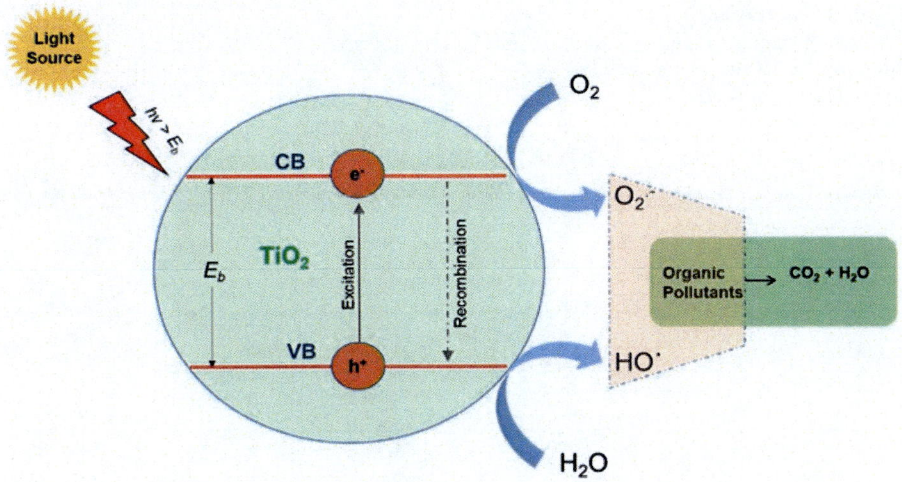

Fig. 3.11 Photocatalysis mechanism of TiO_2 for organic pollutant degradation. Adopted and modified from Chong et al. (2010), Altin and Sökmen (2014)

band react with the water or hydroxide ions to produce hydroxide radicals ($HO^{\cdot}$), on the other hand, photo-generated electrons get reacted with the absorbed oxygen molecules to generate superoxide radicals ($O_2{}^{\cdot}$) which in turn transforms into some highly reactive species like $HO^{\cdot}$, $HO_2{}^{\cdot}$ and H_2O_2. These active radical species play an important role in degradations of harmful organic contaminants and disinfection of pathogenic microorganisms by oxidation–reduction reactions and convert them to harmless substances such as CO_2 and H_2O (Altin and Sökmen 2014).

3.4.2 Crystalline Forms of TiO_2: Anatase and Rutile

The photocatalytic activity of TiO_2 depends on various parameters such as crystallinity, surface area, impurities, and density of hydroxy groups on its surface; however, the most noteworthy factor is its crystalline form. TiO_2 is commonly used as a photocatalyst in two crystal structures as shown in Fig. 3.12: rutile and anatase. Out of which, anatase usually has much greater activity than rutile. More fascinating is the fact that the activity of most common commercial form of TiO_2: P-25 (Degussa) which is frequently used as a benchmark for photocatalyst consists of anatase and rutile (4:1 wt./wt), exceeds the activity of pure anatase in several reaction systems (Kawahara et al. 2002),. It is known that the phase mixture of various polymorphs has a synergistic effect, and increased photocatalytic activity is observed compared to pure phases. However, for pure phases, it is generally accepted that anatase has a higher photocatalytic activity compared to rutile TiO_2 (Luttrell et al. 2015).

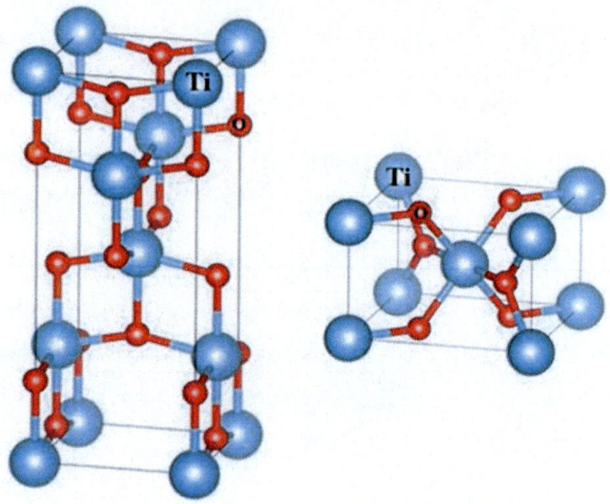

Fig. 3.12 Unit crystal structure of anatase (**a**) and rutile (**b**) and TiO2 (Jia et al. 2016). Copyright © 2016, Springer Nature

Despite rigorous study on TiO_2, there is no fully accepted explanation for this difference in photocatalytic activity of various polymorphs or surface orientations. Although there are certain possible explanations which may project light on it such as; anatase has a larger bandgap (i.e. 3.2 eV) compared to TiO_2 rutile (i.e. 3.0 eV), this reduces the light that can be absorbed, it can increase the maximum of the valence band at higher energy levels than the redox potential of the adsorbed molecules (Luttrell et al. 2015). This increases the oxidation "power" of the electrons and facilitates the transfer of electrons from TiO_2 to adsorbed molecules (Luttrell et al. 2015). This explanation has also been expanded to explain activities dependent on the surface orientation of suggesting that different surfaces have different bandwidth gaps; another possible explanation can be that indirect bandgap shown by anatase is smaller than his direct bandgap. Figure 3.13 shows the two types of bandgaps in semiconductors, i.e. direct bandgap and indirect bandgap and the photoemission process in them. For rutile, on the other hand, its indirect bandgap is remarkably similar to the direct bandgap, resulting in rutile to have its fundamental bandgap as direct bandgap (Luttrell et al. 2015). Indirect bandgap semiconductors generally show longer charge carrier lifetimes than materials with a direct gap. An electron–hole pair life longer in anatase than in rutile would make charge carriers more likely to participate in superficial reactions. A study by Xu et al. (2011) tells that longer charge carriers lifetime in anatase than in rutile comes from transient photoconductivity measurements on single crystalline samples.

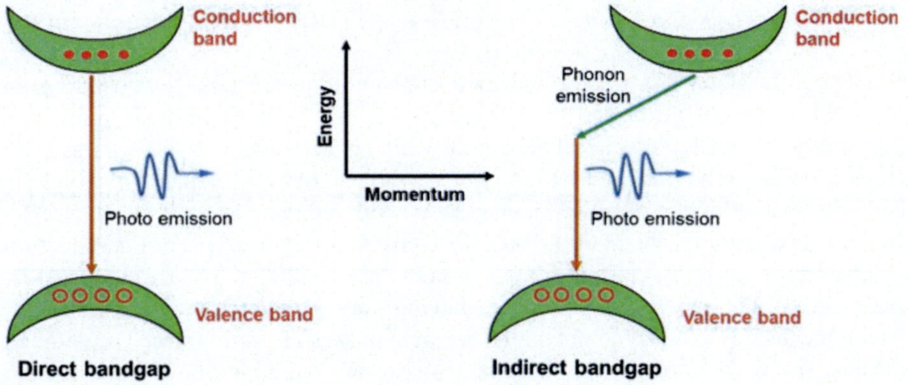

Fig. 3.13 Type of bandgaps in semiconductors. Adopted and modified from Luttrell et al. (2015)

3.5 Magnéli Phase TiO_x: An Electrocatalyst

TiO_2 has been studied in detail globally, and its technological applications have a huge number of varieties ranges such as in the electronic field, medical industry, environmental filed etc. Though stoichiometric TiO_2 has a very low-electrical conductivity near about 10^{-10} Sm^{-1} (Adamaki et al. 2014). Interestingly, the electronic properties of TiO_2 can be radically transformed by establishing oxygen deficiencies in its crystalline lattice which can be created by performing its reduction most commonly by heat-treating the oxide at a high temperature in a reducing atmosphere (Adamaki et al. 2014). The reduction process leads to the formation of new phases of titanium oxides depending on the concentration of oxygen vacancy defects in the crystal lattice. Sub-stoichiometric titanium oxides known as Magnéli phase is a sub-set of those new phases formed with a general chemical composition of Ti_nO_{2n-1}, where n varies from 4 to 9 such as Ti_4O_7, Ti_3O_5 etc. which works as an electrocatalyst (Domaschke et al. 2019).

3.5.1 Synthesis of Magnéli Phases

In order to get reduced titania (oxygen deficient), many preparation methods are being adopted in researches globally which includes electron beam implantation, thermal annealing to high temperatures (over 500 K), vacuum heating TiO_2, UV irradiation, reduced conditions (C, H_2), laser, plasma treating, and high-energy treatments particle (neutron, Ar + , electron, or -ray) bombardment, vacuum activation metal reduction, chemical vapour deposition, electrochemical reduction, etc. (Jayashree and Ashokkumar 2018; Nakajima et al. 2014; Tang et al. 2012). A general reduction sequence of TiO_2 can be shown by the below Eq. (3.1)

$$TiO_2 \rightarrow Ti_nO_{2n-1}(4 < n < 10) \rightarrow Ti_3O_5 \rightarrow Ti_2O_3 \rightarrow TiO \rightarrow Ti_2O \quad (3.1)$$

Hence, the defect formation and their concentration are largely depending on the preparative condition and have a significant effect on its intrinsic properties like structural, optical, electrical behaviour which in turn affect its catalytic activity (Jayashree and Ashokkumar 2018). A typical synthesis of Magnéli phase involves the carbothermal reduction of rutile Titania (TiO_2) with reducing agent PVA, H_2, CaH_2, zirconium, ammonia, etc. at 600–1000 °C (Domaschke et al. 2019) or the reduction of anatase or rutile titania under high temperature (>1200 °C) (Arif et al. 2017). Although all these Magnéli phases are having very analogous lattice parameters, which means it is not easy to obtain the separate phases in a pure form. The detailed mechanism for the formation of these new phases of titanium oxide is like when the O atom is removed, and the nearby Ti–Ti bonds are usually relaxed from the vacancy in order to bond with the rest of the lattice. At the same time, the neighbouring Ti–O bonds displace slightly inward due to the electrostatic pull induced by oxygen vacancies. This internal modification due to oxygen vacancy defects reorganizes atomic positions and reduces the overlap between Ti dangling bonds nearer to the vacancy, thereby reducing the Ti–O bond elongation.

These defects cause the distortion in TiO_6 octahedral structure of TiO_2, which can be transformed into different octahedral packing (e.g. from tetragonal rutile TiO_2 to orthorhombic packing of Ti_4O_7) with increasing concentration of oxygen defects (Fig. 3.14). Hence, with the increase in the concentration of oxygen vacancy defects, ordering of vacancy occurs and leads to the formation of Magnéli phases of titanium oxide because of crystallographic shear.

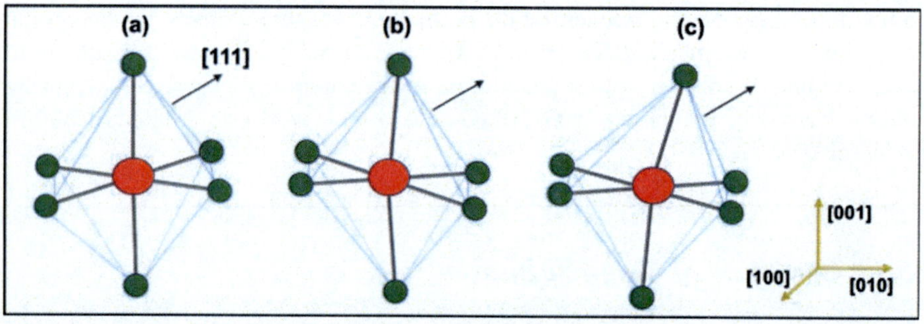

Fig. 3.14 Different types of octahedral packing **a** tetragonal distortion (rutile TiO_2), **b** trigonal distortion (Ti_2O_3), **c** orthorhombic- like distortion (Ti_4O_7). Adopted and modified from Jayashree and Ashokkumar (2018)

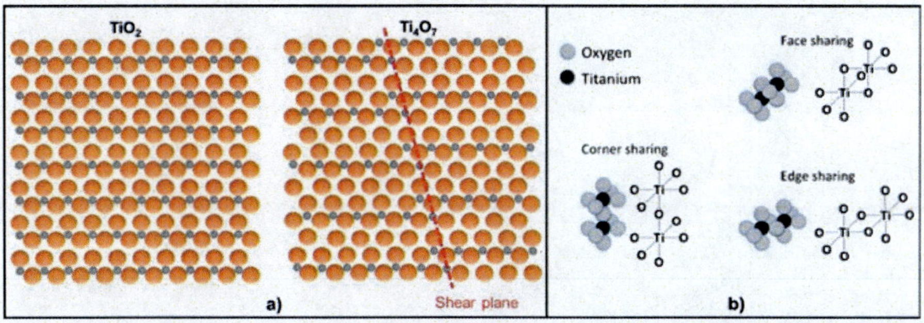

Fig. 3.15 **a** Difference between the crystal structure of TiO_2 and Ti_4O_7 (Magnéli phase) **b** Different type of orientation in TiO_2 octahedra in Magnéli phase. Adopted and modified from Jayashree and Ashokkumar (2018), Copyright © 2010, Elsevier

3.5.2 *Difference Between TiO_2 and Magnéli Phase*

The major difference between the two lies in their catalytic activity, TiO_2 is a well-established semi-conductor which has been widely used for its photocatalytic activities. Whereas, Magnéli phase is a reduced form of TiO_2 having oxygen vacancy in its crystal structure, which leads to an exceptionally good electronic nature and hence makes Magnéli phase an electrocatalyst. The difference can also be seen clearly in their crystalline lattice as shown in Fig. 3.15, the crystal structure of TiO_2 is considered as a 2D chain of TiO_6 octahedra having edge-sharing with individual chains linked by a corner oxygen atom to form the 3D lattice (Jayashree and Ashokkumar 2018). Each octahedron is composed of a titanium atom at its centre and oxygen at each corner as shown in Fig. 3.16. Whereas Magnéli-phase has a crystal structure made up of TiO_2 octahedra blocks with an oxygen deficiency for each nth layer, resulting in the shear planes where the octahedral 2D chains become face sharing to address the deficiency in oxygen (You et al. 2016). This organization eventually reduces the symmetry of the crystal system from tetragonal (i.e. for TiO_2) to triclinic (i.e. for Magnéli phases) as the unit cell size increases.

Also, it has been found in various studies that the Fermi level of Ti_4O_7 at room temperature is located within the conduction band of oxide (Domaschke et al. 2019) because of this reason Ti_4O_7 has a high-electronic conductivity comparable to metals.

Figure 3.16a and b shows the SEM images of P25 TiO_2 and Ti_4O_7 made by thermochemical reduction of TiO_2 at 1050 °C from the study done by Lee et al. (2018). TiO_2 nanoparticles had a small size of 25 nm having highly aggregated morphology. Whereas Ti_4O_7 was sintered and grew to a few micrometres. Also, He et al. (2015) have performed a study where they showed the XPS spectra of Ti-2p (Fig. 3.16c) and O-1 s (Fig. 3.16d) of Ti_8O_{15} (Magnéli phase with n = 8) and TiO_2 powder (anatase). Ti_8O_{15} was fabricated directly on a Ti substrate by an easy one-step evaporation-deposition synthesis procedure under the H_2 atmosphere. In the XPS spectra of Ti-2p, two peaks at binding energies 458.6 and 464.5 eV are there in both the samples, which correspond to the Ti-2p$_{1/2}$ and Ti-2p$_{3/2}$ peaks of Ti^{4+}. However,

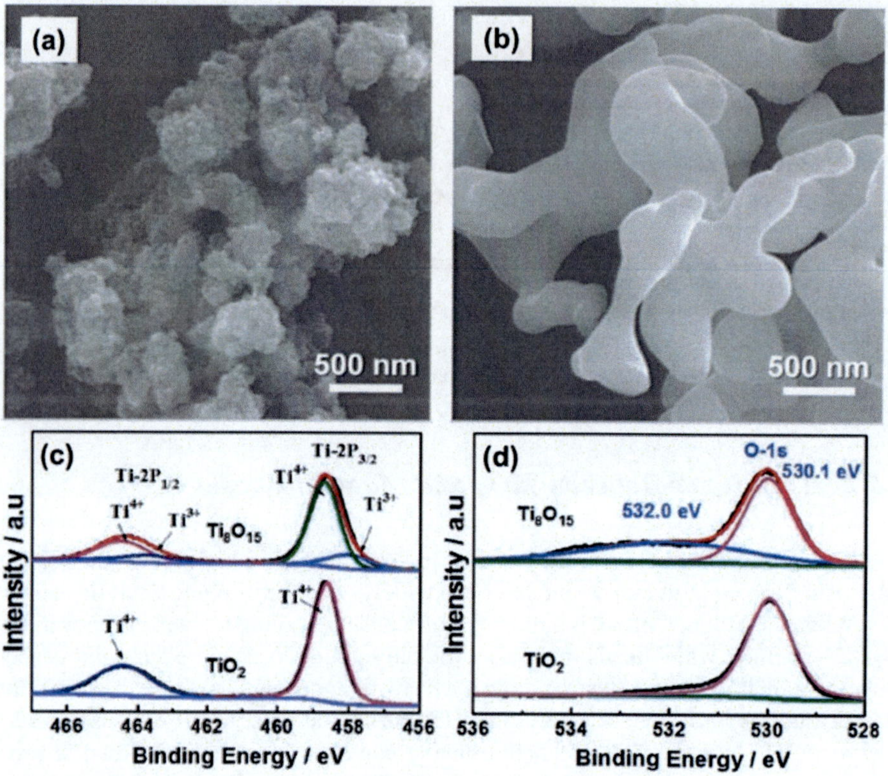

Fig. 3.16 SEM images of P25 TiO2 (**a**) and Ti4O7 made by thermochemical reduction of TiO2 at 1050 °C (**b**) (Lee et al. 2018); (**c** and **d**) Ti-2p and O-1 s XPS spectra of Ti8O15 and TiO2 (He et al. 2015), Copyright © 2018, American Chemical Society; copyright © 2015, Royal Society of Chemistry

it can be clearly seen that there are two extra peaks for Ti_8O_{15} with binding energies at 463.7 and 458.0 eV, which are attributed to the Ti-2$p_{1/2}$ and Ti-2$p_{3/2}$ peaks of Ti^{3+}. And for the XPS spectra of O-1 s, there is a strong additional peak at 532.0 eV in Ti_8O_{15} (related to Ti–OH bond) other than that at 530.1 eV (related to Ti–O-Ti bond in both the case) (He et al. 2015).

Recently, Magnéli phase has gained a lot of attention in the field of conductive materials because the shear planes in Magnéli phase provides a pathway for electron transport and hence, making it an excellent electrical conductor (Jayashree and Ashokkumar 2018). This distinctive structure leads to a combination of exceptional electrical conductivity close to that of metals, a high-corrosion resistance close to that of ceramic materials (You et al. 2016) and durable in electrochemically oxidizing environments (Arif et al. 2017). Also, in Magnéli phases Ti_nO_{2n-1}, the electrical conductivity decreases with increasing n values, and phases with $n = 4–6$ are superior electrically conductive (Zhang et al. 2013), making them more attractive than other phases. Specifically, Ti_4O_7 is an extraordinary member of Magnéli phase with

$n = 4$, which exhibits super electrical conductivity (i.e. 1035 σ/S cm^{-1}) (Walsh and Wills 2010) and electrochemical stability and optical properties (Zhang et al. 2013). The attractive properties are since Ti_4O_7 has one 3d electron for two cationic sites and two possible valence states, $Ti^{3+}(3d^1)$ and $Ti^{4+}(3d^0)$, for the cations (Heckel et al. 2015).

Due to their high-electrical conductivity and chemical resistance, Magnéli phase is of interest for a variety of applications, including cathodic protection, batteries, catalyst support for fuel cells as well as their potential use in the treatment of aqueous waste and contaminated water (Adamaki et al. 2014).

3.6 TiO_2 Doped Graphene

TiO_2 has shown extraordinary photocatalytic activity in many types of research as well as in industries globally. However, TiO_2 in its pure form whether in TiO_2 nanoparticles (most common is P25), in TiO_2 nanorods or in TiO_2 nanotubes (TNTs) etc. experiences some major problem limiting its further application and the most important problem is TiO_2 have a wide bandgap (i.e. 3.2 eV) causing its photocatalytic activity in the UV region, and the terrestrial solar spectrum contains about 5% of UV radiation only (Bhanvase et al. 2017). Therefore, it has limited photocatalytic activity in the visible range of the solar spectrum; and the other problem is easy recombination of its photo-generated electron-holes pairs (Zhao et al. 2015). In order to solve this problem, various methods have been performed to improve their photocatalytic activity, including the loading of noble metals, the doping of non-metals, the doping of transition metal ions, the sensitization of dyes, the co-doping of metals and non-metals, etc. However, the results are still not up to scratch (Zhao et al. 2015).

One of the solution researchers have discovered to bring its reactivity in visible light is by using graphene-TiO_2 composites as graphene owns very unique electronic and photo-electronic properties (Zhao et al. 2015) which improves electron transportation and the recombination of electron–hole pairs get suppressed. Moreover, it is possible that with the use of the graphene-TiO_2 composite, the agglomeration of TiO_2 can be avoided in order to maintain a greater surface and the composite will be having more active sites for the degradation of pollutants and this improves the degradation rate of pollutants with the use of a smaller quantity of photocatalyst based on composite graphene-TiO_2 (Bhanvase et al. 2017; Zhang et al. 2012). In addition, graphene has excellent thermal conductivity, a larger specific area, better charge carrier mobility and good mechanical resistance. These electronic and photonic properties of graphene make it a candidate material for the enhancement of TiO_2 photo-reactivity. Graphene-TiO_2 nanocomposites also have improved photo-activity resulting from greater absorption of extended visible light and a strong affinity for other organic materials (Bhanvase et al. 2017). Figure 3.17 showing the mechanism of TiO_2 incorporated on rGO due to which the bandgap of TiO_2 got lower by some

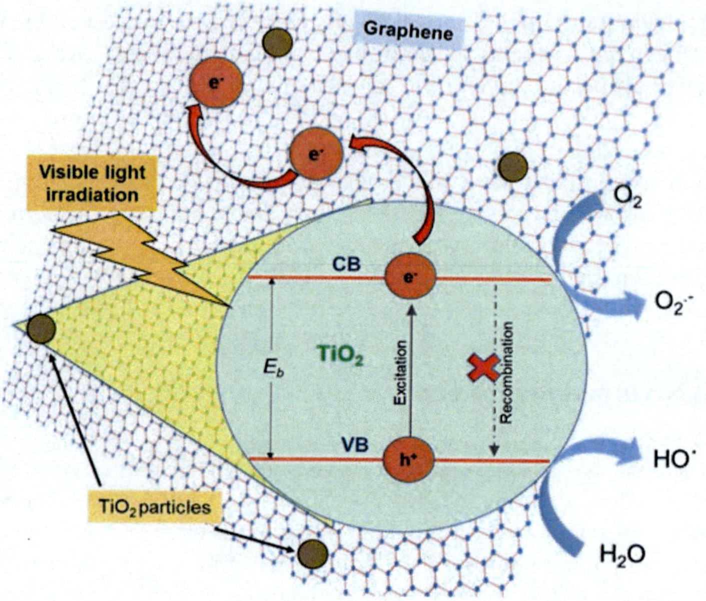

Fig. 3.17 Mechanism of TiO_2 graphene composite. Adopted and modified from Tan et al. (2013)

value resulting in the photo-reactivity of TiO_2 under visible range and hence better and efficient degradation of contaminants in visible light irradiation.

Zhao et al. (2015) in their study have fabricated graphene/TNTs nanocomposites (GTNCs) made up of rGO and TiO_2 nanotubes (TNTs) using hydrothermal process. They showed the advantage of this unique nanostructure in providing sufficient active sites as well as it supplied electron-transport path, which eventually increased the photocatalytic activity of nanocomposite. They checked the photocatalytic performance of GTHCs through the methyl orange (MO) degradation under UV light irradiation and found outstanding photocatalytic activity much better than that of traditional TiO_2 nanotubes. Also, in another study done by Zhang et al. (2010) where they obtained chemically bonded TiO_2-graphene nanocomposite photocatalyst with P25 and GO using hydrothermal method resulting in the simultaneous reduction of GO into rGO and loading on P25 in it. In their study, they showed the enhancement in the photo-degradation of methylene blue under both UV and visible light irradiation using P25-graphene as compared to bare P25and P25-CNTs with the same carbon content. And the reason being was its giant 2-D planar structure, which facilitated an improved platform for adsorption of dyes and charge transportation.

3.7 Conclusion

Graphene is a noble carbon-based nanomaterial having some extraordinary properties such as high specific surface area, high thermal and electron mobility as well as extremely high-mechanical strength. Having such extraordinary properties, graphene and graphene-based nanocomposites can be used for environmental application as photocatalytic, biocidal, electroactive, and adsorbent materials. Laser-induced graphene is a newly invented facile, one-step, cost-effective, and chemical-free method for the synthesis of good quality graphene and graphene-based surfaces such as electrodes. As compared to GO and rGO, good quality LIG can also be easily produced at a large scale-up by roll-to-roll method. LIG can be printed on various carbonaceous materials such as cloth, paper, wood, cork coconut shells, and even on edible items like bread and potato by just changing the laser setting and atmosphere.

Nevertheless, graphene itself does not poses efficient catalytic properties leading to a demand for graphene-based nanocomposites by incorporation of heteroatom into graphene. For the purpose, titanium oxide TiO_x (i.e. TiO_2 as a photocatalyst and Magnéli phases Ti_nO_{n-1} as an electrocatalyst) can be considered as a noble catalyst for doping. The catalysis reaction of TiO_x results in some environmentally friendly by-products such as CO_2 and H_2O in most of the cases. The incorporation of TiO_x with graphene enhances the catalysis reaction as this composite cause the working of TiO_2 more efficiently in solar light and provides a free pathway for electron movement enhancing the electrocatalytic property of Magnéli phase-graphene composite.

References

Adamaki V et al (2014) Manufacturing and characterization of Magnéli phase conductive fibres. J Mater Chem A 2(22):8328–8333. https://doi.org/10.1039/c4ta00685b

Ago H (2015) CVD Growth of High-Quality Single-Layer Graphene. In: Matsumoto K (eds) Frontiers of Graphene and Carbon Nanotubes. Springer, Tokyo. https://doi.org/10.1007/978-4-431-55372-4_1

Al-Khateeb LA, Almotiry S, Salam MA (2014) Adsorption of pharmaceutical pollutants onto graphene nanoplatelets. Chem Eng J 248:191–199. https://doi.org/10.1016/j.cej.2014.03.023

Aldrich R (2019) Laser : Fundamentals—Introduction, pp 1–8

Allmen MV, Blatter A (1995) Physical principles and applications. In: Laser-Beam Interaction and Materials. Springer, vol 196(12). https://doi.org/10.1007/978-3-642-57813-7

Altin I, Sökmen M (2014) Preparation of TiO2-polystyrene photocatalyst from waste material and its usability for removal of various pollutants. Appl Catal B 144:694–701. https://doi.org/10.1016/j.apcatb.2013.06.014

Arif AF et al (2017) Highly conductive nano-sized Magnéli phases titanium oxide (TiO x). Sci Rep 7(1):1–9. https://doi.org/10.1038/s41598-017-03509-y

Bandow S et al (1998) Effect of the growth temperature on the diameter distribution and chirality of single-wall carbon nanotubes. Phys Rev Lett. https://doi.org/10.1103/PhysRevLett.80.3779

Bhanvase BA, Shende TP, Sonawane SH (2017) A review on graphene–TiO2 and doped graphene–TiO2 nanocomposite photocatalyst for water and wastewater treatment. Environ Technol Rev 6(1):1–14. https://doi.org/10.1080/21622515.2016.1264489

Bhuyan MSA, Uddin MN, Islam MM et al (2016) Synthesis of graphene abstract synthesis of graphene | springerlink synthesis of graphene mother of all graphene forms . Graphene is a 2D building material for carbon material of all other dimensionalities. It can be wrapped up into 0D buckyballs, rolle. Int Nano lett 6(2):65–83. https://doi.org/10.1007/s40089-015-0176-1

Brady-Estévez AS, Kang S, Elimelech M (2008) A single-walled-carbon-nanotube filter for removal of viral and bacterial pathogens. Small. https://doi.org/10.1002/smll.200700863

Chandra V et al (2010) Water-dispersible magnetite-reduced graphene oxide composites for arsenic removal. https://doi.org/10.1021/nn1008897

Chong MN et al (2010) Recent developments in photocatalytic water treatment technology: a review. Water Res 44(10):2997–3027. https://doi.org/10.1016/j.watres.2010.02.039

Chowdhury ZZ et al (2018) Electrochemically active carbon nanotube (CNT) membrane filter for desalination and water purification. In: Emerging technologies for sustainable desalination handbook. https://doi.org/10.1016/B978-0-12-815818-0.00010-2

Chyan Y et al (2018) Laser-induced graphene by multiple lasing: toward electronics on cloth, paper, and food. ACS Nano 12(3):2176–2183. https://doi.org/10.1021/acsnano.7b08539

Dasari BL et al (2017) Graphene and derivatives—Synthesis techniques, properties and their energy applications. Energy 140(September):766–778. https://doi.org/10.1016/j.energy.2017.08.048

Davim JP et al (2008) Some experimental studies on CO2 laser cutting quality of polymeric materials. J Mater Process Technol. https://doi.org/10.1016/j.jmatprotec.2007.06.056

Devrim Y, Arıca ED, Albostan A (2018) Graphene based catalyst supports for high temperature PEM fuel cell application. Int J Hydrogen Energy 43(26):11820–11829. https://doi.org/10.1016/j.ijhydene.2018.03.047

Domaschke M et al (2019) Magnéli-phases in anatase strongly promote cocatalyst-free photocatalytic hydrogen evolution. ACS Catal 9(4):3627–3632. https://doi.org/10.1021/acscatal.9b00578

Dong F et al (2014) Nanomaterials for environmental applications. J Nanomater. https://doi.org/10.1155/2014/276467

Drever RWP et al (1993) Laser phase and frequency stabilization using an optical resonator. Appl Phys B 31:97–105

Dubey AK, Yadava V (2008) Laser beam machining—a review. Int J Mach Tools Manuf. https://doi.org/10.1016/j.ijmachtools.2007.10.017

Franklin AD (2015) Carbon nanotube electronics. In: Emerging nanoelectronic devices. https://doi.org/10.1002/9781118958254.ch16

Furmaniak S, Terzyk AP, Gauden PA (2010) Simple model of adsorption on external surface of carbon nanotubes—a new analytical approach basing on molecular simulation data. Adsorption 16:197–213. https://doi.org/10.1007/s10450-010-9223-4

Gao Y et al (2012) Combustion synthesis of graphene oxide-TiO_2 hybrid materials for photodegradation of methyl orange. Carbon 50(11):4093–4101. https://doi.org/10.1016/j.carbon.2012.04.057

Geim A, Novoselov K (2007) The rise of graphene. Nature Mater 6:183–191. https://doi.org/10.1038/nmat1849

Gu Y, Xing M, Zhang J (2014) Synthesis and photocatalytic activity of graphene based doped TiO_2 nanocomposites. Appl Surf Sci 319(1):8–15. https://doi.org/10.1016/j.apsusc.2014.04.182

Gupta A et al (2019) Silver-doped laser-induced graphene for potent surface antibacterial activity and anti-biofilm action. Chem Communi 55(48):6890–6893. https://doi.org/10.1039/c9cc02415h

Ham HT, Choi YS, Chung IJ (2005) An explanation of dispersion states of single-walled carbon nanotubes in solvents and aqueous surfactant solutions using solubility parameters. J Colloid Interface Sci 286:216–223

Han X et al (2018) Laser-induced graphene from wood impregnated with metal salts and use in electrocatalysis. ACS Appl Mater Interfaces 1(9):5053–5061. https://doi.org/10.1021/acsanm.8b01163

He C et al (2015) Direct synthesis of pure single-crystalline Magnéli phase Ti8O15 nanowires as conductive carbon-free materials for electrocatalysis. Nanoscale 7(7):2856–2861. https://doi.org/10.1039/c4nr05806b

Heckel W et al (2015) How the aggregation of oxygen vacancies in rutile-based TiO2-δ phases causes memristive behavior. Phys Rev B—Condens Matter Mater Phys 92(21):1–14. https://doi.org/10.1103/PhysRevB.92.214104

Huang Q et al (2013) Enhanced photocatalytic activity of chemically bonded TiO_2/graphene composites based on the effective interfacial charge transfer through the C-Ti bond. ACS Catal 3(7):1477–1485. https://doi.org/10.1021/cs400080w

Jayashree S, Ashokkumar M (2018) Switchable intrinsic defect chemistry of titania for catalytic applications. Catalysts 8(12):1–25. https://doi.org/10.3390/catal8120601

Jia J et al (2016) On the crystal structural control of sputtered TiO_2 thin films. Nanoscale Res Lett 11. https://doi.org/10.1186/s11671-016-1531-5

Joshi RK et al (2014) Precise and ultrafast molecular sieving through graphene oxide membranes. Science. https://doi.org/10.1126/science.1245711

Julsgaard B et al (2007) Understanding laser stabilization using spectral hole burning. In: Optics infobase conference papers. https://doi.org/10.1364/oe.15.011444

Kamat PV, Haria M, Hotchandani S (2004) C60 Cluster as an electron shuttle in a Ru(II)-polypyridyl sensitizer-based photochemical solar cell. J Phys Chem B 108:5166–5170. https://doi.org/10.1021/jp0496699

Kawahara T et al (2002) A patterned TiO_2(anatase)/TiO_2(rutile) bilayer-type photocatalyst: effect of the anatase/rutile junction on the photocatalytic activity. Angew Chem Int Ed Engl 2 41(15):2811–2813. https://doi.org/10.1002/1521-3773(20020802)41:15<2811::AID-ANIE2811>3.0.CO;2-#

Keblinski P et al (2002) Charge distribution and stability of charged carbon nanotubes. Phys Rev Lett 89:255503. https://doi.org/10.1103/PhysRevLett.89.255503

Leary R, Westwood A (2011) Carbonaceous nanomaterials for the enhancement of TiO_2 photocatalysis. Carbon 49(3):741–772. https://doi.org/10.1016/j.carbon.2010.10.010

Lee J et al (2007) Photochemical production of reactive oxygen species by C60 in the aqueous phase during UV irradiation. Environ Sci Technol 41:2529–2535. https://doi.org/10.1021/es062066l

Lee J et al (2013) Graphene oxide nanoplatelets composite membrane with hydrophilic and antifouling properties for wastewater. J Membr Sci 448:223–230. https://doi.org/10.1016/j.memsci.2013.08.017

Lee S et al (2018) Magnéli-phase Ti4O7 nanosphere electrocatalyst support for carbon-free oxygen electrodes in lithium-oxygen batteries. ACS Catal 8(3):2601–2610. https://doi.org/10.1021/acscatal.7b03741

Li Y et al (2015) In situ fabrication of Mn 3O4 decorated graphene oxide as a synergistic catalyst for degradation of methylene blue. Appl Catal B 162:268–274. https://doi.org/10.1016/j.apcatb.2014.06.058

Li Y et al (2017) Laser-Induced graphene in controlled atmospheres: from superhydrophilic to superhydrophobic surfaces. Adv Mater 29(27):1–8. https://doi.org/10.1002/adma.201700496

Lin J et al (2014) Laser-induced porous graphene films from commercial polymers. Nature Commun 5:5714 (1–8). https://doi.org/10.1038/ncomms6714

Luong DX et al (2019) Laser-Induced Graphene Composites as Multifunctional Surfaces. ACS Nano 13(2):2579–2586. https://doi.org/10.1021/acsnano.8b09626

Luttrell T et al (2015) Why is anatase a better photocatalyst than rutile?—model studies on epitaxial TiO2 films. Sci Rep 4:1–8. https://doi.org/10.1038/srep04043

Mauter MS, Elimelech M (2008) Critical review environmental applications of carbon-based nanomaterials. Am Chem Soc 5843–5859. https://doi.org/10.1016/j.jvs.2008.11.060

McGinnis RL, McCutcheon JR, Elimelech M (2007) A novel ammonia-carbon dioxide osmotic heat engine for power generation. J Membr Sci. https://doi.org/10.1016/j.memsci.2007.08.027

Nair RR et al (2012) Unimpeded permeation of water through helium-leak-tight graphene-based membranes. Science. https://doi.org/10.1126/science.1211694

Nakajima T et al (2014) Rapid formation of black titania photoanodes: Pulsed laser-induced oxygen release and enhanced solar water splitting efficiency. J Mater Chem A 2:6762. https://doi.org/10.1039/C4TA00557K

Oblak R et al (2018) Alternative support materials for titania photocatalyst towards degradation of organic pollutants. J Water Proc Eng 23(March):142–150. https://doi.org/10.1016/j.jwpe.2018.03.015

Park S et al (2011) Hydrazine-reduction of graphite-and graphene oxide. Carbon 49(9):3019–3023. https://doi.org/10.1016/j.carbon.2011.02.071

Perreault F et al (2015a) Antimicrobial properties of graphene oxide nanosheets: Why size matters. ACS Nano 9(7):7226–7236. https://doi.org/10.1021/acsnano.5b02067

Perreault F, Fonseca de Faria A, Elimelech M (2015b) Environmental applications of graphene-based nanomaterials. Chem Soc Rev 44(16):5861–5896. https://doi.org/10.1039/C5CS00021A

Ramesha GK et al (2011) Graphene and graphene oxide as effective adsorbents toward anionic and cationic dyes. J Colloid Interface Sci 361(1):270–277. https://doi.org/10.1016/j.jcis.2011.05.050

Rao CN et al (2009) Graphene: the new two-dimensional nanomaterial. Angew Chem Int Ed Engl 48(42):7752–7777. https://doi.org/10.1002/anie.200901678

Rathinam K et al (2017) Polyimide derived laser-induced graphene as adsorbent for cationic and anionic dyes. Carbon 124:515–524. https://doi.org/10.1016/j.carbon.2017.08.079

Ruparelia JP et al (2008) Potential of carbon nanomaterials for removal of heavy metals from water. Desalination. https://doi.org/10.1016/j.desal.2007.08.023

Sellin PB et al (2001) Laser stabilization at 1536 nm using regenerative spectral hole burning. Phys Rev B 63:155111

Shenderova OA, Zhirnov VV, Brenner DW (2002) Carbon nanostructures. Crit Rev Solid State Mater Sci. https://doi.org/10.1080/10408430208500497

Singh S, Mahalingam H, Singh PK (2013a) Polymer-supported titanium dioxide photocatalysts for environmental remediation: a review. Appl Catal A: Gen 462–463:178–195. https://doi.org/10.1016/j.apcata.2013.04.039

Singh S, Mahalingam H, Singh PK (2013b) Polymer-supported titanium dioxide photocatalysts for environmental remediation: a review. Appl Catal A: Gen 462–463:pp. 178–195. https://doi.org/10.1016/j.apcata.2013.04.039

Singh SP et al (2013) Impact of addition of amendments on the degradation of DDT and its residues partitioned on soil. Chemosphere 92(7):811–820. https://doi.org/10.1016/j.chemosphere.2013.04.025

Singh SP et al (2017) Laser-induced graphene layers and electrodes prevents microbial fouling and exerts antimicrobial action. ACS Appl Mater Interfaces 9(21):18238–18247. https://doi.org/10.1021/acsami.7b04863

Singh SP et al (2018a) Laser-induced graphene biofilm inhibition: texture does matter. ACS Appl Nano Mater 1(4):1713–1720. https://doi.org/10.1021/acsanm.8b00175

Singh SP et al (2018b) Sulfur-doped laser-induced porous graphene derived from polysulfone-class polymers and membranes. ACS Nano 12(1):289–297. https://doi.org/10.1021/acsnano.7b06263

Singh, SP, Bose P (2015) Degradation of soil-adsorbed DDT and its residues by NZVI addition. RSC Adv 5(114):94418–94425. https://doi.org/10.1039/c5ra18282d

Singh SP, Bose P (2016) Degradation kinetics of endosulfan isomers by micron-and nano-sized zero valent iron particles (MZVI and NZVI). J Chem Technol Biotechnol 91(8):2313–2321. https://doi.org/10.1002/jctb.4818

Singh SP, Bose P (2017) Reductive dechlorination of endosulfan isomers and its metabolites by zero-valent metals: reaction mechanism and degradation products. RSC Adv 7(44):27668–27677. https://doi.org/10.1039/c7ra02430d

Sint K, Wang B, Král P (2008) Selective ion passage through functionalized graphene nanopores. J Am Chem Soc. https://doi.org/10.1021/ja804409f

Smith SC, Rodrigues DF (2015) Carbon-based nanomaterials for removal of chemical and biological contaminants from water: a review of mechanisms and applications. Carbon. https://doi.org/10.1016/j.carbon.2015.04.043

Snakenborg D, Klank H, Kutter JP (2004) Microstructure fabrication with a Co2 laser system. J Micromech Microeng. https://doi.org/10.1088/0960-1317/14/2/003

Soler F, Manuel V (2014) Fabrication and characterization of macroscopic graphene layers on metallic substrates. https://www.researchgate.net/publication/266652632

Strickland NM et al (2000) Laser frequency stabilization using regenerative spectral hole burning. Phys Rev B 62:1473–1476

Sun YP et al (2002) Functionalized carbon nanotubes: properties and applications. Acc Chem Res 35(12):1096–1104. https://doi.org/10.1021/ar010160v

Tan LL et al (2013) Reduced graphene oxide-TiO2 nanocomposite as a promising visible-light-active photocatalyst for the conversion of carbon dioxide. Nanoscale Res Lett 8(1):1. https://doi.org/10.1186/1556-276X-8-465

Tang C, Zhou D, Zhang Q (2012) Synthesis and characterization of Magnéli phases: Reduction of TiO_2 in a decomposed NH3 atmosphere. Mater Lett 79:42–44

Thamaraiselvan C et al (2020) Laser-induced graphene and carbon nanotubes as conductive carbon-based materials in environmental technology. Mater Today 115–131. https://doi.org/10.1016/j.mattod.2019.08.014

Vashisth A et al (2020) ReaxFF Simulations of LIG Formation for Multifunctional Polymer Nanocomposites. ACS App Nano Mat 3(2):1881–1890. https://doi.org/10.1021/acsanm.9b02524

Walsh FC, Wills RGA (2010) The continuing development of Magnéli phase titanium sub-oxides and Ebonex® electrodes. Electrochim Acta 55:6342–6351

Walther JH et al (2004) Hydrophobic hydration of C_{60} and carbon nanotubes in water. Carbon 2004(42):1185–1194

Weber WJ, McGinley PM, Katz LE (1991) Sorption phenomena in subsurface systems: concepts, models and effects on contaminant fate and transport. Water Res. https://doi.org/10.1016/0043-1354(91)90125-A

Wu T et al (2013) Three-dimensional graphene-based aerogels prepared by a self-assembly process and its excellent catalytic and absorbing performance. J Mater Chem A 1(26):7612–7621. https://doi.org/10.1039/c3ta10989e

Xu M et al (2011) Photocatalytic activity of bulk TiO_2 anatase and rutile single crystals using infrared absorption spectroscopy. Phys Rev Lett 106(13):138302. https://doi.org/10.1103/PhysRevLett.106.138302

Yaparatne S, Tripp CP, Amirbahman A (2018) Photodegradation of taste and odor compounds in water in the presence of immobilized TiO 2 -SiO 2 photocatalysts. J Hazard Mater 346:208–217. https://doi.org/10.1016/j.jhazmat.2017.12.029

Ye R, James DK, Tour JM (2018) Laser-induced. Account Chem Res 51(7):1609–1620. https://doi.org/10.1021/acs.accounts.8b00084

Yin R et al (2017) Heteroatoms doped graphene for catalytic ozonation of sulfamethoxazole by metal-free catalysis: performances and mechanisms. Chem Eng J 317:632–639. https://doi.org/10.1016/j.cej.2017.01.038

You S et al (2016) Monolithic porous magnéli-phase Ti4O7 for electro-oxidation treatment of industrial wastewater. Electrochim Acta 214:326–335. https://doi.org/10.1016/j.electacta.2016.08.037

Zhang H et al (2010) P25-graphene composite as a high performance photocatalyst. ACS Nano 4(1):380–386. https://doi.org/10.1021/nn901221k

Zhang K, Kemp KC, Chandra V (2012) Homogeneous anchoring of TiO2 nanoparticles on graphene sheets for waste water treatment. Mater Lett 81:127–130. https://doi.org/10.1016/j.matlet.2012.05.002

Zhang X et al (2013) Fabrication and characterisation of Magneli phase Ti4O 7 nanoparticles. Micro Nano Lett 8(5):251–253. https://doi.org/10.1049/mnl.2012.0963

Zhao F et al (2015) A three-dimensional graphene-TiO2 nanotube nanocomposite with exceptional photocatalytic activity for dye degradation. Appl Surf Sci 351:303–308. https://doi.org/10.1016/j.apsusc.2015.05.121

Zhao G et al (2011) Few-layered graphene oxide nanosheets as superior sorbents for heavy metal ion pollution management. Environ Sci Technol. https://doi.org/10.1021/es203439v

Zhao G et al (2012) Synthesis of graphene-based nanomaterials and their application in energy-related and environmental-related areas. RSC Adv 2(25):9286–9303. https://doi.org/10.1039/c2ra20990j

Zhu Y, Murali S, Cai W (2010) Graphene and graphene oxide: synthesis, properties, and applications. Adv Mater 22(35):3906–3924. https://doi.org/10.1002/adma.201001068

Chapter 4
Metal and Carbon-Based Nanomaterials for the Water Disinfection

Nandini Dixit, Amritanshu Shriwastav, and Swatantra P. Singh

Abstract Disinfection is an indispensable part of a water treatment system to provide high-quality water. Various disinfection-based technologies have been implemented for the same, including UV, chlorination, and ozonation, but possess limitations in terms of operations, cost, and carcinogenic by-products formation; therefore, demand such a system that can compensate for these challenges. Electrochemical disinfection is an appealing technology due to its environmentally amicable and cost-effective usage. Due to the intervention of nanomaterial sciences, this technology has shown constructive outcomes for water disinfection. Nanomaterials like carbon and metal-based nanomaterials have shown fabulous results in the field of water disinfection and can enhance the antimicrobial activity via their tunable physicochemical properties. Laser-Induced Graphene (LIG), a novel carbon-based nanomaterial has outstanding surface properties that inhibit bacterial growth and can behave as an antimicrobial surface in the presence of electricity with the simultaneous generation of hydrogen peroxide. Additionally, Metal-based nanoparticles are promising antimicrobial agents as they generate reactive oxygen species (ROS) and creates oxidative stress in the microbes. This chapter articulates the role of carbon and metal-based nanomaterials in disinfection processes and how they can overcome the limitations faced by conventional materials by providing better electrocatalytic and surface properties to escalate the disinfection process.

Keywords Electrochemical disinfection · Laser-induced graphene (LIG) · Reactive oxygen species (ROS) · Antimicrobial surface · Metal nanoparticles

4.1 Introduction

In this era of the developing world, more than 5 million people die every year due to unsafe quality of water and poor hygiene (WHO 2007). This pressing issue of poor cleaning of water needs consideration in terms of social, equitable, and economic

N. Dixit · A. Shriwastav · S. P. Singh (✉)
Environmental Science and Engineering Department, IIT Bombay, Mumbai, India
e-mail: swatantra@iitb.ac.in

S. P. Singh et al. (eds.), *Nanomaterials and Nanocomposites for Environmental Remediation*, Energy, Environment, and Sustainability,
https://doi.org/10.1007/978-981-16-3256-3_4

means. All these factors need to be encompassed for the possible best solution to deal with the bacteriological safety of the water. Chlorination is the oldest and one of the widely used technology to date. Still, due to its excessive use, the generation of harmful disinfection-by-product is increasing, which can cause various ailments, and could lead to cancer (Korshin and Yan 2018). Electrochemical disinfection, which was first used in the nineteenth century, is gaining attention in terms of its chemical-free usage but faces problems due to poor and affordable electrode material (Kraft 2008). The electro-generation of oxidative species and the electric field both create havoc to the microbes leading them to inactivate. Various metal-based electrodes have been investigated and employed for many years. However, they had shown problems like charge reversibility, corrosion and lower current density, etc. and therefore, foster the need for developing low cost, highly conductive engineered materials whose properties can be modified according to a particular application (Wang et al. 2019a, b).

Nanomaterials, especially carbon-based nanomaterials, have exerted their effect in the field of water disinfection for a decade. These materials, due to their unique properties, are versatile in many forms. Membrane-based solutions, when equipped with low-cost nanomaterials, could give an economically viable method for water disinfection. Laser-induced graphene, a 3-D porous carbon nanomaterial which when investigated as an electrochemical membrane filter, has shown good results for disinfection at an applied voltage (Singh et al. 2018a, b), and can further be investigated for its long term usage in the water cleansing technology. Its surface, as well as electrochemical properties, makes it an appealing material for water disinfection. The hydrophilic surface of the material promotes non-adhesion to the microbes. With electricity, this material can reduce oxygen to hydrogen peroxide, which can lead to the generation of reactive oxygen species (ROS). The ROS being toxic to the microbes causes cell disruption and ultimately deactivates them (Buonocore et al. 2010).

The metal-based nanomaterials have shown strong biocidal activity and could be employed in the form of composites with carbon-based nanomaterials to enhance their antimicrobial action. Metal-based nanoparticles, due to their modifiable physicochemical properties, provide a broad spectrum for its synthesis to increase its bioavailability (Wang et al. 2019b). These can be prepared in different shapes and sizes for effective action over a gamut of microbes. The nanomaterial enters the cell via endocytosis, diffuses, and encapsulates the ions which interact with different cellular components depending upon its bond forming ability (Dayem et al. 2017). They are oxidative stress inducers as they generate reactive oxygen species (ROS) inside the cells by damaging the cellular defense system. This could lead to DNA damage, protein rupturing, cell apoptosis, lipid peroxidation, etc. which ultimately destroys the cell (Singh and Dubey 2018).

This book chapter focuses on the limitations that conventional disinfection treatments are facing. Along with this, the role of electrochemical disinfection and the deficiencies faced by them in terms of electrode material is being articulated. The purpose of carbon nanoparticles and metal-based nanoparticles, mainly silver and copper, are being highlighted, which could be employed as composites or on membranes for ecclesiastical disinfection performance. We have also provided some insights into disinfection processes used in spacecraft and International Space Station.

4.2 Overview of Disinfection

Over the last few decades, water disinfection based researches have shown many improvements in terms of materials and technology. There are various technologies used for the disinfection processes, e.g., chlorination, electrochemical disinfection, iodination, ultraviolet radiation method, metal incorporating based filters, etc. Along with this, the results of various studies and experiments have manifested the effectualness of carbon-based nanomaterials as potential disinfecting agents. Whether it is graphene, carbon nanotubes, carbon nanodots or quantum dots and laser-induced graphene, etc. all have significantly strutted their effectiveness against a broad spectrum of microbes (Wang et al. 2019b).

Disinfection is the partial killing of the microbes. Four categories of human enteric organisms of most significant consequences in producing diseases are bacteria, protozoan, virus, oocysts, cysts, and helminths (Metcalf 2003). These microorganisms, if not removed or killed, may cause a plethora of waterborne diseases like cholera, typhoid, diarrhea, etc. to humans. The properties of disinfecting agents are (Simon et al. 2018)

1. It should be readily available.
2. It should not interact with extraneous material (organic matter) other than bacterial cells.
3. It should be non-corrosive, non-toxic, and non-staining.
4. It should be stable and soluble at ambient temperature.

The following section describes the conventional technologies which are currently being employed with the limitations possessed by them. The major disinfection processes used in water treatment are given in Fig. 4.1.

Fig. 4.1 Various disinfection processes used in water treatment

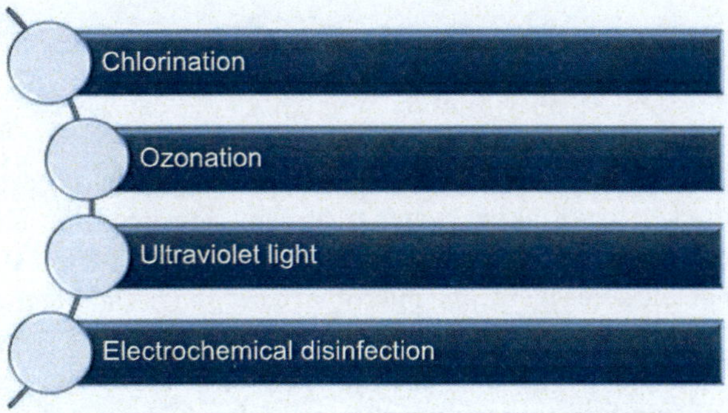

4.2.1 Conventional Disinfection Processes

4.2.1.1 Disinfection by Chlorination

Disinfection by chlorination is one of the most considerable methods used for disinfection in water and wastewater treatment facilities. Chlorine gas, hypochlorite solution (sodium or potassium hypochlorite), chlorine dioxide, and calcium hypochlorite are being employed as a disinfecting agent. Chlorine gas on hydrolysis gives hydrochloric acid and hypochlorous acid. Hypochlorous acid on further hydrolysis gives hypochlorite ion (OCl^-) ion, which is a pH-dependent species (Martínez-Huitle and Brillas 2008). Hypochlorite ion and hypochlorous acid concentrations are termed as free chlorine or active chlorine, which is responsible for the deactivation of microbes by inhibiting the enzymes responsible for their growth and metabolism. Chlorine dioxide is a yellow-green colored pungent gas characterized by high oxidizing power and has a robust germicidal action towards pathogenic microbes. The stronger action than hypochlorite and chlorine gas make it interact with extraneous materials like organic matter or other ions, which decreases the efficacy of maintaining the microbiological quality of water (Thorn et al. 2012).

$$Cl_2 + H_2O \rightarrow HOCl + HCl \tag{4.1}$$

$$HOCl \rightarrow H^+ + ClO^- \tag{4.2}$$

For distribution purposes, residual chlorine is an essential element to safeguard against microbial contamination. Generally, the range up to which residual concentration is provided lies between 0.5 and 1.0 mg/L, which is quite low up to any physiological relevance towards mammals (Metcalf 2003); however, higher concentration could be dangerous. The major problem with chlorine disinfection arises with the generation of disinfection by-products like chloroform, trihalomethanes, etc. which are potential carcinogens (Richardson et al. 2007). Additionally, the corrosiveness and toxicity of this gas is a primary issue for its operation and transportation. Due to its high oxidative power, high doses of chlorine used in the system can cause immunosuppression, infertility, cancer, etc. (Thornton and Campbell 2001).

4.2.1.2 Disinfection by UV

Ultraviolet light is one of the most commonly used technologies available in the market for drinking water treatment. In many households, it is used in combination with reverse osmosis water purification systems. UV radiation interacts with nucleic acids inside the microbial cells and permanently damages the DNA of the cell (Metcalf 2003). The nucleic acid readily absorbs UV radiation (250–280 nm) in the DNA. Most bacteria and viruses require lower doses of UV, while protozoan and cysts are significantly resistant to even higher UV doses. The inability to provide residual effect is the main disadvantage of the UV; therefore, chances of recontamination occur during distribution or storage. Along with this photoreactivation, phenomena may occur in which damaged bacterial cells regrow themselves when exposed to light (Wolfe 1990). Also, UV treated water and wastewater were found to contain antibiotic resistance genes (ABGs) and antibiotic resistance microbes (ABRs) (McKinney and Pruden 2012). The main advantage of it lies in no harmful by-product formation and easy applicability. Still, its high cost and presence of high level of suspended solids and turbidity in the water can cause unsuitability for adopting the technology for long term purposes.

4.2.1.3 Disinfection by Ozonation

Ozone is an unstable molecule that is composed of three oxygen atoms and is produced via a two-step reaction. The first step includes the dissociation of oxygen molecules into oxygen atoms in the presence of light or any other high energy source, and the second step includes the collision between atoms to form ozone. Ozone is highly unstable and has an extremely short lifetime; therefore, signified as highly reactive. Its oxidation strength is high enough to oxidize a vast domain of species (Fig. 4.2). It is usually employed by generating it in situ at the sites for immediate treatment due to its high toxicity and short-lived nature. Conventionally ozone is produced by treatment with high energy ultraviolet radiation photolytic reactions, radiochemical reactions, or via electrochemical generation through a corona discharge (Kraft 2008). Ozone produces hydroxide radicals which are responsible for disrupting proliferation and killing microbes by attacking the cell wall. Ozone is also used with many other combinations such as ozone with the ultraviolet light (UV), and

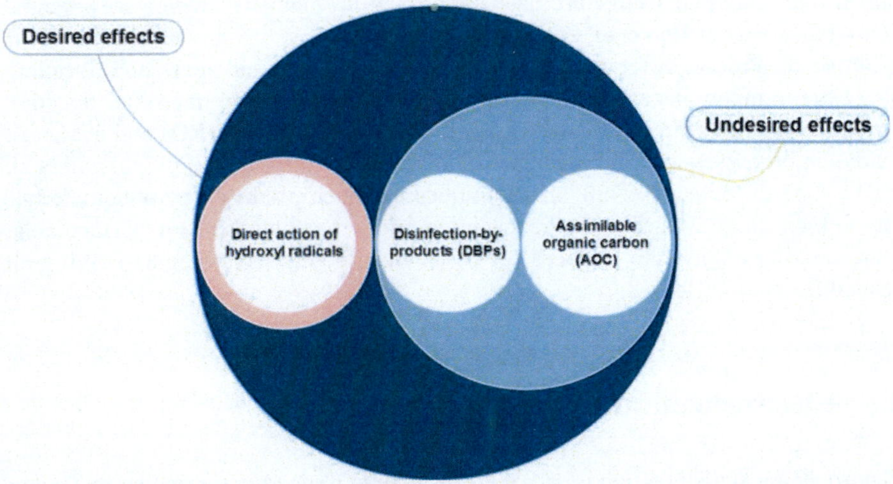

Fig. 4.2 A diagrammatic representation showing desired and undesired effects of ozone disinfection

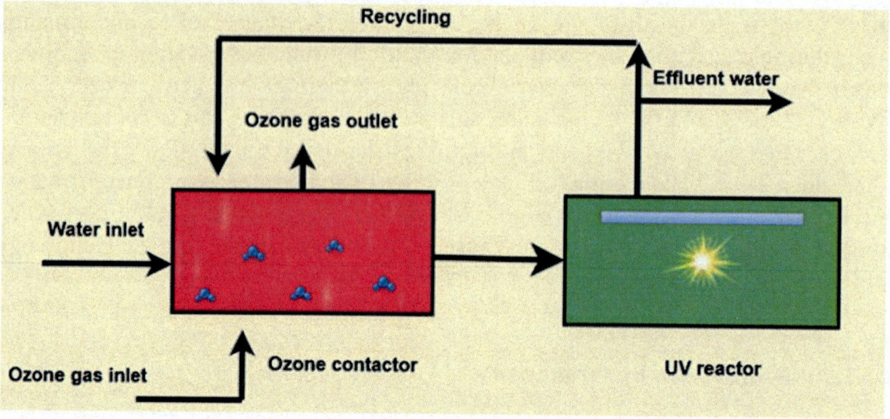

Fig. 4.3 Integrated ozone-UV based reactor. Adapted and modified from (Oh et al. 2007)

ozone with hydrogen peroxide, etc. in wastewater reuse and reclamation technology (Metcalf 2003). The integration of ozone and the UV method leads to the formation of radical reactions that are activated by the photochemical reactions between the UV and ozone. An Integrated ozone-UV disinfection reactor is shown in Fig. 4.3. The combination of the UV with ozone converts the unstable ozone to stable oxygen and generates hydrogen peroxide, which exacerbates the hydroxyl radical formation in the presence of UV light (Collivignarelli et al. 2018). The major problem associated with ozonation is the formation of assimilable organic carbon (AOC), which allows the microbes to regrowth in the water system. Therefore, biologically active filters are required for the post-treatment to ozonation, which enhances the overall cost of operation, leading to its unsuitability for smaller utilities (Wolfe 1990). Ozonation also produces DBPs like bromate and bromonitromethanes, which are potential genotoxins (Richardson et al. 2007).

Other than these, different methods, such as rapid sand and slow sand filtration, also exist in many places. Some facility uses membrane-based microbial rejection processes like ultrafiltration, nanofiltration, or reverse osmosis (RO) and advanced oxidation processes with hydrogen peroxide or ozonation, etc. (Collivignarelli et al. 2018). One of the state-of-the-art techniques has been produced by nanomaterials due to their unique tunable physicochemical properties and the large surface area; thus, creating engineered surfaces for better disinfecting technologies is lying all around (Li et al. 2015).

4.3 Electrochemical Disinfection

Electrochemical disinfection is one of the prominent methods used to disinfect water by maintaining the aesthetics and odor in a non-hostile and affordable manner. In any disinfection methodology, there are two steps, the one being primary disinfection, and the other one is the residual effect, which needs consideration during the distribution of water. Most of the electrochemical water disinfection systems rely on the electrogeneration of mixed-oxidants, which have reliable properties of inactivation and killing microbes. Along with this, techniques like electroporation, electrocution, and electrosorption have been explored (Martínez-Huitle and Brillas 2008). The primary mechanism for the inactivation of microbes occurs due to:

1. Oxidative stress due to Reactive oxygen species (ROS) or Reactive chlorine species (RCS) production which is either produced by extraneous interaction of microbial cells with toxic components such as metal nanoparticles or some toxic chemical present in the system or by the generation of radicals caused by electrochemical reactions (Suresh et al. 2013).
2. Oxidation of essential cellular components on account of induced electric field exposure which inhibits the basic functioning of the cell (Huo et al. 2017).
3. Irreversible permeabilization of cell membrane on exposure to electricity leading to leaking of cellular components or up taking of toxic materials (Ghernaout 2017).

4.3.1 Electrochlorination

The chlorine-based disinfection technique is the historical method till date and the one which is most commonly used. Electrochlorination method incorporates chloride-containing water, which can either be natural chloride or externally added chlorine in insignificant quantities lesser than the permitted concentration of chlorine in drinking water. Drinking water disinfection is generally performed via in-line or on-site electrolysis of water. The pH of the reaction mixture is of utmost importance as chloride ion changes its behavior at different pH. At pH < 4, no chloride ion would be there, but at pH > 10, chloride ion would be prominent. The pH of the disinfected water is around 6.5–8, so the overall process works between distributed pH (Thorn et al. 2012). Inactivation of microbes by electrochlorination occurs via two methods, one involves the use of brine in electrolyzer to produce free chlorine while other electrolyzes the water to produces oxidants like $OH^{\cdot}$, singlet oxygen, hydrogen peroxide or ozone (Martínez-Huitle and Brillas 2008).

Chlorine dioxide is one of the disinfection products formed during electrolysis and generates different by-products at distributed pH with different ions such as nitrates or sulfates. The presence of these ions could also produce oxidants like peroxydisulfate, peroxydiphosphate due to their preferential oxidation (Martínez-Huitle and Brillas 2008). In electrochemical chlorination, the allowed concentration of chloride ion is very low; therefore, it becomes necessary to control the diffusion of active chlorine formation into the bulk solution. In addition to this, Cl^- ions also act as a catalyst to embrace the formation of oxidants or free radicals like OH^- by involving in chain reactions thereby, increasing the activity of the radicals for cell destruction (Kapałka and Fóti 2010).

Active chlorine which is a key oxidant in electrochlorination process forms on the anode in the following manner:

$$2Cl^- \rightarrow Cl_2 + 2e^- \tag{4.3}$$

$$Cl_2 + 2OH^- \rightarrow H_2O + OCl^- + Cl^- \tag{4.4}$$

$$Cl_2 + 4H_2O \rightarrow 2ClO_2 + 8H^+ + 8e^- \tag{4.5}$$

4.3.2 Electrochemical Disinfection by Hydrogen Peroxide

Electrochemical disinfection via hydrogen peroxide involves the electro-generation of hydrogen peroxide at the cathode, leading to the inactivation of microbes due to its oxidizing nature. Water gets electrolyzed to form oxygen at the anode; the oxygen can produce hydrogen peroxide at cathode if the electrodes used can compromise a high overvoltage of oxygen to favor the reduction of it. Earlier metal electrodes, including platinum, ruthenium, iridium, etc. were used, but nowadays people are using carbon-based electrodes for generating hydrogen peroxide due to their greater abundance (Liu et al. 2015b; Wang et al. 2019a). Along with this gas diffusion electrodes are also being employed which have greater efficiency as they use the oxygen in the air surrounding the electrode to directly convert into hydrogen peroxide (Särkkä et al. 2015). These electrodes have a three-phase interface electrode configuration, which can greatly enhance the kinetics and the overall current efficiency of the gas-phase electrochemical process (Poornesh et al. 2010). The hydrogen peroxide has a lower oxidizing potential ($E_o = 2.31$ V) than chlorine and ozone and may require in higher quantities or could be employed for longer disinfection time. However, the main advantage lies in less to no by-product formation (Kraft 2008).

$$2H_2O \rightarrow O_2 + 4H^+ + 4e^- \quad \text{(Anode)} \tag{4.6}$$

$$O_2 + 2H_2O + 2e^- \rightarrow H_2O_2 + 2OH^- \quad \text{(Cathode)} \tag{4.7}$$

4.3.3 Electrocoagulation

In Electrocoagulation, coagulants are formed in-situ by electrochemical reactions on the electrodes, which tend to aggregate the suspended particulates to form flocs. The microorganism suspended in the water gets enmeshed into the flocs leading to their removal via sweep flocculation (Ghernaout et al. 2011). The anode behaves as a sacrificial electrode which loses electrons to make cations which immediately reacts with oxygen to form hydroxides. These hydroxides act as coagulants which neutralize the surface charge of the microbes that occur near the anodic zone followed by sweeping via adsorption. Along with this, depending upon the type of electrode and electrolyte used, the production of various oxidative species can cause additional effects (Ghernaout et al. 2019). At pH = 12, iron hydroxide is formed at anode according to the below reactions.

$$Fe \rightarrow Fe^{2+} + 2e^- \tag{4.8}$$

$$4Fe + 10H_2O + O_2(g) \rightarrow 4Fe(OH)_3(s) + 4H_2(g) \tag{4.9}$$

$$2H_2O + 2e^- \rightarrow 2OH^- + H_2(g) \tag{4.10}$$

Figure 4.4 is highlighting the mechanism of microbial inactivation via Electrocoagulation.

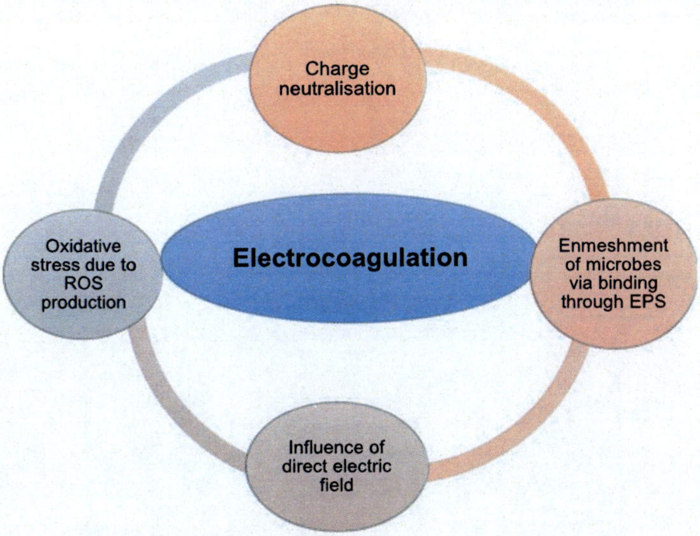

Fig. 4.4 A flowchart representing the mechanism of inactivation of microbes via Electrocoagulation. Adapted and modified from (Ghernaout et al. 2019)

4.3.4 Electro-Fenton Process

The Electro-Fenton process uses a high oxygen overvoltage electrode used for the reduction of water to produce hydrogen peroxide. When a small amount of iron is added to the mixture, iron gets oxidized and yields Fe^{3+} and OH. at pH = 3, which is a powerful oxidant even higher than H_2O_2, as shown in Fig. 4.5 (Brillas et al. 2009). This process efficiently kills a diversity of microbes and can be modified via different means to optimize the pH. In advance oxidation, this method has successfully shown its potential for the degradation of recalcitrant compounds and heavy metals. The incorporation of nanoparticles into this process could lead to enhanced efficiency. Various membrane and light assisted Fenton process has been developed for this with pH control technologies to drive its potential in multiple sectors of wastewater remediation (Liu et al. 2007).

Applications of Electro-Fenton process

1. In one study, carbon nanotubes membrane stack filter was designed for killing microbiota through in-situ Electro-Fenton reagent generation. A carbon nanotube gridded cathode, CNT-COOFe^{2+} cathode for the production of

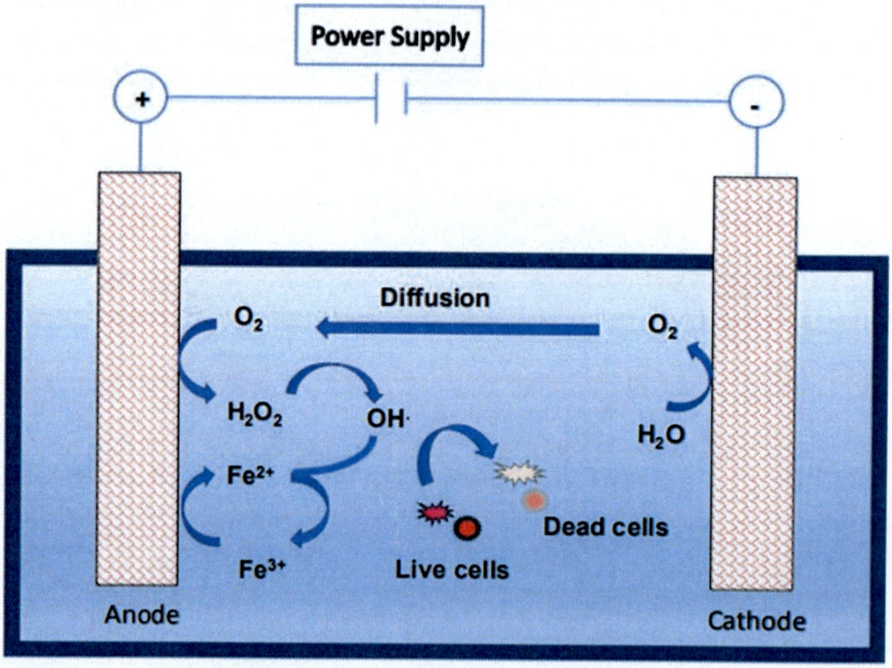

Fig. 4.5 A schematic showing inactivation of microbes in an Electro-Fenton process. Adapted and modified from (Chen et al. 2019)

hydroxyl radical through H_2O_2 and Fe^{2+} in-situ, and a gas diffusion porous PVDF and PTFE separator along with CNT filter anode (Gao et al. 2015).
2. A recently developed decentralized system with the Fenton process is used for the domestic sewage system, consisting of degradation from organics to a wide range of microorganisms in the wastewater. This system is a stacked flow-through reactor consists of graphite felt cathode and DSA mesh anode employed at voltage around 2.5 V and flow rate 30 mL/sec. The overall concentration of bacteria was reached to 4 log reduction (Ren et al. 2020).
3. A pH-responsive membrane-based filtration assembly is being designed in which a two-stage stacked configuration system is produced. The upper assembly generates hydrogen peroxide, and the lower one is incorporated with iron nanoparticles with a pH-responsive membrane for filtration (Li et al. 2015). This type of membrane is prepared by brushing pH-sensitive chemicals or grafting polymers whose side chains are sensitive to a particular pH (Wandera et al. 2010).

4.3.5 Problems Related to Conventional Electrode Usage

The electrodes of various materials have been employed in electrochemical disinfection processes including IrO_2, RuO_2, or combined IrO_2/RuO_2, boron-doped diamond, diamond thin-film electrodes, graphite, platinum electrodes, iron electrode, dimensionally stable electrodes or mixed metal electrodes generally coated with titanium, etc. (Gonzalez-Rivas et al. 2019). These materials faced a variety of problems, which led to the possible usage of novel materials, which showed enhanced efficiency of the overall process. The major disadvantages include:

1. **Low current efficiency**: Lower current efficiency affects both the electrogeneration of the oxidants and the working capacity of the electrode. This hinders the electrode usage for long term applications (Thorn et al. 2012). Due to lower current efficiency, the amount of hydrogen peroxide produced would be lesser than the theoretically calculated amount, which ultimately promotes lesser antimicrobial effect. Also, lower current efficiency would significantly increase the specific cost of an electrolyzer (Särkkä et al. 2015).
2. **Polarity reversal**: Many metal electrodes undergo polarity reversal due to the accumulation of species on it, which causes the functioning of the electrode to change. It is a major problem when the scaling of electrodes occurs due to the presence of calcium or magnesium salts in the water. This type of process was earlier used to clean the electrode, but nowadays, methods like ultrasonication, etc. are under employment (Kerwick et al. 2005). As a result of polarity reversal, the life expectancy of an electrode decreases.
3. **Low oxygen overvoltage**: Most of the metal electrodes have shown lower oxygen overvoltage. Overvoltage is the voltage that is applied over the equilibrium voltage. In a redox reaction at a certain voltage, ions have to move from the bulk to the interface of the electrode, where it encounters resistance from the

neighboring ions. Therefore, for removing the resistance, some extra potential is provided, which is termed as overpotential or overvoltage. This could be solved by using a catalyst that can help in providing that extra voltage. The problem of lowered catalytic activity of the electrode for oxygen reduction reduces the rate of production of oxidative species like hydrogen peroxide or RCS etc. (Kraft et al. 2006).

4.4 Applications of Disinfection in Space

In space missions, the chances of the contamination may arise from different assemblies, which create an environment conducive for microbes to thrive. Various strains of bacteria and fungi have been found in the spacecraft, including Staphylococcus, Micrococcus, *Bacillus, Aspergillus, Penicillium, Cladosporium.* These strains were found in the air in a study done by the USA space shuttle program. The average pre-flight level of bacterial concentration should be 300 CFU/100 cm^2, while fungal concentrations are 100 CFU/100 cm^2. Bacterial level in potable water is generally very low, and *Burkholderia cepacia* is the most commonly cultured bacterial species. Different techniques of disinfection are being employed for various spacecraft missions depending upon the duration, which provides better habitability to the crew members in the flight (Pierson 2019).

A wide variety of disinfection techniques exists, but only some of them are useful for manned spacecraft missions. Historically, disinfection was performed by chlorination, iodination, fuel cells, water condensate, and ozone, but the choice of disinfectant depends upon journey to journey (Gerald and Richard 1987). Long-time missions could employ technologies which involve fewer chemicals and are energy-intensive while shorter missions could employ chemical-based technologies due to lesser storage facility instead of energy-intensive ones.

ATV, i.e., Automated Transport Vehicle, is a program launched by the European Space Agency for transporting and delivering potable water to International Space Station (ISS) (Lobascio et al. 2004). There are two different types of potable water standards defined by Russian and US agencies which are:

1. The Russian water is mineralized and disinfected with silver.
2. The US water is minimally mineralized and disinfected with iodine.

The use of iodine was found to be reliable than chlorine for disinfection because chlorine has higher vapor pressure, which tends to make its delivery complex. Due to its corrosive nature, its handling becomes difficult, and reaction with organic matter to produce DBPs makes it unsuitable to be used in microgravity. While iodine has a lower vapor pressure, and its handling is easier. Additionally, iodine is solid at room temperature while chlorine is a gas; therefore, iodine formed through electrolytic method would not pass through the gas to the valve, which can happen with chlorine. Therefore, the problem of outgassing was not found with I_2 (Punyani et al. 2006).

4.4.1 Disinfection via Iodine

Iodine was firstly used to control microbial contamination in the First World War to disinfect drinking water for troops. The US Army in the Second World War used iodine in the form of globulin (tetraglycine hydroperiodide). Iodine has been employed in various ways comprising of organic iodine compounds, iodophors, and iodine incorporated resins (Backer and Hollowell 2000).

NASA used the iodine-polyvinyl pyrrolidone complex for the controlled release of iodine on its flights at the levels of 2–3 μg/mL of molecular iodine I_2 and 0.5–1.5 μg/mL of iodide. The complexation of molecular iodine with ligands enhances its bioavailability, provides complexation sites for biological activity, and changes its morphology (Moulay 2013). Although the main biocidal effect is caused by iodine as iodide has no biocidal property, and it just goes inside the body and gets absorbed by the blood. A Microbial Check Valve (MCV), a flow-through device containing an iodinated polymer, was made, which imparts a bacteriostatic residual concentration up to 2 mg/L (Little et al. 2020).

An enzyme-based iodine disinfectant system is discovered, which included a two-stage iodine resin filter. Firstly, microfiltration is employed to remove cryptosporidium oocysts that are resistant to halogen disinfection, and the next stage includes granulated activated charcoal to further reduce the concentration of iodine in effluent water. A pentaiodide resin was investigated by NASA for the US space mission, which contains more than 70% of elemental iodine. This resin has a low residual effect but can provide effective disinfection by killing bacteria up to 10^9 CFU/mL with the contact time of 10 s with the resin (Punyani et al. 2006).

Disadvantages of iodine usage

NASA has employed various iodine-based composites for disinfection; however, significant drawbacks of its use had found on the health of crew members. Iodine species accumulate in the thyroid and produce a noticeable change in the taste of drinking water. Additionally, it may also result in the iodine disinfection products, which could be either allergic, cytotoxic, or genotoxic (Thorstenson et al. 1987). Thus, it becomes necessary to remove the bulk of the iodine before consumption. An activated carbon exchange iodine removal unit (ACTEX) is used for removing the iodine, which increases the consumable cost and burdens up the Equivalent System Mass (ESM) of the iodine system thus, makes the system prohibitive for extended duration missions (Roberts et al. 2007).

4.4.2 Disinfection by Silver

Silver is found to be an effective biocide and provides a potential advantage over iodine and could be safely consumed by the crew members if released in a controlled way. It has been employed by the Russian Space Agency for disinfection purposes.

The agency uses silver via an electrochemical method for releasing the silver ion in the form of ionic silver or colloidal silver (Li et al. 2018). The precipitate AgCl formed is removed through microfiltration, and the water is transported to a container coated with silver. The material compatibility is one of the issues, and wetted materials of construction are required for maintaining the biocidal concentration of silver to adequate levels (Roberts et al. 2007).

Silver is a potent antimicrobial agent and, once taken in the body, can subsequently cross biological barriers and accumulate in the tissues, including lungs, kidneys, stomach, brain, and blood. In spacecraft, oral ingestion can be the main route for silver exposure. Thus, to take care of the harmful effects of silver accumulation, strict regulations have been set up by the space agencies. The maximum allowable concentration of silver in a 100 days mission is 0.36 ppm/day (Lobascio et al. 2004).

A silver-coated filter media has been used in which silver was coated on the sand, diatomaceous, and clay-like material by AgNO3 solution followed by heating. The increase in contact time was achieved by keeping the disinfected water in a container containing silver nitrate solution. The concentration of the silver is maintained to be in the range of 10–50 ppb. Another method involves the use of activated carbon doped with the silver atom. Ion-exchange resin has also been used in several flights, such as silver zirconium resins (Ag-Zr-P), which is a silver substituted hexagonal zirconium phosphate (Li et al. 2018). A recently developed technique by NASA for the active release of silver in the solution is being developed. Silver nanoparticles incorporated polyurethane foam has found to be an effective biocide along with the controlled release of silver ion to prevent the accumulation of it in the body, as shown in Fig. 4.6 (Lee et al. 2020). This method compatibility with various materials has been checked for its suitability for long term space missions. Polyurethane foam is found to be a compatible material that is widely used for packaging and does not have any stability issues. This method was proved to be a fruitful way to deliver silver in a controlled manner; however, more testing on its suitability needs to be performed (Lee et al. 2020).

4.4.3 Electrochemical Systems for Disinfection

Currently, the US and Russian space agency are relying on iodine and silver-based disinfection. However, a secondary method for disinfection has to be investigated for a future space mission. NASA has examined electrochemical systems generating ozone, hydrogen peroxide, peracetic acid, and sodium hypochlorite (Rodriguez et al. 2013).

An electrochemical hydrogen peroxide generator was invented by using a gas-diffusion anode and a platinized titanium cathode, which are separated by a cation-exchange membrane. The process produced high purity H_2O_2 as there was no use of any chemicals in the medium, and only 2-electron oxygen reduction happens on the cathode. The catalyst mix contains a quaternary ammonium compound, which enabled enhanced production and current efficiency of the process (Tennakoon et al.

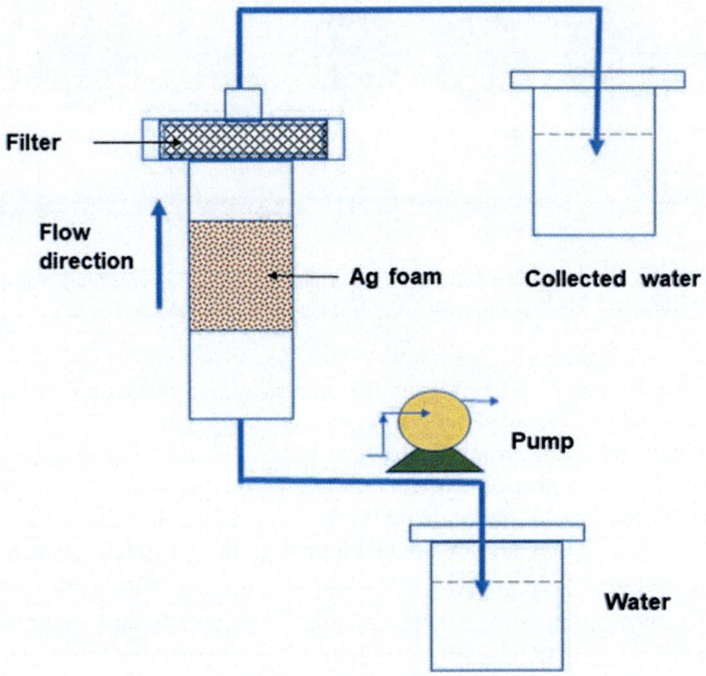

Fig. 4.6 A schematic showing polyurethane foam incorporated silver ion disinfection process in space. Adapted and modified from (Lee et al. 2020)

2010). Another method for generating ozone for short-term missions was examined in which cathode and anode were separated by a proton exchange membrane, which was made up of perfluorinated sulfonic acid polymer: the O_3 and O_2 gases generated at the anode while H_2 is generated at the cathode. The composition of ozone in the mixture of gases was separated by choice of electrolyte, which produces high overvoltage for oxygen generation. The gases were collected, diffused in the water, and sent for the ozone consumption process. Also, a refrigeration subsystem was there, which removes the heat generated by the electrolyzer and maintains temperature around 35 °C to enhance the dissolution of ozone for a longer time (Johnson 2003).

4.5 Nanocarbon Materials for Water Disinfection

Carbon-based nanomaterials have revolutionized the whole era of water disinfection system by engaging it's unique, state-of-the-art properties towards bacterial inhibition (Wang et al. 2019b). Various types of carbon-based nanomaterials have been evolved and studied for a gamut of applications ranging from biological applications to electronic media. However, the interactions of nanocarbon materials with human cells, i.e., the toxicity issues, constitute a significant concern (Li et al. 2015). Along

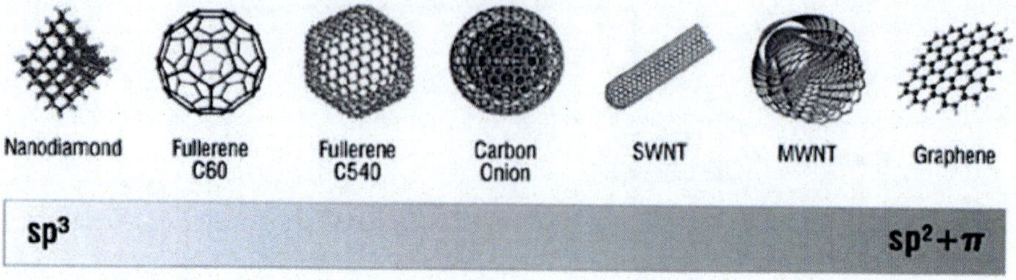

Fig. 4.7 Different types of carbon-based nanomaterials. Reprinted (adapted) with permission from (Mauter and Elimelech 2008). Copyright (2008) American Chemical Society

with this, its cost, recyclability, reusability, and sustainability issues still need to be revamped (Jin and Maduraiveeran 2018).

The various types of carbon-based nanomaterials include graphene, graphene oxide (GO), reduced graphene oxide (rGO), carbon nanotubes, nanodiamonds, fullerenes, carbon dots, graphene quantum dots and laser-induced graphene (LIG) as shown in Fig. 4.7. The physicochemical, thermal, or photocatalytic transmogrifying effects may ameliorate the properties of these materials for effective action.

4.5.1 Fullerenes

Fullerene is one of the allotropes of carbon, and the most common one is Buckminsterfullerene, i.e., C60. The other one includes C540, C70, etc. having structure similitudes to a football containing hexagonal and pentagonal rings. Fullerenes are inert and do not exhibit any antimicrobial property instead when chemically modulated with chemical reagents. They promote the light-induced generation of reactive oxygen species (ROS) which causes lipid peroxidation, cellular component leakage, DNA damage, etc. (Wang et al. 2019b).

4.5.2 Nanodiamonds

Nanodiamonds are nano-sized diamonds having a size range of 2–10 nm comprising sp^3 hybridized carbon core with an sp^2 hybridized carbon shell prepared by the detonation of explosive materials (Mochalin et al. 2012). The antibacterial property of the material could be obtained by modifying the surface chemistry. The functionalization of the surface is an essential aspect of the antimicrobial action of the materials. Chemically reconstructed nanodiamonds, more specifically partially oxidized negative surfaces, show antimicrobial properties (Mochalin et al. 2012; Wang et al. 2019b). Glycan modified nanodiamonds show enhanced antibacterial activity with comparatively less cytotoxicity. A copolymer functionalized nanodiamond creates hydrophilic effects via cations embedment, creates repulsive behavior in the material, leading it to break in exceedingly small particles ranging from 1 to 10 nm, respectively. This type of technique has elucidated the way of nanomaterial size miniaturization, which helps in driving the physicochemical properties for intended applications (Barras et al. 2013).

4.5.3 Carbon Dots

Carbon dots or carbon quantum dots are semiconducting materials used in various optoelectronic or biomedical applications due to photoluminescent, highly conducting, and large surface area properties. Various studies have depicted their antimicrobial properties against various species. Carbon dots do not exhibit toxicity to cells until modified by some capping agent and show comparatively less cytotoxicity than other carbon precursors (Liu et al. 2015a). Their structure resembles a conjugated center with some oxygenated functional groups like carboxyl, hydroxyl, and aldehyde. Carbon dots are a mixture of both graphene quantum dots and carbon quantum dots but exhibit differences in terms of crystallography (Scheerschmidt and Werner). Graphene quantum dots are sp^2 hybridized and crystalline, while carbon quantum dots are sp^3 and amorphous. Along with this, their morphologies might differ depending upon the nature of the precursor used, which leads to a different surface to volume ratio, hydrodynamic diameter, and biocompatibility. They can be prepared by either top-down or bottom-up approach, but the most preferable one is the bottom-up approach because it provides the flexibility of the precursor used and creates less to almost no wastage than top-down approach (Wang et al. 2019a).

A recent study showed bacterial inhibition against gram-positive bacteria comprising *Staphylococcus aureus*, Methicillin-resistant *Staphylococcus aureus* (MRSA), *Staphylococcus epidermis*, *Enterococcus faecalis*. In this study, quaternary ammonium carbon dots were prepared by 2,3-epoxytrimethylammonium chloride and diallyl dimethylammonium chloride via a green synthesis (Zhao et al. 2020). Along with this, various studies have been performed with doped photocatalysts to heighten the antimicrobial or electronic properties. The multiple factors which affect the properties of carbon dots include surface charge, size, and surface curvature of the substance, i.e., the surface curvature of the material should match with bacteria for proper association (Wang et al. 2019b).

4.5.4 Graphene and Laser-Induced Graphene

Graphene is prepared by either exfoliation of graphite or by a thermo-chemical bottom-up approach using carbon precursors. Graphene has been extensively studied for its antimicrobial properties due to its sharp edges, which, when perpendicularly interacts with the cell membrane damages it and leaks out the cellular components present inside it (Pham et al. 2015). Graphene oxide, a derivative of graphene with the oxidized functional group present on the basal plane and the edges of it. The functionalization with oxygen-containing atoms makes it hydrophilic, which engages its property against biofouling and microbial attachment. In some cases, doping with various metal nanoparticles can lead to the generation of ROS, which disrupts the antioxidant defense system along with lipid peroxidation or DNA damage, as shown in Fig. 4.8. A wide variety of studies have been carried out on functionalization or reconstruction of graphene structure with various doped heteroatoms or photocatalysts or by using environmentally non-hostile precursors, which makes it a versatile agent for deactivating a broad range of microbial species (Moosavi et al. 2015).

Laser-induced graphene a 3-D porous carbon-based nanomaterial that is prepared by scribing laser on any carbonaceous surface (Ye et al. 2019). It has explicitly shown remediating properties for environmental systems, whether it is water or air (Singh et al. 2017; Stanford et al. 2019). LIG based electrochemical filtration technique had shown a great replacement of metal-based electrochemical disinfection method due to its high surface area and conductivity (Singh et al. 2017). The tuneable property of the nanomaterial could make it super hydrophilic, i.e., contact angle = 0 °C, which hinders the biofilm formation over the surface (Li et al. 2017; Singh et al. 2018b). These properties provide an ideal solution over various chemicals and energy-intensive disinfection processes, which have numerous disadvantages in terms of energy, cost, and suitability.

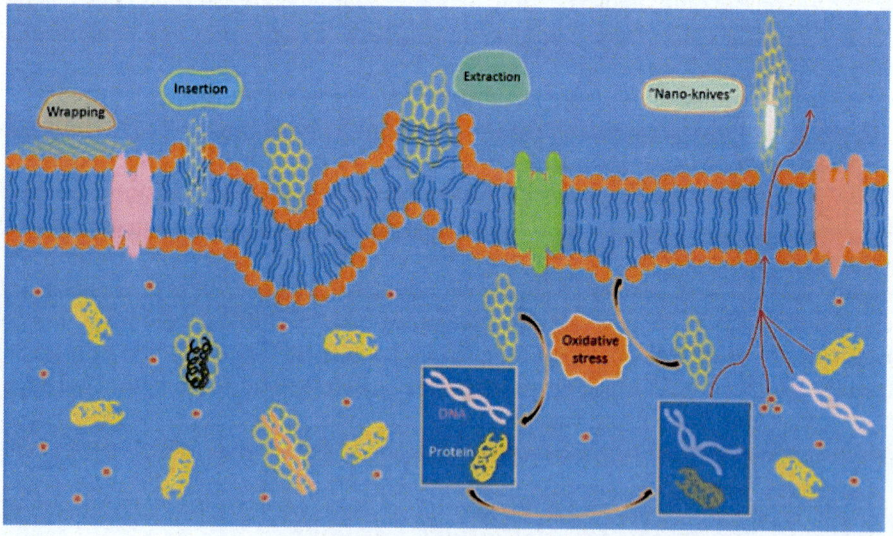

Fig. 4.8 Antimicrobial action of graphene-based nanoparticles. Reprinted (adapted) with permission from (Zou et al. 2016). Copyright (2016) American Chemical Society

4.6 Mechanism of Electrochemical Disinfection by LIG

Carbon materials have proven to be a propitious solution for the electrosynthesis of hydrogen peroxide due to their high ampleness, electro-reduction activity, and low cost. Different carbon materials like graphite, carbon fibers, N-doped porous carbon fibers, and carbon nanotubes, etc. have shown electro-reduction activity for the synthesis of hydrogen peroxide (You et al. 2012).

Laser-induced graphene is a carbon-based nanomaterial whose excellent surface property inhibits bacterial attachment onto it. The surface charge or zeta potential of the LIG creates electrostatic repulsion to the bacteria, thereby hindering the adherence of it. Along with this, when it is used as an electrode, it can act as an electro-catalyst to generate hydrogen peroxide (Singh et al. 2017). Its high surface area, large porous volume, and good electrical conductivity make it an ideal material to behave as electrode surfaces. The electrochemical generation of hydrogen peroxide can occur either via a two-electron transfer oxygen reduction reaction or a four-electron transfer water reduction (Das et al. 2020).

The H_2O_2 generation reaction scheme is complex because of the simultaneous formation of consecutive reactions, all of which are thermodynamically favorable and exothermic. Among these, the unwanted reactions are—(a) the formation of water (Reaction 4.3), (b) the decomposition of hydrogen peroxide (Reaction 4.1), and (c) the reduction of H_2O_2 (Reaction 4.2) as shown in Fig. 4.9. Each of these reactions can be favored depending upon the type of catalyst used, the promoters, or additives in the reaction medium.

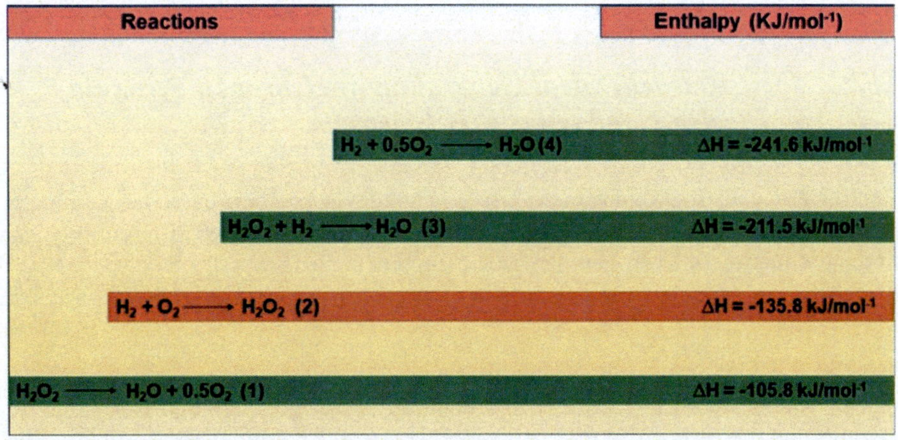

Fig. 4.9 A scheme of reactions involving in synthesis, decomposition, and reduction of hydrogen peroxide. Adapted and modified from (Das et al. 2020)

4.6.1 *Advantages of Carbon-Based Electrode*

Carbon-based electrodes reduce the overpotential of oxygen, thereby favoring the reduction of oxygen, thus, increases the selectivity of the reaction. The use of carbon-based materials facilitates oxygen transfer on the electro-catalyst at atmospheric pressure by providing high surface area. These types of electrodes promote specific sites at which hydrogen peroxide generation occurs. Also, it has been shown that the oxidation treatment of nanostructured carbon generates abundant oxygen functional groups, which may tailor the electronic structure of carbon material significantly to modulate their O_2 reduction activity. The calculations shown by density functional theory depicts that carbon atoms which are adjacent to the oxygen functional group like -COO and C-O-C are the active sites for oxygen reduction (Lu et al. 2018).

One of the studies on carbon nanotubes found that these graphene folded tubes reduced the overpotential of the oxygen reduction reaction up to 0.06 V. The over-potential of this reaction mainly occurs due to hydrogenation of oxygen or reduction of OOH^* to form H_2O_2, as shown in the reactions (4.11–4.13) (Lu et al. 2018).

$$O_2 + H_2O + e^- \rightarrow OOH^* + OH^- \tag{4.11}$$

$$OOH^* + e^- \rightarrow HO_2^- \tag{4.12}$$

$$HO_2^- + e^- \rightarrow H_2O_2 \tag{4.13}$$

4.6.2 *The Electrochemical Generation of Hydrogen Peroxide on Electro-Conductive LIG Electrodes*

Electrochemically hydrogen peroxide is generated when carbon-based laser-induced graphene electrodes are used. The background electrolyte causes different impacts on the production of hydrogen peroxide (Singh et al. 2017).

1. **NaCl as electrolyte**

H_2O_2 production is less when NaCl is present. Since the reduction potential of OH_2^- ions are higher than Cl^- therefore, preferentially hydrogen peroxide is formed via reduction on the cathode. While on the anode reactive chlorine species (RCS) like dichloride radical anion (Cl_2), hypochlorous acid (HOCl) and hypochlorite (OCl^-) get oxidatively generated (Lim and Hoffmann 2019).

$$\textbf{Anode}: -2\,Cl^-(aq) \rightarrow Cl_2(g) + 2e^- \tag{4.14}$$

$$\textbf{Cathode}: -O_2(g) + 2H^+(aq) + 2e^- \rightarrow H_2O_2 \quad (4.15)$$

Despite the oxidation of chlorine at the anode, no chlorine gas or reactive chlorine species was detected. Reactive chlorine species (RCS) contribute to the decomposition of hydrogen peroxide. The bimolecular rate constant for the reaction of OCl^- with H_2O_2 at $Pk_a = 7.6$ is 7.5×10^3 $M^{-1}S^{-1}$, which is very high (Barazesh et al. 2015).

Along with low H_2O_2 production, the concentration of reactive oxygen species is too low to be detected. It might have occurred that RCS generated on the anodic surface gets reduced back to chloride on the cathode since cathodic chloride reduction is competitive with the reduction of O_2, leading to H_2O_2 production. Reactions 4.16–4.21 are showing different ways by which cathodic reduction of RCS can occur.

$$H_2O_2 + Cl^{\cdot} \rightarrow HOO^{\cdot} + H^+ + Cl^- \quad (4.16)$$

$$H_2O_2 + Cl_2^{\cdot -} \rightarrow HOO^{\cdot} \cdot + H^+ + 2Cl^- \quad (4.17)$$

$$H_2O_2 + HClO \rightarrow H_2O + O_2 + H^+ + Cl^- \quad (4.18)$$

$$H_2O_2 + OCl^- \rightarrow Cl^- + O_2 + H_2O \quad (4.19)$$

$$HClO + H^+ + 2e^- \rightarrow Cl^- + H_2O \quad (4.20)$$

$$OCl^- + H_2O + 2e^- \rightarrow Cl^- + 2OH^- \quad (4.21)$$

Therefore, low concentrations of hydrogen peroxide generation in NaCl electrolyte could occur due to reaction with reactive chlorine species and competitive reduction between oxygen and chloride (Yu 2004).

2. **Na_2SO_4as electrolyte**

It was observed that hydrogen peroxide generation was significantly increased when Na_2SO_4 is used as an electrolyte. It may be assumed that the anodic oxidation of sulfate ($SO_4{}^{2-}$) may have been occurred to form ($SO_4{\cdot}^-$). Also, less to no decomposition of hydrogen peroxide might have occurred because it is decomposable only in alkaline conditions. Above pH 5–6, the decomposition of the H_2O_2 increases sharply. Although the anodic decomposition of hydrogen peroxide and sulfate ion is competitive due to the high oxidation potential of H_2O_2 than sulfate (Lim and Hoffmann 2019).

4.6.3 Electroporation

Another mechanism that could happen is electroporation. Electroporation is a phenomenon in which destabilization of specific regions in the cell membrane occurs when the induced transmembrane potential becomes greater than the critical transmembrane potential or threshold membrane potential under the influence of direct electric field as shown in Fig. 4.10 (Liu et al. 2013). The destabilization causes the nano-sized pore to expand either reversibly or irreversibly depending upon the strength of the electric field, and the range lies between 0.5 and 25 mV in case of reversible electroporation while for irreversible electroporation it can go up to 1 kV (Calvin and Hanawalt 1988). In reversible electroporation, the pore gets resealed after a certain time when the electric field is removed, unlike irreversible electroporation in which the pore expands, and complete deformation occurs, which leads to cell lysis or cell death (Kotnik et al. 2015).

Factors depend upon electroporation (Chang et al. 1991):

(a) The geometry of the cell
(b) Conductivity
(c) Permittivity
(d) External medium
(e) Applied electric field
(f) Type of electrode
(g) No. of pulses
(h) Time duration applied for an electric field.

Electroporation causes the realignment of phospholipids present in the cell membrane. The lipid bilayer configuration gets changed and separated by a pore, which permits the passage of extracellular material to pass through the membrane. Disruption of the metabolic process and change in the shape of cells occurs along with the production of toxic products, which eventually leads to death. As a result of the induction of the electric field, the nano-sized pore expands exponentially in size

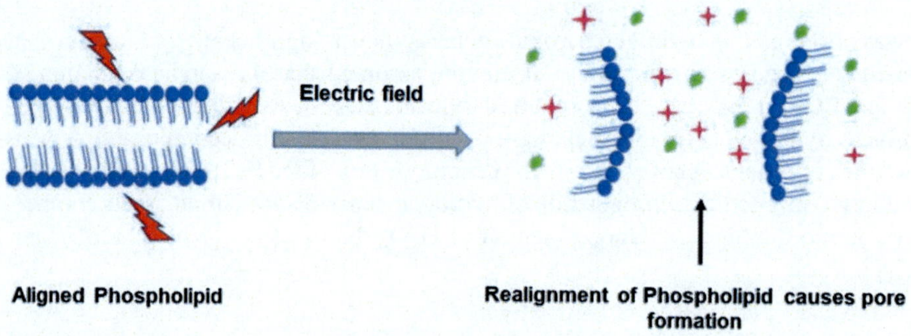

Fig. 4.10 Schematic showing electroporation in the cell membrane under the influence of the electric field

and number and thereby increases the permeability of exogenous materials inside the cell membrane, which is known as electropermeabilization (Aslam et al. 2019).

4.6.4 Mechanism of Hydrogen Peroxide Interaction with the Cells

Hydrogen peroxide is a widely known disinfecting agent used for sterilizing various items such as food, medical devices, water, etc. In biology, it is known as the primary transmitter of redox signals (Särkkä et al. 2015). It reacts poorly with most of the biological molecules due to its low activation energy. However, it's one-electron—reduction product, i.e., hydroxyl radical, is highly oxidative ($E_o = 2.31$ V) and have high reactivity due to its low activation energy. The oxidation reactions occur mostly at close to the diffusion-controlled rate (Winterbourn 2013). The major attribute to biological damage by H_2O_2 is due to the generation of hydroxyl radicals, reactions with transition metal centers, and redox signaling to oxidation of thiol proteins in regulating pathways (Imlay 2003).

Many studies have shown that it is reactive with proteins, lipids, and nucleic acids, which are the vital components in a cell. Metals like iron and copper cleave the O-O bond, which is electrophilic to form hydroxyl radicals or activated metal complexes that oxidize better than H_2O_2 like low molecular weight chelates, heme peroxidases or redox-active metalloproteins like iron/sulfur proteins. The activity of the protein is rendered by the oxidizing modified amino acids at the binding sites. Low molecular weight thiols like cysteine can directly be targeted by H_2O_2. Thiols get oxidized to sulfenic acids, which again can be oxidized to sulfinic acids, as shown in Fig. 4.11. Sulfenic acid acts as an oxidative stress mediator, creates oxidative stress in the

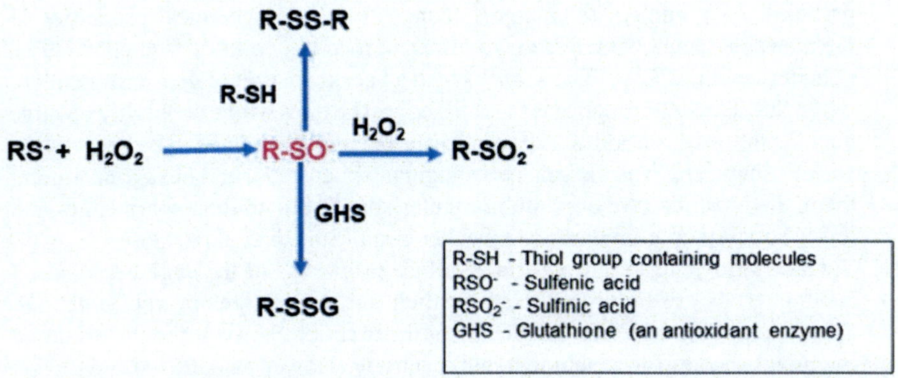

Fig. 4.11 A schematic showing reaction of hydrogen peroxide with thiol- group to form sulfenic acid, a main precursor for inducing oxidative stress. Adapted and modified from (Winterbourn 2013)

microbes leading them to stop proliferation and, ultimately, death even in mill molar concentrations, which is beyond physiological relevance (Imlay 2008).

4.7 Metal Nanoparticles as Antimicrobial Agents

Metal nanoparticles have been vastly used as inhibitory agents for a broad spectrum of microbes ranging from bacteria, viruses, protozoa, and fungi. Nanotechnology has given various techniques to tune or transform the properties of these metals on the nanoscale, which evinced their felicitousness for antimicrobial activity. Many studies have been published for silver nanomaterials as an effective antimicrobial agent. Numerous transitions, metal, metal nanoparticles, metal oxides, or different metal-based complexes have shown potent disinfectant properties for plenty of applications. Silver-based nanomaterials have been extensively researched for long years and proved their versatility in various fields such as biomedicine, biosensor, disinfectant, cancer therapies, orthopedic infections, etc. (Foldbjerg et al. 2011; Tugulea et al. 2014). Many other transition metals such as copper, nickel, zinc, manganese, etc. are also effective against different microbial species and currently being further researched for implementation (Li et al. 2019).

Factors affecting metal nanoparticles biocidal activity

The tuneable physiochemical characteristics of metal-based nanomaterials depend upon various factors that influence their bioavailability and biocidal activity towards microorganisms. The different factors include:

1. **Size and shape**: The size of the particle depicts the surface to volume ratio, which gives the maximum available surface area to which the particle can contact. Smaller particles give a larger surface area as compared to larger particles. As particle size changes, changes in physicochemical properties of the material occurs, which leads to different reactivity and surface chemistries (Chatterjee et al. 2014). The shape is also a necessary trait to demonstrate reactivity. Surfaces with rough sides and edges are the ones which are highly reactive than the one with smoother surfaces. Different nanoparticles can be prepared in various shapes such as the square, triangular, sphere, rod, and hexagon. Among them, the most reactive ones are triangular-shaped due to their sharp edges and lower bonding coordination, i.e., weaker bond (Suresh et al. 2013).
2. **Surface charge or zeta potential**: Surface properties of the engineered based nanomaterials are highly affected by their surface charges or zeta potential. Surface charges influence the electrostatic attraction between the material and microbial species for attachment and reactivity (Howes et al. 2014).
3. **Method of preparation**: Preparation methodologies also affect the biocidal activity of nanoparticles. Non-toxic nanomaterials could be toxic to species if prepared by toxic solvents such as formaldehyde, etc. The formulation of the

material and the remnants of solvents could affect cytotoxicity in the material's property (Khezerlou et al. 2018).

4. **Surface coating**: Nanoparticles are always surrounded by some covering agent or envelope, which affects the reactivity of the metal towards species by influencing its surface charge. Different factors such as pH, type of solvent, electrochemical potential, and the presence of other metal ions impinge the chemical reactivity. The surface coating also determines the fate and transport of the nanomaterial by determining the overall reactivity and bioavailability of it (Khan and Javed 2018).

4.7.1 Role of Reactive Oxygen Species (ROS) and Genotoxicity

Reactive oxygen species are a group of species that includes oxygen radicals and some non-radical species which are oxidizing and can form radicals. Non-radical species include HOCl, HOBr, O_3, $ONOO^-$, O_2, H_2O_2, etc. ROS are produced by the reduction of water in the cells. ROS triggers the antioxidant defense system leading to cell apoptosis. ROS production is not harmful until it gets accumulated and starts disrupting the enzymes required for the proper functioning of the system (Buonocore et al. 2010). In the mitochondrial membrane, the oxygen reductant is ubisemiquinone, which is responsible for ROS production. Different types of ROS are produced endogenously involving mainly superoxide ion, hydroxyl ion, or singlet oxygen. The exogenous production of ROS could occur through applications of different nanomaterials-based chemicals, harmful metals, and radiations, etc. (Simon et al. 2000).

4.7.1.1 Production of Reactive Oxygen Species (ROS)

The main sources of intracellular generation of ROS are mitochondria, endoplasmic reticulum, peroxisomes, microsomes, and NOX complexes. During the formation of ATP and water via mitochondria, a small amount of oxygen is produced. The complex I NADH (ubiquinone oxidoreductase) and complex III (co-enzyme Q, bc1 complex and ubiquinone or cytochrome c-reductase) produces superoxide anion, which is the first step of ROS generation. The anion is formed in between the mitochondrial matrix and intermembrane space, respectively. The metal ions released by the nanoparticles act as catalysts and convert the superoxide ion into a more stable hydrogen peroxide form (Dayem et al. 2017). Nanoparticles enter the cell via encapsulation within vesicles by endocytosis (Iversen et al. 2011). Different factors like interaction with mitochondrial component, free radical reaction, activation of growth factors, and NOX depend upon the formation of ROS. Extracellular generation of ROS can also occur via xenobiotic components, radiations, and metal nanoparticles. Metal nanoparticles such as iron (Fe), chromium (Cr), cobalt (Co), and vanadium

(Va) react with hydrogen peroxide to form OH radical via Haber–Weiss reaction in which metal nanoparticles react with hydrogen peroxide to form hydroxyl radicals (Valko et al. 2006).

4.7.2 *Biocidal Activity of Reactive Oxygen Species (ROS)*

The singlet oxygen, a potent ROS damages the DNA by directly oxidizing the proteins, lipids, and nucleic acids. Mitochondrial dysfunction is one of the main impacts of ROS on cell membranes and occurs when oxidative stress increases (Rezvani et al. 2019). Nanoparticles also release endogenous NO by nitric oxide synthase enzymes, which may lead to the association of genomic instability. A small amount of superoxide is released during oxidative phosphorylation and is balanced by the antioxidant enzymes such as glutathione or superoxide dismutase and catalase. Under the attack, the mitochondrial functioning gets disrupted due to exogenous or endogenous production of ROS, leading to misbalancing the immune system. Also, studies have revealed that the metal ion release in the system by nanoparticles increases the amount of ROS production. Some studies say that the increase in intracellular Ca^{2+} showed oxidative stress in the cells (Bhattacharya and Mukherjee 2008). This effect is associated with oxidative injury and up-regulation of some genes involved in oxidative stress conditions and apoptosis. The release of these species can cause inflammatory actions. Figure 4.12 is depicting the effect of metal nanoparticles on microbial inactivation.

ROS toxicity can also be amplified by the presence of free metal ions such as iron, copper, and manganese from either different metalloprotein complexes or exogenous entry via different sources. The harmful effects of ROS include lipid peroxidation in the cell membrane, DNA mutation, oxidative damage to proteins, and activation to pro-cell death factors (Suresh et al. 2013). Since the property of ROS production influenced by metal has a huge impact on disinfection applications and many studies, have been proven powerful against different microorganisms.

4.7.3 *Silver Nanoparticles as an Antibacterial Agent*

Silver nanoparticles are particles of size in the range of 1–100 nm, which are capped or surrounded by some agent for stability. Silver nanoparticle's toxic effect is caused by the release of Ag^+ ion in the environment. This nanosilver, which, when reacts with oxygen, gets oxidized to silver ion. In the process of releasing ions, particles first tend to displace the coated agent and become unstable, and to maintain stability, they either agglomerate or react with the oxygen to form Ag^+ ion (Bhattacharya and Mukherjee 2008).

Silver has a strong affinity for sulfur and sulfur-containing ligands, which tends to make strong bonds with sulfur in the form of silver sulfide ($ksp = 10^{-52}$) (Rezvani

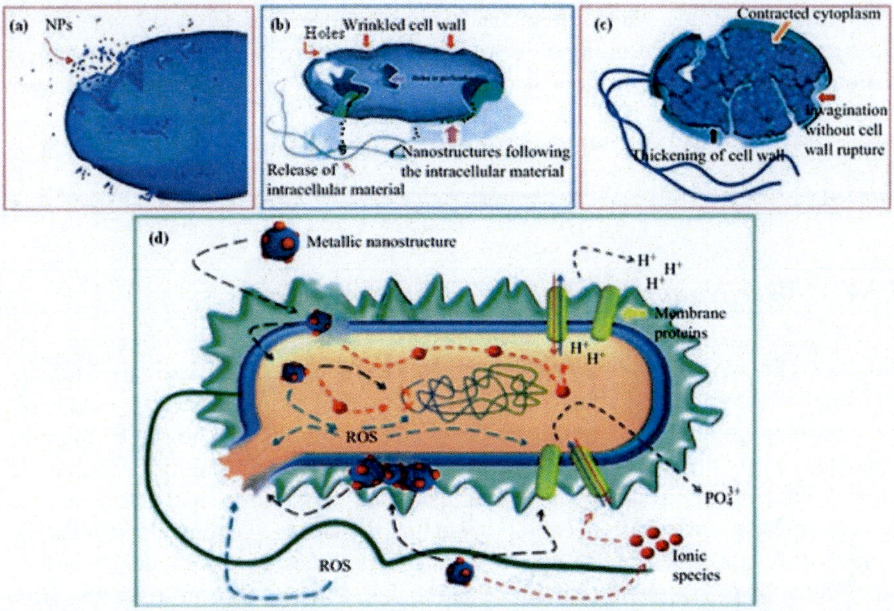

Fig. 4.12 Nanoparticles biocidal action mechanism. **a** Nanoparticle enters through endocytosis by forming a bud, **b** cutting of cell wall is shown along with releasing of intracellular components, **c** interaction of nanoparticles within the cytoplasm, **d** generation of ROS and interaction with membrane proteins. Reprinted (adapted) from (Singh and Dubey 2018). Copyright (2018) American Chemical Society

et al. 2019). The silver is first up taken by the cells through diffusion and subsequently reacts with the thiol groups in the cellular membrane leading to mitochondrial dysfunction and apoptosis (Pulit-Prociak et al. 2014). Nanosilver can react with the cell proteins activating the signaling pathways to cause antiproliferation of the cells. They accumulate in the mitochondria and cause direct damage to the cells leading to disruption of respiratory chain enzymes and generation of oxidative species. Some of the experiments have shown that the damage in the fibroblasts membrane allows calcium influx and causes intracellular calcium overload, which further exaggerates ROS production (Marambio-Jones and Hoek 2010).

Nanosilver interacts with the cell by adsorbing on the surface of the cell. The surface chemistry plays a vital role in the interaction of nanosilver with the extracellular proteins in the cell. pH, potential difference, and NaCl concentration are crucial for surface interaction. The cellular uptake of nanosilver also depends upon the size and shape of the cell. The cytotoxicity of the nanosilver is also responsible due to the production of ROS in the cells, which triggers the cellular antioxidant mechanism in the system. When ROS generation exceeds the capacity of the cellular antioxidant defense system, it depletes glutathione and protein-bound sulfhydryl groups leading them to cause apoptosis (Durán et al. 2016).

Nanosilver binding with proteins and amino acids causes DNA damage. It can cause protein corona, unfolding of the protein, leading to protein disruption. Conformational changes in the protein structure occur due to the breaking of hydrogen bonds between the alpha-helices and an increased number of beta-pleated sheets. Direct interaction of DNA with nanosilver can cause genotoxicity (MiclĂuş et al. 2016).

4.7.4 Silver Nanoparticles as an Antiviral Agent

Viruses are known to cause various health infections to humans and animals. They are very small-sized particles having unique characteristics. As they do not have enzymes for synthesizing proteins, they hijack the protein synthesis machinery of the host organism and produce necessary proteins that they require to multiply. Viruses are intracellular parasites and show tissue tropism, i.e., they have a specific host range. The viruses have a protein coat called capsid that protects the genome present inside; it may either composed of RNA or DNA. The genomes are contained in virions whose function is to protect the genome and to deliver it into cells in which the virus can replicate (Nelson and Cox 2004). The virus binds to the cell either by spikes, which are tissue-specific sites for the host or by the fusion through the envelope. The genome enters the cytoplasm by release through the capsid. Then transcription of viral genome occurs, and viral mRNA directs protein synthesis. Finally, genomes replicate together with viral structural proteins and assemble new virions, which are then released from the cell (Khan and Javed 2018).

As silver has a considerable affinity to bind with sulfur and sulfur-containing molecules, it tends to bind with the glycoproteins present on the spikes, which tend to prohibit the entry and attachment with the host cell. It also tends to bind with phosphates, which directly attaches with the DNA molecules and could stop the replication, thereby decreasing the infectivity. Since the biological interactions are multivalent, the interplay between the microbes and the host cell requires different copies of receptors and ligands, which binds in a coordinated manner (Khandelwal et al. 2014). The microbes are unlikely to develop resistance by the silver as silver acts on a broad range of targets in the microbial species. For example, in HIV, the most prominent target of silver is the sulfide-bearing glycoprotein knobs. The inactivation of bacteriophage MNV-1 occurred by the interaction of silver with capsid proteins.

4.7.5 Copper Nanomaterial as Antimicrobial Agent

Copper is an essential micronutrient for the body used for various metabolic activities. It is an abundant metal found in a variety of rocks and minerals. Copper is a transition metal having an electronic configuration of a completely filled d shell. It is quite stable and unreactive towards mammalian cells but could lead to toxicity if it can be oxidized to form a cuprous or cupric ion. Copper nanoparticle toxicity is related to the decapsulation of the surface coating leading to the ingression of the copper ion into the solution. The nanoparticles released are too reactive that it may either be agglomerate with other nanoparticles to form clusters or can react with something at first place depending upon the affinity of the particle to coordinate or form a bond. There are three main processes by which conversion and mobility of the nanoparticle is being influenced, i.e., precipitation, complexation and adsorption which determines its bioavailability, fate, and transport (Flemming and Trevors 1989).

There are not many studies available to reveal the actual toxicity mechanism of copper nanoparticles, but few of them are known and widely accepted. These involve; (1) accumulation of nanoparticles in the cell membrane, which leads to change in membrane permeability and leakage of intracellular enzymes, lipopolysaccharides proteins, and intracellular biomolecules, (2) Formation of reactive oxygen species with corresponding oxidation of cellular components, (3) Ingestion of metal ions followed by interaction with DNA or proteins to prevent DNA replication and ATP synthesis (Grass et al. 2011). One of the studies had contemplated that the linoleic acid capped copper nanoparticle inactivates the enzymes in the cells by generating hydrogen peroxide. In another study, the copper and -SH group bonding was hypothesized, which led to protein denaturation. It has been elucidated that the copper nanoparticles strive on cell membranes because of their affinity for phosphates, carboxyl, sulfur, and amino groups (Ingle et al. 2014). Some studies have shown that copper can act on respiratory enzymes in the cellular membrane, which was evidenced while recovery of injured *E. coli,* which started following fermentative pathways due to reduced oxygen levels (Thurman and Gerba 1989). Copper nanoparticles inhibit viral proteins by the generation of ROS, which causes capsid oxidation (Rai et al. 2014).

4.8 Conclusion

Halogens like chlorine and iodine, due to their significant oxidation potential, are being employed as disinfecting agents. Chlorination is one of the oldest, yet most widely used technology for water disinfection but possesses various limitations in terms of handling and excessive exposure. Due to outgassing problems, it has been replaced by iodine in microgravity. Electrochemical disinfection has certain advantages over conventional disinfection methodologies, including its eco-friendly operation, low cost, and chemical-free usage but suffers drawbacks like charge reversibility, overvoltage, and low current efficiencies. These obstacles could be rescued by using electrode material, having better electrocatalytic properties. Carbon-based nanomaterials have found to be excellent material for replacing conventional metal-based electrodes, and one of them being is Laser-Induced Graphene (LIG). It is a novel material that can be prepared easily by laser scribing on any carbonaceous surface. Its excellent surface property inhibits microbial growth and makes it an antimicrobial surface when employed under the electric field as membranes. Additionally, metal-based nanoparticles due to their potential biocidal effect generate oxidative stress in the microbes and acts as antimicrobial surfaces. However, the biocidal activity can be modified on the nanoscale by varying their physicochemical properties. These materials exert their action via ion release followed by interaction with various components inside the cell depending upon their affinity to form bonds and could be employed with different nanoparticles for upsurging the microbial inactivation process. The impeccable properties of these nanomaterials could be applied in the form of either composites or hybrids for various point-of-use or large-scale water disinfection systems for enhanced antimicrobial action.

References

Aslam MA, Riaz K, Mahmood MQ, Zubair M (2019) Hybrid analytical-numerical approach for investigation of differential effects in normal and cancer cells under electroporation. RSC Adv 9:41518–41530. https://doi.org/10.1039/c9ra07428g

Backer H, Hollowell J (2000) Use of iodine for water disinfection: iodine toxicity and maximum recommended dose. Environ Health Perspect 108:679–684. https://doi.org/10.1289/ehp.00108679

Barazesh JM, Hennebel T, Jasper JT, Sedlak DL (2015) Modular advanced oxidation process enabled by cathodic hydrogen peroxide production. Environ Sci Technol 49:7391–7399. https://doi.org/10.1021/acs.est.5b01254

Barras A, Martin FA, Bande O et al (2013) Glycan-functionalized diamond nanoparticles as potent E. coli anti-adhesives. Nanoscale 5:2307–2316. https://doi.org/10.1039/c3nr33826f

Bhattacharya R, Mukherjee P (2008) Biological properties of "naked" metal nanoparticles. Adv Drug Deliv Rev 60:1289–1306. https://doi.org/10.1016/j.addr.2008.03.013

Brillas E, Sirés I, Oturan MA (2009) Electro-fenton process and related electrochemical technologies based on Fenton's reaction chemistry. Chem Rev 109:6570–6631. https://doi.org/10.1021/cr900136g

Buonocore G, Perrone S, Tataranno ML (2010) Oxygen toxicity: chemistry and biology of reactive oxygen species. Semin Fetal Neonatal Med 15:186–190. https://doi.org/10.1016/j.siny.2010.04.003

Calvin NM, Hanawalt PC (1988) High-efficiency transformation of bacterial cells by electroporation. J Bacteriol 170:2796–2801. https://doi.org/10.1128/jb.170.6.2796-2801.1988

Chang DC, Saunders JA, Chassy BM, Sowers AE (1991) Guide to electroporation and electrofusion

Chatterjee AK, Chakraborty R, Basu T (2014) Mechanism of antibacterial activity of copper nanoparticles. Nanotechnology 25. https://doi.org/10.1088/0957-4484/25/13/135101

Chen L, Pinto A, Alshawabkeh AN (2019) Activated carbon as a cathode for water disinfection through the electro-fenton process. Catalysts 9. https://doi.org/10.3390/catal9070601

Collivignarelli MC, Abbà A, Benigna I et al (2018) Overview of the main disinfection processes for wastewater and drinking water treatment plants. Sustain 10:1–21. https://doi.org/10.3390/su10010086

Gerald CV, Richard SL (1987) Review of water disinfection techniques. In: Intersociety conference on environmental systems. SAE Publication Divisions, Seattle, Washington, p 8

Das S, Mishra A, Ghangrekar MM (2020) Production of hydrogen peroxide using various metal-based catalysts in electrochemical and bioelectrochemical systems: mini review. J Hazardous Toxic Radioact Waste 24:1–6. https://doi.org/10.1061/(ASCE)HZ.2153-5515.0000498

Dayem AA, Hossain MK, Bin LS et al (2017) The role of reactive oxygen species (ROS) in the biological activities of metallic nanoparticles. Int J Mol Sci 18:1–21. https://doi.org/10.3390/ijms18010120

Durán N, Durán M, de Jesus MB et al (2016) Silver nanoparticles: a new view on mechanistic aspects on antimicrobial activity. Nanomed Nanotechnol Biol Med 12:789–799. https://doi.org/10.1016/j.nano.2015.11.016

Flemming CA, Trevors JT (1989) Copper toxicity and chemistry in the environment: a review. Water Air Soil Pollut 44:143–158. https://doi.org/10.1007/BF00228784

Foldbjerg R, Dang DA, Autrup H (2011) Cytotoxicity and genotoxicity of silver nanoparticles in the human lung cancer cell line, A549. Arch Toxicol 85:743–750. https://doi.org/10.1007/s00204-010-0545-5

Gao G, Zhang Q, Hao Z, Vecitis CD (2015) Carbon nanotube membrane stack for flow-through sequential regenerative electro-fenton. https://doi.org/10.1021/es505679e

Ghernaout D (2017) Microorganisms' electrochemical disinfection phenomena. EC Microbiol 9:160–169

Ghernaout D, Naceur MW, Aouabed A (2011) On the dependence of chlorine by-products generated species formation of the electrode material and applied charge during electrochemical water treatment. Desalination 270:9–22. https://doi.org/10.1016/j.desal.2011.01.010

Ghernaout D, Touahmia M, Aichouni M (2019) Disinfecting water: electrocoagulation as an efficient process. Appl Eng 3:1–12. https://doi.org/10.11648/j.ae.20190301.11

Gonzalez-Rivas N, Reyes-Pérez H, Barrera-Díaz CE (2019) Recent advances in water and wastewater electrodisinfection. ChemElectroChem 6:1978–1983. https://doi.org/10.1002/celc.201801746

Grass G, Rensing C, Solioz M (2011) Metallic copper as an antimicrobial surface. Appl Environ Microbiol 77:1541–1547. https://doi.org/10.1128/AEM.02766-10

Howes PD, Chandrawati R, Stevens MM (2014) Colloidal nanoparticles as advanced biological sensors. Science (80)346. https://doi.org/10.1126/science.1247390

Huo ZY, Luo Y, Xie X et al (2017) Carbon-nanotube sponges enabling highly efficient and reliable cell inactivation by low-voltage electroporation. Environ Sci Nano 4:2010–2017. https://doi.org/10.1039/c7en00558j

Imlay JA (2003) Pathways of oxidative damage. Annu Rev Microbiol 57:395–418. https://doi.org/10.1146/annurev.micro.57.030502.090938

Imlay JA (2008) Cellular defenses against superoxide and hydrogen peroxide. Annu Rev Biochem 77:755–776. https://doi.org/10.1146/annurev.biochem.77.061606.161055

Ingle AP, Duran N, Rai M (2014) Bioactivity, mechanism of action, and cytotoxicity of copper-based nanoparticles: a review. Appl Microbiol Biotechnol 98:1001–1009. https://doi.org/10.1007/s00253-013-5422-8

Iversen TG, Skotland T, Sandvig K (2011) Endocytosis and intracellular transport of nanoparticles: present knowledge and need for future studies. Nano Today 6:176–185. https://doi.org/10.1016/j.nantod.2011.02.003

Jin W, Maduraiveeran G (2018) Nanomaterial-based environmental sensing platforms using state-of-the-art electroanalytical strategies. J Anal Sci Technol 9. https://doi.org/10.1186/s40543-018-0150-4

Johnson LB (2003) Alternative electrochemical systems for ozonation of water hydrogen gas, ozone gas, and ozonated water can be delivered under pressure. In: Interferometer for measuring displacement to within 20 pm, pp 24–25

Kapałka A, Fóti G (2010) Electrochemistry for the environment, 1st edn. Springer, New York

Kerwick MI, Reddy SM, Chamberlain AHL, Holt DM (2005) Electrochemical disinfection, an environmentally acceptable method of drinking water disinfection? Electrochim Acta 50:5270–5277. https://doi.org/10.1016/j.electacta.2005.02.074

Khan K, Javed S (2018) Functionalization of inorganic nanoparticles to augment antimicrobial efficiency: a critical analysis. Curr Pharm Biotechnol 19:523–536. https://doi.org/10.2174/1389201019666180731121401

Khandelwal N, Kaur G, Kumar N, Tiwari A (2014) Application of silver nanoparticles in viral inhibition: a new hope for antivirals. Dig J Nanomater Biostruct 9:175–186

Khezerlou A, Alizadeh-Sani M, Azizi-Lalabadi M, Ehsani A (2018) Nanoparticles and their antimicrobial properties against pathogens including bacteria, fungi, parasites and viruses. Microb Pathog 123:505–526. https://doi.org/10.1016/j.micpath.2018.08.008

Korshin G, Yan M (2018) Electrochemical dehalogenation of disinfection by-products and iodine-containing contrast media: a review. Environ Eng Res 23:345–353. https://doi.org/10.4491/eer.2018.054

Kotnik T, Frey W, Sack M et al (2015) Electroporation-based applications in biotechnology. Trends Biotechnol 33:480–488. https://doi.org/10.1016/j.tibtech.2015.06.002

Kraft A, Stadelmann M, Wünsche M, Blaschke M (2006) Electrochemical ozone production using diamond anodes and a solid polymer electrolyte. Electrochem Commun 8:883–886. https://doi.org/10.1016/j.elecom.2006.02.013

Kraft BA (2008) Electrochemical water disinfection: a short review. 177–185. https://doi.org/10.1595/147106708X329273

Tennakoon CLK, Singh W, Anderson K (2010) Electrochemical hydrogen peroxide generator. NASA Tech Briefs

Lee S, Speight JG, Lee S, Speight JG (2020) Water systems. Environ Technol Handb 93–117. https://doi.org/10.1201/9780367813390-6

Li P, Li J, Feng X et al (2019) Metal-organic frameworks with photocatalytic bactericidal activity for integrated air cleaning. Nat Commun 10:1–10. https://doi.org/10.1038/s41467-019-10218-9

Li R, Zhang L, Wang P (2015) Rational design of nanomaterials for water treatment. Nanoscale 7:17167–17194

Li W, Calle LM, Hanford AJ et al (2018) Investigation of silver biocide as a disinfection technology for spacecraft—an early literature review. In: 48th international conference on environmental systems (ICES)—2018, p 82

Li Y, Luong DX, Zhang J et al (2017) Laser-induced graphene in controlled atmospheres: from superhydrophilic to superhydrophobic surfaces. Adv Mater 29:1–8. https://doi.org/10.1002/adma.201700496

Lim J, Hoffmann MR (2019) Substrate oxidation enhances the electrochemical production of hydrogen peroxide. Chem Eng J 374:958–964. https://doi.org/10.1016/j.cej.2019.05.165

Little GA, Nemeth JE, Keck DE et al (2020) Life systems. JHC

Liu C, Xie X, Zhao W et al (2013) Conducting nanosponge electroporation for affordable and high-efficiency disinfection of bacteria and viruses in water. Nano Lett 13:4288–4293. https://doi.org/10.1021/nl402053z

Liu H, Wang C, Li X et al (2007) A novel electro-Fenton process for water treatment: reaction-controlled pH adjustment and performance assessment. Environ Sci Technol 41:2937–2942. https://doi.org/10.1021/es0622195

Liu W, Yao J, Jin J et al (2015) Microbial toxicity of a type of carbon dots to escherichia coli. Arch Environ Contam Toxicol 69:506–514. https://doi.org/10.1007/s00244-015-0222-z

Liu Y, Quan X, Fan X et al (2015) High-yield electrosynthesis of hydrogen peroxide from oxygen reduction by hierarchically porous carbon. Angew Chemie Int Ed 54:6837–6841. https://doi.org/10.1002/anie.201502396

Lobascio C, Bruno G, Grizzaffi L et al (2004) Quality of ATV potable water for ISS crew consumption. SAE Tech Pap 113:1122–1129. https://doi.org/10.4271/2004-01-2491

Lu Z, Chen G, Siahrostami S et al (2018) High-efficiency oxygen reduction to hydrogen peroxide catalysed by oxidized carbon materials. Nat Catal 1:156–162. https://doi.org/10.1038/s41929-017-0017-x

Marambio-Jones C, Hoek EMV (2010) A review of the antibacterial effects of silver nanomaterials and potential implications for human health and the environment. J Nanoparticle Res 12:1531–1551. https://doi.org/10.1007/s11051-010-9900-y

Martínez-Huitle CA, Brillas E (2008) Electrochemical alternatives for drinking water disinfection. Angew Chemie Int Ed 47:1998–2005. https://doi.org/10.1002/anie.200703621

Mauter MS, Elimelech M (2008) Enviromental application of cacbon-based nanomaterial. Environ Sci Technol 42:5843–5859

McKinney CW, Pruden A (2012) Ultraviolet disinfection of antibiotic resistant bacteria and their antibiotic resistance genes in water and wastewater. Environ Sci Technol 46:13393–13400. https://doi.org/10.1021/es303652q

Metcalf E (2003) Wastewater engineering teatment and reuse, 4th ed. Metcalf & Eddy, Inc. J Wastewater Eng

MiclĂuş T, Beer C, Chevallier J et al (2016) Dynamic protein coronas revealed as a modulator of silver nanoparticle sulphidation in vitro. Nat Commun 7:1–10. https://doi.org/10.1038/ncomms11770

Mochalin VN, Shenderova O, Ho D, Gogotsi Y (2012) The properties and applications of nanodiamonds. Nat Nanotechnol 7:11–23. https://doi.org/10.1038/nnano.2011.209

Moosavi R, Ramanathan S, Lee YY et al (2015) Synthesis of antibacterial and magnetic nanocomposites by decorating graphene oxide surface with metal nanoparticles. RSC Adv 5:76442–76450. https://doi.org/10.1039/c5ra15578a

Moulay S (2013) Molecular iodine/polymer complexes. J Polym Eng 33:389–443. https://doi.org/10.1515/polyeng-2012-0122

Nelson D, Cox M (2004) Lehninger principles of biochemistry (4th ed.). https://www.researchgate.net/publication/48376766_Lehninger_Principles_of_Biochemistry

Oh BS, Park SJ, Jung YJ et al (2007) Disinfection and oxidation of sewage effluent water using ozone and UV technologies. Water Sci Technol 55:299–306. https://doi.org/10.2166/wst.2007.036

Pham VTH, Truong VK, Quinn MDJ et al (2015) Graphene induces formation of pores that kill spherical and rod-shaped bacteria. 8458–8467. https://doi.org/10.1021/acsnano.5b03368

Pierson DL (2019) Microbial contamination of spacecraft. NASA Johnson Space Center, Houston, TX

Poornesh KK, Cho C, Kim DY, Tak Y (2010) Effect of gas-diffusion electrode material heterogeneity on the structural integrity of polymer electrolyte fuel cell. Energy 35:5241–5249. https://doi.org/10.1016/j.energy.2010.07.041

Pulit-Prociak J, Stokłosa K, Banach M (2014) Nanosilver products and toxicity. Environ Chem Lett 13:59–68. https://doi.org/10.1007/s10311-014-0490-2

Punyani S, Narayana P, Singh H, Vasudevan P (2006) Iodine based water disinfection: a review. J Sci Ind Res (India) 65:116–120

Rai M, Deshmukh SD, Ingle AP et al (2014) Metal nanoparticles: the protective nanoshield against virus infection. 7828:1–11. https://doi.org/10.3109/1040841X.2013.879849

Ren G, Zhou M, Zhang Q et al (2020) A novel stacked flow-through electro-Fenton reactor as decentralized system for the simultaneous removal of pollutants (COD, NH_3-N and TP) and disinfection from domestic sewage containing chloride ions. Chem Eng J 387:124037. https://doi.org/10.1016/j.cej.2020.124037

Rezvani E, Rafferty A, McGuinness C, Kennedy J (2019) Adverse effects of nanosilver on human health and the environment. Acta Biomater 94:145–159. https://doi.org/10.1016/j.actbio.2019.05.042

Richardson SD, Plewa MJ, Wagner ED et al (2007) Occurrence, genotoxicity, and carcinogenicity of regulated and emerging disinfection by-products in drinking water: a review and roadmap for research. Mutat Res Rev Mutat Res 636:178–242. https://doi.org/10.1016/j.mrrev.2007.09.001

Roberts MS, Hummerick ME, Edney SL et al (2007) Assessment of silver based disinfection technology for CEV and future US spacecraft: microbial efficacy. SAE Tech Pap 116:481–491. https://doi.org/10.4271/2007-01-3142

Rodriguez B, Shindo D, Montgomery E (2013) Electrochemical disinfection feasibility assessment materials evaluation for the International Space Station. In: 43rd international conference on environmental systems, pp 1–9. https://doi.org/10.2514/6.2013-3471

Särkkä H, Bhatnagar A, Sillanpää M (2015) Recent developments of electro-oxidation in water treatment—a review. J Electroanal Chem 754:46–56. https://doi.org/10.1016/j.jelechem.2015.06.016

Scheerschmidt K, Werner P Characterization of structure and composition of quantum dots by transmission electron microscopy

Simon HU, Haj-Yehia A, Levi-Schaffer F (2000) Role of reactive oxygen species (ROS) in apoptosis induction. Apoptosis 5:415–418. https://doi.org/10.1023/A:1009616228304
Simon RG, Stöckl M, Becker D et al (2018) Current to clean water—electrochemical solutions for groundwater, water, and wastewater treatment. Chem-Ing-Tech 90:1832–1854. https://doi.org/10.1002/cite.201800081
Singh A, Dubey AK (2018) Various biomaterials and techniques for improving antibacterial response. ACS Appl Bio Mater 1:3–20. https://doi.org/10.1021/acsabm.8b00033
Singh SP, Li Y, Be'Er A et al (2017) Laser-induced graphene layers and electrodes prevents microbial fouling and exerts antimicrobial action. ACS Appl Mater Interfaces 9:18238–18247. https://doi.org/10.1021/acsami.7b04863
Singh SP, Li Y, Zhang J et al (2018) Sulfur-doped laser-induced porous graphene derived from polysulfone-class polymers and membranes. ACS Nano 12:289–297. https://doi.org/10.1021/acsnano.7b06263
Singh SP, Ramanan S, Kaufman Y, Arnusch CJ (2018) Laser-induced graphene biofilm inhibition: texture does matter. ACS Appl Nano Mater 1:1713–1720. https://doi.org/10.1021/acsanm.8b00175
Stanford MG, Li JT, Chen Y et al (2019) Self-sterilizing laser-induced graphene bacterial air filter. ACS Nano 13:11912–11920. https://doi.org/10.1021/acsnano.9b05983
Suresh AK, Pelletier DA, Doktycz MJ (2013) Relating nanomaterial properties and microbial toxicity. Nanoscale 5:463–474. https://doi.org/10.1039/c2nr32447d
Thorn RMS, Lee SWH, Robinson GM et al (2012) Electrochemically activated solutions: evidence for antimicrobial efficacy and applications in healthcare environments. Eur J Clin Microbiol Infect Dis 31:641–653. https://doi.org/10.1007/s10096-011-1369-9
Thornton J, Campbell D (2001) Pandora's poison: chlorine, health and a new environmental strategy. BMJ 322:497. https://doi.org/10.1136/bmj.322.7284.497
Thorstenson YR, Sauer RL, Janik DS (1987) Effects of iodine disinfection products in spacecraft water. SAE Tech Papers
Thurman RB, Gerba CP (1989) The molecular mechanisms of copper and silver ion disinfection of bacteria and viruses. Crit Rev Environ Control 18:295–315. https://doi.org/10.1080/10643388909388351
Tugulea AM, Bérubé D, Giddings M et al (2014) Nano-silver in drinking water and drinking water sources: stability and influences on disinfection by-product formation. Environ Sci Pollut Res 21:11823–11831. https://doi.org/10.1007/s11356-014-2508-5
Valko M, Rhodes CJ, Moncol J et al (2006) Free radicals, metals and antioxidants in oxidative stress-induced cancer. Chem Biol Interact 160:1–40. https://doi.org/10.1016/j.cbi.2005.12.009
Wandera D, Wickramasinghe SR, Husson SM (2010) Stimuli-responsive membranes. J Memb Sci 357:6–35. https://doi.org/10.1016/j.memsci.2010.03.046
Wang D, Saleh NB, Sun W et al (2019) Next-generation multifunctional carbon-metal nanohybrids for energy and environmental applications. Environ Sci Technol 53:7265–7287. https://doi.org/10.1021/acs.est.9b01453
Wang L, Yuan Z, Karahan HE et al (2019) Nanocarbon materials in water disinfection: state-of-the-art and future directions. Nanoscale 11:9819–9839. https://doi.org/10.1039/c9nr02007a
WHO (2007) Combating waterborne disease at the household level
Winterbourn CC (2013) The biological chemistry of hydrogen peroxide, 1st ed. Elsevier Inc
Wolfe RL (1990) Ultraviolet disinfection of potable water: current technology and research needs. Environ Sci Technol 24:768–773. https://doi.org/10.1021/es00076a001
Ye R, James DK, Tour JM (2019) Laser-induced graphene: from discovery to translation. Adv Mater 31:1–15. https://doi.org/10.1002/adma.201803621
You JM, Kim D, Jeon S (2012) Electrocatalytic reduction of H_2O_2 by Pt nanoparticles covalently bonded to thiolated carbon nanostructures. Electrochim Acta 65:288–293. https://doi.org/10.1016/j.electacta.2012.01.070

Yu XY (2004) Critical evaluation of rate constants and equilibrium constants of hydrogen peroxide photolysis in acidic aqueous solutions containing chloride ions. J Phys Chem Ref Data 33:747–763. https://doi.org/10.1063/1.1695414

Zhao C, Wu L, Wang X et al (2020) Quaternary ammonium carbon quantum dots as an antimicrobial agent against gram-positive bacteria for the treatment of MRSA-infected pneumonia in mice. Carbon 163:70–84. https://doi.org/10.1016/j.carbon.2020.03.009

Zou X, Zhang L, Wang Z et al (2016) Mechanisms of the antimicrobial activities of graphene materials mechanisms of the antimicrobial activities of graphene materials. https://doi.org/10.1021/jacs.5b11411

Chapter 5
Application of g-C_3N_4-based Materials for the Efficient Removal and Degradation of Pollutants in Water and Wastewater Treatment

Karthik Rathinam, Meenakshi M. Nara, Ibrahim M. A. ElSherbiny, Imran Ali, and Stefan Panglisch

Abstract The existence of the several pollutants in water resources has led the water unfit for consumption or reuse, and eventually becoming one of the greatest threats to the environment and human health. Pollutants in water are not desirable even if they have less toxicity, therefore, they must be either removed or degraded. Despite several materials reported, polymeric materials have been widely suggested and used nowadays in water treatment applications to get rid-off the multiple pollutants. In this context, graphitic carbon nitride (g-C_3N_4), a two-dimensional organic polymeric material is emerging as effective adsorbent and photo-catalyst for the rapid removal and degradation of various pollutants from water. Therefore, this chapter is devoted to summarizing the latest applications of g-C_3N_4-based materials in water and wastewater treatment processes. Moreover, the challenges and future perspectives of the g-C_3N_4 is also discussed in this chapter.

Keywords Pollutants · Carbon nitride · Adsorption technology · Photocatalytic degradation · Water treatment

5.1 Introduction

Water pollution by various organic and inorganic pollutants is a challenging problem globally. The adverse effect of the pollutants on the environment has become one of the greatest threats worldwide. Primary sources for such pollutants are attributed to geological processes and anthropogenic activities. Pharmaceuticals and personal

K. Rathinam (✉) · M. M. Nara · I. M. A. ElSherbiny · S. Panglisch
Chair for Mechanical Process Engineering and Water Technology, University of Duisburg-Essen, 47057 Duisburg, Germany

I. Ali
Department of Chemistry, College of Sciences, Taibah University, Al-Medina Al-Munawara 41477, Saudi Arabia

Department of Chemistry, Jamia Millia Islamia (Central University), New Delhi 11025, India

S. P. Singh et al. (eds.), *Nanomaterials and Nanocomposites for Environmental Remediation*, Energy, Environment, and Sustainability,
https://doi.org/10.1007/978-981-16-3256-3_5

care products (PPCPs) have been widely used today. Due to the massive production and consumption, PPCPs have been emerging as a major group of micro-pollutants in water. Nowadays, more than 200 pharmaceuticals have been identified in river waters globally. Though PPCPs adverse effects on human beings have not been revealed yet these has been proven on aquatic systems, especially on fish.

The presence of heavy metals such as lead (Pb), arsenic (As), chromium (Cr), mercury (Hg), cadmium (Cd), etc., in water may result in carcinogenic and mutagenic effects to both human and aquatic life systems. Moreover, these heavy metals are non-biodegradable and highly toxic. Upon intake, heavy metals tend to accumulate in enzymes, and they act as strong inhibitors to numerous physiological and biochemical processes in living organisms. In recent years, human exposure to heavy metals has been raised dramatically because of the exponential increase in the use of heavy metals in industries such as mining, coal and petroleum consumption, dyes and pigments, chrome plating, battery, etc.

With the continuous growth in population, the production of food items has also become intensive. Meanwhile to ensure more crop production, the usage of pesticides, insecticides and fertilizers are also increased substantially in agricultural sectors. It is obvious that these extensive activities resulting in contamination of groundwater and surface water bodies and posing a serious problem to all the living species. Eutrophication is one of the major problems encountered in surface water due to the increased amount of nutrients in water. Furthermore, excessive levels of nitrate in drinking water lead to methemoglobinemia in human beings.

Water pollution by crude oil is yet another factor that harms the environment especially terrestrial and aquatic ecosystems. The primary causes are including oil spillage accidents, oil leakages, and oily wastewater discharge to the water. The crude oil consists of mixtures of aliphatic, aromatic, and heterocyclic hydrocarbons. Until now, several oil spillage accidents have been reported across the world (Atlas and Hazen 2011; Fayad 1986; Horn and Neal 1981; Islam et al. 2018, Patton et al. 1981).

Dyes have found a wide range of applications in textile, paint, paper, cosmetics, food processing, etc. According to Yagub et al., the textile industries across the world consume more than 10,000 tons/year of dyes and discharge approximately 100 tons/year of dyes to the water bodies (Yagub et al. 2014). Inappropriate disposal of dyes to the water streams could certainly affect the water qualities and make water bodies unfit for usage. Therefore, dyes are also considered as one of the important water pollutants that must be treated properly.

To address these critical issues, polymeric materials have been widely employed to treat contaminated water sources (Rathinam et al. 2019, 2018; Rathinam and Singh 2020). In this context, graphitic carbon nitride (g-C_3N_4), a two-dimensional structured organic polymeric material has been receiving more attention today and found quite a number of applications in various disciplines of science. Berzelius and Liebig have first discovered the carbon nitride polymer in 1834 and named as “melon” (Liebig 1834). g-C_3N_4 is the most stable allotrope of carbon nitride under ambient conditions. g-C_3N_4 polymer is consisting of covalently linked, sp^2 hybridized carbon and nitrogen atoms besides some impurity hydrogen atoms, which are connected via tri-s-triazine framework. In addition, it has basic surface functionalities, high

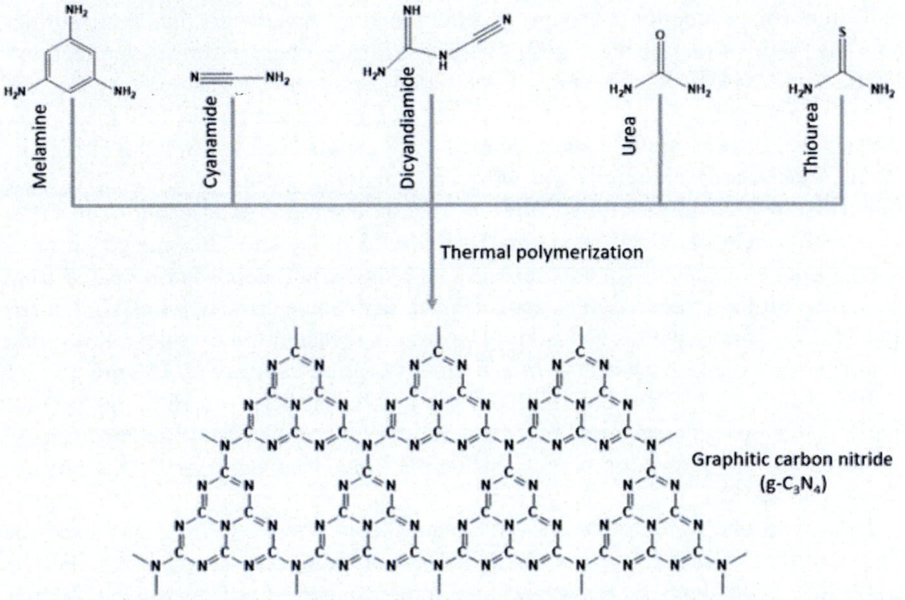

Fig. 5.1 Synthesis of g-C_3N_4 polymer from various precursors

electron-rich properties, H-bonding motifs due to the presence of N and H atoms, and high thermal stability either in the presence of O_2, N_2 or air atmosphere. Furthermore, it has high chemical stability in various solvents including (water, alcohols, dimethylformamide, tetrahydrofuran, diethyl ether and toluene (Zhu et al. 2014) and also possessed high thermal conductivity and tensile strength as well (Mortazavi et al. 2015).

The g-C_3N_4 can be prepared from various nitrogen-rich sources such as melamine, urea, cyanamide, dicyandiamide, etc. The typical synthesis of g-C_3N_4 polymer from various precursors via thermal polymerization (i.e. pyrolysis) is shown in Fig. 5.1. Despite g-C_3N_4 has versatile applications in solar cells, gas sensors, CO_2 storage, metal-free catalysis, energy storage and conversion, energy production, etc., this overview focuses on the potential applications of g-C_3N_4 as adsorbent and photocatalyst in water treatment applications.

5.2 Removal of Pollutants from Water Using g-C_3N_4-based Adsorbents

Owing to the similar analog with the graphene, g-C_3N_4 has also been widely recognized as an adsorbent in water treatment processes for the removal of various pollutants including micropollutants, dyes, heavy metals, etc. (Yazdankish et al. 2020). The built-in nitrogen functional groups, excellent textural parameters, moderate surface area and porous structure make g-C_3N_4, a powerful adsorbent in various applications. For instance, Xiao et al., prepared g-C_3N_4 nanosheets from the guanidine hydrochloride with a thickness less than 1.6 nm and specific surface area of 111.2 $m^2\ g^{-1}$ for the removal of both anionic and cationic heavy metals such as Cr(VI), Cd(II), and Pb(II). The tri-s-triazine units and nitrogen containing groups of the g-C_3N_4 were found to be responsible for the adsorption process. The reported adsorption capacities of g-C_3N_4 for Cr(VI), Cd(II), and Pb(II) were 684.5, 123.2 and 136.6 mg g^{-1}, respectively (Xiao et al. 2019). In another work, g-C_3N_4 nanosheets were prepared from melamine for the remediation of actinides and lanthanides (Eu(III), La(III), Nd(III) and Th(IV)) from aqueous solutions. The results revealed the excellent adsorption performance of g-C_3N_4 with the maximum adsorption capacity of 155 mg g^{-1} for Eu(III), 122.3 mg $g^{-}1$ for La(III), 132.5 mg g^{-1} for Nd(III), and 185.6 mg g^{-1} for Th(IV). Moreover, it was found that the ionic strength of the adsorption medium did not influence the adsorption process, while pH influenced significantly the process (Liao et al. 2018b).

Exfoliated and un-exfoliated g-C_3N_4 nanosheets were developed and used for the removal of various heavy metals including Cu(II), Mn(II), Zn(II), Pb(II), Fe(III) and Cd(II) from synthetic water and acid mine drainage from the Sungun copper mine in Iran. The exfoliated g-C_3N_4 showed remarkable adsorption capacities for the heavy metals than the un-exfoliated g-C_3N_4, due to the increase in active surface area (Dehkharghani 2019).

g-C_3N_4 derived from urea was also tested for the removal of Pb(II) and aniline from water through batch experiments. Although g-C_3N_4 showed high adsorption performance, the removal of Pb(II) and aniline was strongly dependent on pH, contact time, ionic strength, and temperature. At pH < 7.0, the adsorption of Pb(II) was dominated by ion-exchange or outer sphere surface complexation, whereas at pH $>$ 7.0, the inner sphere complexation has played a major role. In the case of aniline, at pH < 5.0, the adsorption was attributed to the electrostatic interactions, while at pH > 5.0 the adsorption was attributed to π-π electron donor–acceptor interactions (Hu et al. 2015).

Furthermore, the preparation conditions were found to affect the adsorption capacity of g-C_3N_4 for various pollutants. For example, Cai et al. obtained g-C_3N_4 by pyrolysis of urea and investigated the influence of preparation conditions on the recovery of Cd(II) and methylene blue (MB) from contaminated water. It was seen that g-C_3N_4 prepared at 550 °C showed the best adsorption capacity for Cd(II) and MB (Cai et al. 2017). Additionally, Pb(II), Cu(II), Cd(II), and Ni(II) containing water was treated using g-C_3N_4 synthesized from melamine using a salt melt method. It was reported that the adsorption capacities of g-C_3N_4 for Cd(II) and Ni(II) were relatively low compared to Pb(II) and Cu(II). It was ascribed that the adsorption of heavy metals on g-C_3N_4 was driven by the Lewis acid-base interaction followed by a strong surface complexation mechanism (Shen et al. 2015).

Mesoporous carbon nitride (MCN) has been developed by different chemical methods and used for water treatment applications. In a study by Yan et al., mesoporous silica SBA-15 was used as a template for preparing MCN for the adsorption of perfluorooctane sulfonate (PFOS). It was revealed that the adsorption capacity of MCN for PFOS was increased from 433.7 to 625 mg g^{-1} when the calcination temperature was decreased from 873 to 673 K. Also, the extent of removal of PFOS was observed to decrease with increasing pH value. Hydrophobic and electrostatic interactions between PFOS and the basic groups of MCN controlled the PFOS adsorption onto MCN (Yan et al. 2013).

To make use of the basic sites present in MCN, selective removal of the slightly acidic phenol was also performed. The removal of phenol by MCN was compared with the commercially available activated carbon and CMK-3-150. The adsorption capacity of MCN was found to be 2.7 and 1.29 times higher than that of activated carbon and CMK-3-150, respectively. It was concluded that -NH and -NH_2 groups in MCN could effectively contribute to the removal of toxic phenol (Haque et al. 2010).

In parallel, MCN was prepared from dicyandiamide using silica nanoparticles as a hard template and its adsorption capacity for the removal of Ni(II) ions was investigated. The removal of the silica template after the synthesis helped in generating more pore structures on the g-C_3N_4 surface and hence promoted the adsorption capacity for Ni(II) ions (Xin et al. 2016). Besides, Huang et al., developed a protonated MCN from cyanamide and used for the removal of microcystins (MCs) such as MC-LR and MC-RR. It was observed that the protonation not only enhanced the adsorption capacities of MCN for MCs but also accelerated the MCs removal rate (Huang et al. 2015).

g-C_3N_4 was also subjected to several chemical modifications to improve its affinity towards a variety of pollutants (Xie et al. 2020a). For instance, g-C_3N_4 was oxidized using a ternary mixture of H_2SO_4/HNO_3/H_2O_2 to obtain Ox-g-C_3N_4, which was then decorated onto polyaniline nanofibers (PANI-NF) to produce Ox-g-C_3N_4/PANI-NF. Due to the positive surface charge (+21 mV), Ox-g-C_3N_4/PANI-NF showed a higher affinity towards Cr(VI) rather than Cu(II). Moreover, the Cr(VI) adsorption was highly dependent on solution pH, temperature and concentration of Cr(VI) (Kumar et al. 2017). PANI/oxidation etching g-C_3N_4 was also applied for the removal of radioactive uranium(VI) from uranium-containing wastewater (Liu et al. 2019a). In another study, Yousefi et al., developed an oxidized g-C_3N_4 by exfoliation in the liquid phase through a combination of sonication and oxidation. It was demonstrated as a promising adsorbent for the remediation of antibiotic tetracycline drug, cationic dyes (MB, methylene green, rose Bengal and basic fuchsin) and anionic dyes (rhodamine B, and methyl orange). Owing to the improved dispersibility in water, increased surface area and abundance of the surface functional groups, the oxidized g-C_3N_4 possessed higher adsorption capacity over the unoxidized g-C_3N_4 (Yousefi et al. 2019).

Furthermore, sodium alginate (SA), a biopolymer was used as a modifier to functionalize g-C_3N_4 for the effective removal of heavy metals (Pb(II), Ni(II) and Cu(II)) (Shen et al. 2020). Chitosan was also incorporated into g-C_3N_4 and used as a metal-free heterogeneous adsorbent and catalyst for the removal of chlorpyrifos (Vigneshwaran et al. 2019). Multifunctional g-C_3N_4/β-cyclodextrin was synthesized and applied for the adsorptive remediation of MB and Pb(II) from aqueous solution.

Density functional theory test revealed that Pb(II) removal by g-C_3N_4/β-cyclodextrin was dominated by surface complexation and electrostatic attractions, while MB removal was driven by hydrogen bonds and π-π stack interactions (Zou et al. 2016). Moreover, Guo et al., developed a nanoscale zero-valent iron modified potassium doped g-C_3N_4 composites and applied as an adsorbent for the treatment of Cr(VI) contaminated water. The surface area and pore volume measurements showed that the developed composite possessed 4.7 and 3.7 times higher values than the pure g-C_3N_4 and thus it showed higher adsorption capacity for Cr(VI) ions (Guo et al. 2020). Recently, phosphate and Pb(II) removal by adsorption onto protonated g-C_3N_4/acid-activated montmorillonite composite (g-C_3N_4/Mt) was demonstrated. The maximum adsorption capacity for phosphate and Pb(II) was achieved by 2%-g-C_3N_4/Mt, owing to the synergistic effects of the individual counterparts (Wan et al. 2019).

Besides, it was believed that carbon doping on g-C_3N_4 could result in the enhancement of the π-conjugation system and surface area. Ren et al., investigated the adsorption of MB using a carbon doped g-C_3N_4 as an adsorbent. It was found that the surface area and adsorption capacity of g-C_3N_4 for MB was enhanced after carbon doping (Ren et al. 2018). Likewise, S-block metals doped g-C_3N_4 (e.g., lithium, sodium, potassium, magnesium, calcium, strontium, and barium) was prepared from cyanamide for the remediation of water pollutants. A substantial increment in the adsorption capacity of g-C_3N_4 towards the removal of MB and Cu(II) ions was noted after doping with s-block metals (Fronczak et al. 2018, 2017). Furthermore, phosphorus doped g-C_3N_4 was developed and applied for the adsorption of MB from aqueous solution (Chegeni and Dehghan 2020). Recently, S-doped g-C_3N_4/graphene oxide framework was developed by Li and co-workers to promote the Hg(II) removal from the desulfurization slurry. It was deduced from the results that the presence of S groups and tubular geometry in S-doped g-C_3N_4 had strong impacts on recovering the Hg(II) through enhanced electrostatic forces and co-ordination bonding (Li et al. 2020c).

Also, it is very important to separate the exhausted adsorbent after the adsorption process to avoid any adverse effects on the environment. Due to the very small size of g-C_3N_4-based materials, the complete separation of g-C_3N_4 from aqueous solution is still posing a significant challenge to the researchers. To facilitate the separation, magnetic properties have been incorporated into the g-C_3N_4 nanosheets. For instance, Yan et al., introduced magnetic nanoparticles to the MCN (MMCN) for the potential removal of PFOS and perfluorooctanoic acid (PFOA) with the consideration of simple adsorbent recovery by applying an external magnetic field. The adsorption study revealed that MMCN had a higher affinity for PFOS adsorption (454.6 mg g^{-1}) than PFOA (370.4 mg g^{-1}). Moreover, when the temperature was increased from 278 to 318 K, adsorption of PFOS and PFOA on MMCN was also found to be increased. The removal of PFOS and PFOA by MMCN was mainly controlled by the combination of electrostatic and hydrophobic interactions (Yan et al. 2014). Likewise, Guo et al., reported one-step preparation of magnetic g-C_3N_4 for the removal of Zn(II), Pb(II) and Cd(II). The magnetic g-C_3N_4 showed improved removal of Zn(II), Pb(II) and Cd(II) with the increase of the solution pH (Guo et al. 2018a).

A novel three-dimensional (3D) magnetic nanocomposite (Fe_3O_4@g-C_3N_4) was also prepared and tested for the potential adsorption of Pb(II) from water. By comparing the adsorption capacities of pristine g-C_3N_4, Fe_3O_4, and Fe_3O_4@g-C_3N_4, the binary composite showed higher adsorption capacity due to the addition of more layer spacing (Guo et al. 2018b). Besides, Sahoo and co-workers synthesized magnetically separable and reusable graphene oxide/g-C_3N_4 decorated Fe_3O_4 nano adsorbent for the elimination of tetracycline and MB. The synthesized nano adsorbent showed excellent adsorption capacity due to more π-π and hydrogen bonding interactions. In addition, the elimination of the tetracycline and MB was significantly affected by the solution pH (Sahoo et al. 2020).

An efficient adsorbent for adsorption and separating crude oil from water was additionally prepared by combining g-C_3N_4 along with magnetic $FeNi_3$ and fatty acid functionalization. The prepared magnetic hydrophobic/oleophilic g-C_3N_4 exhibited excellent oil sorption performance (Talukdar et al. 2019).

An overview of the literature studying the application of g-C_3N_4-based materials for the adsorptive removal of various environmental pollutants is listed in Table 5.1.

5.3 Photodegradation of Pollutants Using g-C_3N_4-based Photocatalysts for Wastewater Treatment Application

In recent years, photodegradation of pollutants has been turned out to be one of the effective environmental cleaning-up methods. Among several photocatalysts, g-C_3N_4-based materials have gained much attention as versatile photocatalysts for the environmental remediation of various pollutants. This is mainly because of its unique properties, i.e., low cost, eco-friendly, metal-free catalyst, visible light response, and appealing electronic bandgaps (Ong et al. 2016; Wen et al. 2017).

Additionally, the photocatalytic activity of g-C_3N_4 could be further enhanced by exposing g-C_3N_4 into various chemical modifications (Zhang et al. 2019). Nevertheless, the photocatalytic activity of the catalysts depends certainly on the preparation conditions. For instance, charcoal/g-C_3N_4 composite visible light photocatalyst was developed under different atmospheric conditions and tested their photoactivity towards the degradation of MB and Cr(VI). The results indicated that charcoal/g-C_3N_4 composites prepared under the air atmosphere showed higher photocatalytic activity for MB, while charcoal/g-C_3N_4 composites prepared under the oxygen atmosphere exhibited good photoreduction of Cr(VI) (Lamkhao et al. 2018). Hu and co-workers developed g-C_3N_4 based catalysts co-doped with iron, phosphorus, and sulfur for the efficient photodegradation of various dyes and phenols (Hu et al. 2018, 2014). The sulfur self-doped g-C_3N_4 prepared at different temperatures (450–575 °C) were shown to have different photocatalytic activity behavior for MB. For instance, sulfur self-doped g-C_3N_4 prepared at 550 °C showed optimal visible light photocatalytic activity performance, which might be due to the unique layered structure with more surface area and reduced band gap induced sulfur-doping (Cao et al. 2015).

Table 5.1 Overview of the literature studying the application of g-C_3N_4-based materials explored for the adsorptive removal of various environmental pollutants

Adsorbent	Target pollutant(/s)	Reference
g-C_3N_4 assembled with Mg-Co-Al-layered double hydroxide	Iodine 131	Yazdankish et al. (2020)
g-C_3N_4	Cr(VI), Cd(II), Pb(II)	Xiao et al. (2019)
g-C_3N_4	Eu(III), La(III), Nd(III), and Th(IV)	Liao et al. (2018b)
Exfoliated g-C_3N_4	Cu(II), Mn(II), Zn(II), Pb(II), Fe(III), and Cd(II)	Dehkharghani (2019)
g-C_3N_4	Pb(II) and aniline	Hu et al. (2015)
g-C_3N_4	Cd(II) and methylene blue	Cai et al. (2017)
g-C_3N_4	Pb(II), Cu(II), Cd(II), and Ni(II)	Shen et al. (2015)
Mesoporous carbon nitride	PFOS	Yan et al. (2013)
Mesoporous carbon nitride	Phenol	Haque et al. (2010)
Mesoporous g-C_3N_4	Ni(II)	Xin et al. (2016)
Protonated mesoporous g-C_3N_4	Microcystins	Huang et al. (2015)
g-C_3N_4/attapulgite hybrid	Cd(II)	Xie et al. (2020a)
Ox-g-C_3N_4 /PANI-NF	Cr(VI) and Cu(II)	Kumar et al. (2017)
PANI/ oxidation etching g-C_3N_4	U(VI)	Liu et al. (2019a)
Oxidized g-C_3N_4	Tetracycline, methylene blue, methylene green, rose Bengal, basic fuchsin, rhodamine B, and methyl orange	Yousefi et al. (2019)
g-C_3N_4 functionalized by SA biopolymer	Pb(II), Ni(II), and Cu(II)	Shen et al. (2020)
Chitosan incorporated g-C_3N_4	Chlorpyrifos	Vigneshwaran et al. (2019)
g-C_3N_4/β-cyclodextrin	Pb(II)	Zou et al. (2016)
Nanoscale zero valent iron modified potassium-doped g-C_3N_4	Cr(VI)	Guo et al. (2020)
Protonated g-C_3N_4/acid-activated montmorillonite composite	Phosphate and Pb(II)	Wan et al. (2019)
Carbon-doped g-C_3N_4	Methylene blue	Ren et al. (2018)
S-block metals-doped g-C_3N_4	Methylene blue and Cu(II)	Fronczak et al.(2018)
Sodium-doped g-C_3N_4	Methylene blue	Fronczak et al. (2017)
Phosphorus-doped g-C_3N_4	Methylene blue	Chegeni and Dehghan (2020)

(continued)

Table 5.1 (continued)

Adsorbent	Target pollutant(/s)	Reference
S-doped g-C_3N_4	Hg(II)	Li et al. (2020c)
Magnetic nanoparticles incorporated mesoporous carbon nitride	PFOS and PFOA	Yan et al. (2014)
Magnetic g-C_3N_4	Zn(II), Pb(II), and Cd(II)	Guo et al. (2018a)
Fe_3O_4@g-C_3N_4	Pb(II)	Guo et al. (2018b)
Fe_3O_4-g-C_3N_4	Zn(II), Pb(II), and Cd(II)	Guo et al. (2018c)
Graphene oxide/g-C_3N_4 decorated Fe_3O_4	Tetracycline and methylene blue	Sahoo et al. (2020)
g-C_3N_4/ $FeNi_3$	Crude oil	Talukdar et al. (2019)
Carboxylmethyl cellulose/ g-C_3N_4/ZnO	Methyl violet	Sharma et al. (2020)
g-C_3N_4	Aspirin	Chegeni et al. (2020)
g-C_3N_4/Fe_3O_4/ZIF-8 nanocomposite	Malachite green	Zhao et al. (2020c)
V_2O_5/S-doped g-C_3N_4	Methylene blue and phenol	Chegeni et al. (2019)
g-C_3N_4/MgO nanocomposite	Pb(II)	Zhou et al. (2019)
Microporous Fe_2O_3/g-C_3N_4	Phosphate	Gamshadzehi et al. (2019)
Biochar/g-C_3N_4 composite	Methylene blue	Li (2019)
g-C_3N_4	P(V), Cr(VI) and Re(VII)	Liao et al. (2018a)
Polyoxoniobate/g-C_3N_4	Methylene blue	Gan et al. (2018)
$SrCO_3$/g-C_3N_4 nanocomposite	Crystal violet, rhodamine b, and methylene blue	Lu et al. (2018)

In recent days, metal/non-metal doped g-C_3N_4 catalysts were developed to increase the photocatalytic activity of g-C_3N_4 nanosheets by reducing the band gap, hindering the recombination of generated electrons and holes, increasing the surface area, and as well as increasing the visible light absorption ability. For instance, boron doped g-C_3N_4 exhibited superior photocatalytic activity for rhodamine B (RhB), compared to pristine g-C_3N_4 nanosheets due to the improvement in dye adsorption and light absorption of catalyst (Yan et al. 2010). Moreover, it was observed that after incorporating palladium nanocones into g-C_3N_4, recombination of holes and generated electrons were hampered and hence the catalytic reduction of Cr(VI) to Cr(III) was further promoted (Wu et al. 2019b). Besides, carbon dots decorated g-C_3N_4 based photocatalysts were shown to possess higher photocatalytic activities than the pristine g-C_3N_4 towards various pollutants (Fang et al. 2016; Miao et al. 2017; Zhang et al. 2016). The impregnation of carbon dots on the g-C_3N_4 surface not only enhanced the production of photogenerated electron–hole pairs, but also inhibited electron–hole recombination (Zhang et al. 2016).

Vellaichamy and Periakaruppan reported on the synthesis of sulfur and phosphorous co-doped g-C_3N_4 grafted polyaniline for the effective degradation of MB under visible light. A substantial enhancement in the MB degradation was achieved by the developed composite, compared to its counterparts (Vellaichamy and Periakaruppan 2018). Pan et al., reported the facile synthesis of iron (Fe) and cerium (Ce) co-doped g-C_3N_4 (Fe-Ce/g-C_3N_4) and investigated its photocatalytic activity towards the degradation of Cr(VI). The results revealed the superior catalytic performance of Fe-Ce/g-C_3N_4 for the degradation of Cr(VI). In addition, the degradation of Cr(VI) was also tremendously improved in the presence of MO, RhB or 2,4-dichloro phenol mixed pollutant systems (Pan et al. 2020a). The proposed mechanistic pathways for the Cr(VI) reduction and dyes degradation using Fe-Ce/g-C_3N_4 under visible light is given in Fig. 5.2. Likewise, hydrochloric acid and phytic acid co-doped polyaniline (PANI) coupled g-C_3N_4, 3D hierarchical structures with high surface area were also synthesized and exhibited a higher catalytic reduction of Cr(VI) and photodegradation of MB in water. Based on the results, it was assumed that photoexcited charge transfer, electric conductivity and unpaired electrons within the PANI coupled g-C_3N_4 played key roles during the photocatalysis processes (Wu et al. 2019a). In addition, gold nanoparticles stabilized sulfur-doped g-C_3N_4 (Au@S-g-C_3N_4) was used for the catalytic reduction of 4-nitro phenol (4-NP). The results revealed the superior catalytic performance of Au@S-g-C_3N_4 towards 4-NP reduction than in the case of S-g-C_3N_4. Higher reduction of 4-NP was attributed to the synergistic effects of the gold nanoparticles and S-g-C_3N_4 (Balakumar et al. 2020).

Zhao and co-workers prepared different combinations of g-C_3N_4-chitosan beads and investigated the photodegradation of MB. g-C_3N_4-chitosan beads exhibited higher removal of MB than pristine g-C_3N_4 because of the combined effects of the in-situ adsorption and photocatalysis process. It was also found that super oxide

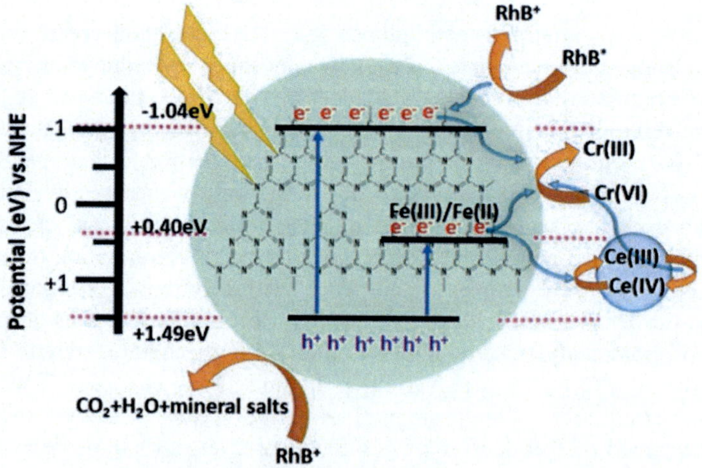

Fig. 5.2 The proposed mechanistic pathways of Cr(VI) reduction and dyes degradation using Fe-Ce/g-C_3N_4 under visible-light irradiation. Modified from (Pan et al. 2020a)

radicals ($^{\bullet}O_2^-$) and holes (h^+) were mainly responsible for the photodegradation of MB (Zhao et al. 2018). Furthermore, the presence of carbon nanotubes in g-C_3N_4 helped to transfer the electron produced during the degradation of RhB and promoted the separation of electrons-holes (Liu et al. 2020a). In another study, superior degradation of Acid red G was attained using $AgFeO_2$/g-C_3N_4 composite as a catalyst, which is ~7.5 and ~10.7 times higher than pristine g-C_3N_4 and $AgFeO_2$, respectively. This enhanced photocatalytic performance was ascribed to the formation of core–shell heterojunction between $AgFeO_2$ and g-C_3N_4, to superior electron–hole separation efficiency, and the superior specific surface area, etc. (Xue et al. 2015). In parallel, nitrogen-doped graphene/g-C_3N_4 composite (NG@g-C_3N_4) was demonstrated to be an ideal photocatalyst for the degradation of methyl orange (MO) from water. It was mainly attributed to the excellent charge separation and reduction in electron-holes recombination (Santha kumar et al. 2019). A binary polymer composite consisting of poly(diphenylbutadiyne) and g-C_3N_4 exhibiting higher visible light photocatalytic activity for the degradation of RhB and phenol than pristine g-C_3N_4 and poly(diphenylbutadiyne) were developed. Notably, the visible light absorption ability and separation efficiency of photo-generated electrons and holes of g-C_3N_4 was significantly enhanced via the modification with poly(diphenylbutadiyne), and thus, higher photocatalytic behavior was achieved (Lei et al. 2017). Similar kinds of results were reported by several researchers for different g-C_3N_4-based photocatalysts (Gawande and Thakare 2012; Ge et al. 2012; Huy et al. 2017; Qian et al. 2018).

Furthermore, photocatalytic degradation of 4-chlorophenol (4-CP) as well as reduction of highly toxic Cr(VI) into less toxic Cr(III) has been achieved using porous g-C_3N_4. An obvious synergistic photocatalytic effect has been observed for both Cr(VI) and 4-CP in a Cr(VI)/4-CP binary system than the individual system under acidic conditions (pH 3). This was mainly attributed to the accelerated redox reaction between $Cr_2O_7^{2-}$ and 4-CP via electron transfer through g-C_3N_4 (Wei et al. 2017). In another study, photo oxidation of bisphenol A and photoreduction of Cr(VI) was investigated in a binary system using g-C_3N_4 prepared under nitrogen gas atmosphere. Yet again, higher degradation efficiency was obtained in the binary system over the individual systems. It was established that the produced H_2O_2 and photogenerated electrons were responsible for the reduction of Cr(VI) into Cr(III), while $^{\bullet}O_2^-$ radicals were responsible for the oxidation of bisphenol A (Wang et al. 2019a). In the work of Wang and co-workers, the pyromellitic diimide doped g-C_3N_4 showed non-adsorptive nature for the oxidative product As(V) formed during the photocatalysis processes. More interestingly, a small amount of Cr(III) adsorbed on g-C_3N_4 surface promoted the oxidation of As(III) during the process (Wang et al. 2019b).

Moreover, g-C_3N_4 nanosheets have also been explored as photocatalyst for the mineralization of various PPCPs from wastewater (Zhu et al. 2016). For example, mineralization of four pharmaceutical compounds (tetracycline, ciprofloxacin, salicylic acid, and ibuprofen) in water was conducted using g-C_3N_4 nanosheets. The capacity of g-C_3N_4 for PPCPs degradation followed the sequence: tetracycline > ciprofloxacin > salicylic acid > ibuprofen. It was also found that the photogenerated holes, H_2O_2 and $^{\bullet}OH$ radicals were responsible for the mineralization of tetracycline (Hernández-Uresti et al. 2016). Also, Xu and co-workers developed ultrathin g-C_3N_4 nanosheets decorated hydroxyapatite (HAp) for the degradation of tetracycline, an antibiotic drug molecule from wastewater. A faster tetracycline degradation rate was noted for the composite over pristine g-C_3N_4 nanosheets and HAp because of the formation of $^{\bullet}O_2^-$ radicals and h^+ holes formed (Xu et al. 2019). Ibuprofen (an anti-inflammatory drug) and 2,4-Dicholorophneoxy acetic acid (herbicide) were also mineralized using sustainable g-C_3N_4-based hybrid photocatalysts (single, binary, and tertiary). Tertiary catalysts exhibited substantial photocatalytic activity over the single and binary catalysts for the mineralization of ibuprofen and 2,4-Dicholorophneoxy acetic acid. This improved performance by tertiary photocatalysts was mainly attributed to high visible light absorption and high charge separation and diffusion (Kumar et al. 2019). Moreover, the crystalline structure in g-C_3N_4 facilitated the O_2 absorption process which in turn produced greater amount of H_2O_2, and thus, generated more $^{\bullet}OH$ radicals to degrade several PPCPs including naproxen (NPX), indomethacin (IDM), diclofenac (DCF), carbamazepine (CBZ), triclosan (TCS), ofloxacin (OFX), enrofloxacin (ENR), and sulfamethoxazole (SME). Also, by considering the complex nature of PPCPs in the wastewater matrices, degradation of mixed PPCPs was investigated. The results showed a degradation efficiency of ~79.5% for NPX and SME, 100% for IDM, DCF and CBZ, 97.2% for TCS, 84.2% for OFX, and 95.1% for ENR (Wang et al. 2020d).

In addition, Wang and co-workers (Wang et al. 2015) and Li and co-workers (Li et al. 2016) synthesized g-C_3N_4 and titanium dioxide (TiO_2)-based hybrid photocatalysts and investigated their photocatalytic activity for the degradation of ciprofloxacin (antibiotic drug) and acyclovir (antiviral drug), respectively. The results indicated the troubles in the complete degradation of acyclovir under normal conditions. Besides, it was found that some of the intermediates formed during the photocatalytic process were more toxic than acyclovir.

On the other hand, Marta Jiménez-Salcedo and co-workers compared the photocatalytic degradation of ibuprofen using g-C_3N_4 and TiO_2 photocatalysts and investigated the intermediates formed during the process. Under the employed conditions, TiO_2 showed better degradation than g-C_3N_4. Additionally, after analyzing the intermediates, it has been found that two of the formed intermediates were higher toxic than ibuprofen (Jiménez-Salcedo et al. 2019). Recently, Li and co-workers developed rare earth metal samarium (Sm)-doped g-C_3N_4 nanosheets and examined its photocatalytic performance towards the degradation of tylosin and RhB. The results showed an increased photocatalytic activity of Sm-doped g-C_3N_4 than pristine g-C_3N_4. The photocatalytic activity of g-C_3N_4 nanosheets found to be increased upon Sm-doping, however, extensive doping of Sm had an adverse impact on the photocatalytic degradation performance towards selected pollutants, since extensive Sm doping has resulted in promoting the recombination of holes and electrons (Li et al. 2020b).

Synergistic photocatalytic activity of persulfate and g-C_3N_4 has been investigated toward the degradation of sulfamethoxazole under simulated sunlight. A substantial increment in the degradation efficiency was solely seen for the system involving persulfate and $g\text{-}C_3N_4$. It was emphasized that in the binary system, $g\text{-}C_3N_4$ induced the sulfate to transfer $SO_4^{-\bullet}$, $S_2O_8^{2-}$ and favored the generation of $O_2^{-\bullet}$, $SO_4^{-\bullet}$, and h^+ which in turn enhanced the degradation of sulfamethoxazole (Song et al. 2020). In parallel, Liu et al., modified $g\text{-}C_3N_4$ with four typical nanocarbons and studied their photocatalytic activities for the degradation of sulfachloropyridazine and dyes. Carbon nanospheres modified $g\text{-}C_3N_4$ exhibited the highest degradation efficiency for sulfachloropyridazine and dyes, which was ascribed to a large number of active sites as well as the increased charge, mass transport and chemical composition (Liu et al. 2020b).

Moreover, the photodegradation of methyl-parabens, ethyl-parabens and propyl-parabens in ultrapure, tap and river water was conducted using $g\text{-}C_3N_4$ as a metal-free photocatalyst. Experiments with single parabens demonstrated that the photocatalytic system was sensitive to the structure of the paraben. In addition, the existence of organic and inorganic ions in water had a significant influence on the degradation of the parabens (Fernandes et al. 2020). Besides, Moreira and co-workers investigated the photochemical degradation of various micropollutants (MPs) including carbamazepine, isoproturon, clopidogrel, diclofenac, atenolol, bezafibrate, tramadol, venlafaxine, and fluoxetine from urban wastewater treatment plant using $g\text{-}C_3N_4$ in both batch and continuous mode. Significant degradation of the selected MPs was achieved in the batch mode, while lower degradation was observed in the continuous mode, which was mainly due to lower residence time (Moreira et al. 2019). A list of literature reporting on $g\text{-}C_3N_4$-based photocatalysts developed for photodegradation of various pollutants is given in Table 5.2.

5.4 Challenges and Future Perspectives

The graphitic carbon nitride is an evolving material used for water treatment in photodegradation and adsorption methods. This material is gaining popularity in water treatment due to its ease of preparation at an economic level from easily available precursors. This material has no toxicity with tunable structure and hence, can be mold easily for getting the desired band gap. These features made this material an ideal candidate for increased solar energy usage and good photocatalytic agent for degrading water pollutants. The lenience to increase the surface area of $g\text{-}C_3N_4$ by generating nanoscale or porous structure not only advances the charge separation but also offers extra accessible locations for the photo-degradation reactions. The formation of a Z-scheme heterojunction of graphitic carbon nitride with other photocatalysts may stimulate charge separation and use of low-energy visible photons without conceding the redox capacity of photo-generated electrons and holes. Of course, graphitic carbon nitride is a good material for photo-degradation but still needs improvement so that it may be used at a large scale in industrial applications. For this purpose, it is required to modify the structure of $g\text{-}C_3N_4$ to get the final goal of the commercial and economic industrialization of $g\text{-}C_3N_4$ in water treatment.

Table 5.2 List of literature reporting on $g\text{-}C_3N_4$-based photocatalysts used for photodegradation of various environmental pollutants

Photocatalyst	Target pollutant(/s)	Reference
$g\text{-}C_3N_4$ modified by carbon nanotubes	Rhodamine B	Liu et al. (2020a)
$Ag/Fe_3O_4/g\text{-}C_3N_4$ composite	Tetracycline	Zhu et al. (2016)
$g\text{-}C_3N_4$	Tetracycline, ciprofloxacin, salicylic acid, and ibuprofen	Hernández-Uresti et al. (2016)
$g\text{-}C_3N_4$ nanosheets decorated hydroxyapatite	Tetracycline	Xu et al. (2019)
$g\text{-}C_3N_4$/ACN/RGO@ Biochar and $g\text{-}C_3N_4$/PANI/RGO @Biochar	Ibuprofen and 2,4-Dicholorophneoxy acetic acid	Kumar et al. (2019)
$g\text{-}C_3N_4$	Naproxen, indomethacin, diclofenac, carbamazepine, triclosan, ofloxacin, enrofloxacin and sulfamethoxazole	Wang et al. (2020d)
$g\text{-}C_3N_4\text{-}TiO_2$ (P25)	Ciprofloxacin	Wang et al. (2015)
$g\text{-}C_3N_4\text{-}TiO_2$	Acyclovir	Li et al. (2016)
$g\text{-}C_3N_4$	Ibuprofen	Jiménez-Salcedo et al. (2019)
Sm-doped $g\text{-}C_3N_4$	Tylosin and rhodamine B	Li et al. (2020b)
Carbon quantum dots modified $g\text{-}C_3N_4$	Sulfachloropyridazine, methylene blue, methyl orange and rhodamine B	Liu et al. (2020b)
Reduced graphene oxide modified $g\text{-}C_3N_4$		
Carbon nanotube modified $g\text{-}C_3N_4$		
Carbon nanospheres modified $g\text{-}C_3N_4$		
$g\text{-}C_3N_4$	Methyl-, ethyl- and propyl-parabens	Fernandes et al. (2020)
Exfoliated $g\text{-}C_3N_4$	Carbamazepine, isoproturon, clopidogrel, diclofenac, atenolol, bezafibrate, tramadol, venlafaxine, and fluoxetine	Moreira et al. (2019)
$Fe_3O_4/g\text{-}C_3N_4$ nanospheres	Methyl orange	Zhou et al. (2013)
$BiOBr\text{-}g\text{-}C_3N_4$ composite	Rhodamine B	Ye et al. (2013)
Fe(III)/graphene $g\text{-}C_3N_4$	Methyl orange	Liu et al. (2016)
$V_2O_5/g\text{-}C_3N_4$	Rhodamine B and tetracycline	Hong et al. (2016)
$g\text{-}C_3N_4/SmVO_4$ composite	Rhodamine B	Li et al. (2013)
$g\text{-}C_3N_4/Bi_2WO_6$ composite	Methyl orange	Ge et al. (2011a)
$g\text{-}C_3N_4/Bi_2MoO_6$ composite	Rhodamine B	Li et al. (2014)

(continued)

Table 5.2 (continued)

Photocatalyst	Target pollutant(/s)	Reference
g-C_3N_4/Ag_3VO_4 composite	Fuchsine	Wang et al. (2014)
Ag/g-C_3N_4	Methyl orange	Ge et al. (2011b)
WO_3-TiO_2@g-C_3N_4	Acteylsalicylate and methylthebromine	Tahir et al. (2019)
ZnO-Ag_2O/porous g-C_3N_4	Ciprofloxacin	Rong et al. (2016)
g-C_3N_4/TiO_2	Rhodamine B	Tong et al. (2015)
Macro/mesoporous g-C_3N_4/TiO_2	Rhodamine B	Hao et al. (2016)
Quaternary magnetic BiOCl/g-C_3N_4/Cu_2O/Fe_3O_4	Sulfamethoxazole	Kumar et al. (2018)
TiO_2@g-C_3N_4	Tetracycline	Wang et al. (2017b)
Carbon dots modified MoO_3/g-C_3N_4	Tetracycline	Xie et al. (2018)
Sn_3O_4/g-C_3N_4	Tetracycline	Li et al. (2018)
Nitrogen doped carbon dots/g-C_3N_4	Indomethacin	Wang et al. (2017a)
Carbon dots and Ag nanoparticles decorated g-C_3N_4	Methyl orange and p-nitrophenol	Dadigala et al. (2017)
g-C_3N_4/Ag_2SO_4 nanocomposite	Rhodamine B, methylene blue and fuchsine	Akhundi and Habibi-Yangjeh (2016)
g-C_3N_4	Rhodamine B and methyl orange	Yu and Wang (2016)
g-C_3N_4/ZnO/AgCl nanocomposite	Rhodamine B	Akhundi and Habibi-Yangjeh (2015)
Nested g-C_3N_4 fiber surrounding graphene oxide	Rhodamine B	Wang et al. (2020e)
g-C_3N_4/CuS	Methyl orange	Wang et al. (2020a)
Carboxymethyl cellulose with Zn-Cu-Mg mixed metal hydroxide/g-C_3N_4 composite	Sulfadiazine	Gholami et al. (2020)
Terephthalic acid functionalized C_3N_4/TiO_2/Fe_2O_3@SiO_2	Ibuprofen, benzophenone, and carbamazepine	Kumar et al. (2020)
Cu(I) coordinated g-C_3N_4 framework	Chlortetracycline	Wang et al. (2020b)
2D graphene/g-C_3N_4	Bisphenol A, sulfamethoxazole, ciprofloxacin, 2-chlorphenol and diphenhydramine	Li et al. (2020a)
Hollow TiO_2@g-C_3N_4/Co_3O_4 core–shell microspheres	Tetracycline and methyl orange	Yu et al. (2020)
Nitrogen rich g-C_3N_4	Cefotaxime	Gao et al. (2020)

(continued)

Table 5.2 (continued)

Photocatalyst	Target pollutant(/s)	Reference
Ultrathin Ag_2WO_4-coated P-doped g-C_3N_4	Indomethacin	Huang et al. (2020)
Mn_3O_4 nanodots loaded g-C_3N_4	4-chlorophenol	Chen et al. (2020a)
Pd/g-C_3N_4	Bezafibrate	Yin et al. (2020)
2D-2D WO_3/g-C_3N_4	Tetracycline	Pan et al. (2020b)
Amorphous goethite nanoparticles decorated g-C_3N_4	Tetracycline	Zhao et al. (2020b)
K-doped g-C_3N_4/BiOBr hybrid	Tetracycline and rhodamine B	Qu et al. (2020)
Covalent triazine-based frameworks/g-C_3N_4	Sulfamethazine	Cao et al. (2020)
AgBr/P-g-C_3N_4	Ephedrine	Chen et al. (2020b)
g-C_3N_4/ZnO nanorods	Methylene blue, rhodamine B, Cr(VI) and eosin	Zhong et al. (2020)
g-C_3N_4/MoS_2	Levofloxacin and methylene blue	He et al. (2020)
g-C_3N_4/TiO_2	Acid orange 7	Liu et al. (2020c)
Amorphous g-C_3N_4/TiO_2	Reactive red 4	Azami et al. (2020)
MoS_2/Ag/g-C_3N_4	Tetracycline	Jin et al. (2020)
Au-SnO_2 quantum dot-anchored g-C_3N_4 nanosheets	Rhodamine B	Babu et al. (2020)
g-C_3N_4/Ag-TiO_2	Rhodamine B	Sui et al. (2020)
Waste sludge–doped g-C_3N_4	Eriochrome Black T	Gu et al. (2020)
Fe^{3+}-doped g-C_3N_4	Sulfadiazine	Ou et al. (2020)
Carbon quantum dots modified g-C_3N_4	Diclofenac	Liu et al. (2019b)
Carbon quantum dots modified tubular g-C_3N_4	Carbamazepine	Zhao et al. (2020a)
g-C_3N_4 supported Fe_3O_4 nanoparticles/ZnO nanorods	Pantoprazole	Raha and Ahmaruzzaman (2020)
Nitrogen defects g-C_3N_4	Meropenem	Wang et al. (2020c)
g-C_3N_4/La_2O_3	Methylene blue, rhodamine B and methyl orange	Xie et al. (2020b)
Sepiolite/c/Pd nanocomposite	Ciprofloxacin	Chuaicham et al. (2020)
$CoWO_4$/ g-C_3N_4 nanocomposite	Norfloxacin	Prabavathi et al. (2019)
Pd/CeO_2/g-C_3N_4 nanocomposite	Cr(VI)	Saravanakumar et al. (2017)

There may be some methods to amend the structure of g-C_3N_4 for excellent performance. Among them, theoretical simulation studies are useful to save time and costly chemicals. The most important tool that may be used are DFT and molecular dynamics simulations. With the help of these simulations, one can module the structure, properties, and photo-catalytic capability of g-C_3N_4. Besides, the guidelines for balanced material design may be obtained before synthesis. It is important to establish the photo-stability, recyclability, and separation of g-C_3N_4 in water for industrial applications. Still, one needs to develop the methods and their accuracies of the application of g-C_3N_4 in complex water *i.e.* industrial and municipal effluents. The most important challenge is to find out the photo-degradation products because, sometimes, a few degradation products may be toxic to the environment and human beings. Certainly, there are some challenges commercialize g-C_3N_4 in water treatment methods. The applications of g-C_3N_4 in the adsorption process are still limited because most of the papers are with batch mode applications. Therefore, there is a great need to modulate this material for column operations. There is a big need to amend the structure of this material in many ways to make it ideal adsorbent and photo-degradation material in column operations. Briefly, there is a lot to work with this material in the future. Certainly, we hope that this material will be a boon in water treatment in the near future.

5.5 Summary and conclusions

Over the years, the unlimited inventions in the field of g-C_3N_4, a two-dimensional structured organic polymeric material has certainly witnessed the novel attractive properties with remarkable applications. Due to its unique properties such as highly porous and π-conjugated structures, high chemical tunability, reduced band gap, and high visible light absorption, g-C_3N_4-based materials are emerging as ideal candidates for variety of environmental applications. In this chapter, the potential environmental applications of g-C_3N_4-based materials in water and wastewater treatment were reviewed.

To date, even though g-C_3N_4-based materials have been revealed as promising candidates for designing and fabricating powerful adsorbents and photocatalysts, there are still many challenges facing the widespread application of g-C_3N_4-based materials. The aforementioned shortcomings can be potentially alleviated by employing the most appropriate precursors and preparation conditions, development of heterojunction nanocomposites, non-metal and metal doping, or copolymerization with other functional polymers. This will open up for further research that might be needed in the area of design and development of g-C_3N_4-based materials with improved performance and recovery. Moreover, pollutants remediation from real wastewater/river water under continuous system using g-C_3N_4-based materials is still regarded as challenges to newly emerged materials. Further research is required in designing, developing, and optimizing the continuous mode application toward the efficient removal of pollutants. Overall, it is believed that this chapter offers a comprehensive and updated overview on the state of the art as well as recent developments and applications of g-C_3N_4-based materials in water and wastewater treatment.

References

Akbari Dehkharghani A (2019) Exfoliated graphitic carbon nitride for the fast adsorption of metal ions from acid mine drainage: a case study from the Sungun copper mine. Mine Water Environ 38(2):335–341

Akhundi A, Habibi-Yangjeh A (2015) Ternary g-C_3N_4/ZnO/AgCl nanocomposites: synergistic collaboration on visible-light-driven activity in photodegradation of an organic pollutant. Appl Surf Sci 358:261–269

Akhundi A, Habibi-Yangjeh A (2016) Novel g-C_3N_4/Ag_2SO_4 nanocomposites: fast microwave-assisted preparation and enhanced photocatalytic performance towards degradation of organic pollutants under visible light. J Colloid Interface Sci 482:165–174

Atlas RM, Hazen TC (2011) Oil biodegradation and bioremediation: a tale of the two worst spills in U.S. History. Environ Sci Technol 45(16):6709–6715

Azami MS, Ismail K, Ishak MAM, Zuliahani A, Hamzah SR, Nawawi WI (2020) Formation of an amorphous carbon nitride/titania composite for photocatalytic degradation of RR4 dye. J Water Process Eng 35:101209

Babu B, Koutavarapu R, Shim J, Yoo K (2020) Enhanced visible-light-driven photoelectrochemical and photocatalytic performance of Au-SnO_2 quantum dot-anchored g-C_3N_4 nanosheets. Sep Purif Technol 240:116652

Balakumar V, Kim H, Ryu JW, Manivannan R, Son Y-A (2020) Uniform assembly of gold nanoparticles on S-doped g-C_3N_4 nanocomposite for effective conversion of 4-nitrophenol by catalytic reduction. J Mater Sci Technol 40:176–184

Cai X, He J, Chen L, Chen K, Li Y, Zhang K, Jin Z, Liu J, Wang C, Wang X, Kong L, Liu J (2017) A 2D-g-C_3N_4 nanosheet as an eco-friendly adsorbent for various environmental pollutants in water. Chemosphere 171:192–201

Cao L, Wang R, Wang D (2015) Synthesis and characterization of sulfur self-doped g-C_3N_4 with efficient visible-light photocatalytic activity. Mater Lett 149:50–53

Cao S, Zhang Y, He N, Wang J, Chen H, Jiang F (2020) Metal-free 2D/2D heterojunction of covalent triazine-based frameworks/graphitic carbon nitride with enhanced interfacial charge separation for highly efficient photocatalytic elimination of antibiotic pollutants. J Hazardous Mater 391:122204

Chegeni M, Dehghan N (2020) Preparation of phosphorus doped graphitic carbon nitride using a simple method and its application for removing methylene blue. Phys Chem Res 8(1):31–44

Chegeni M, Goudarzi F, Soleymani M (2019) Synthesis, characterization and application of V_2O_5/S-doped graphitic carbon nitride nanocomposite for removing of organic pollutants. ChemistrySelect 4(46):13736–13745

Chegeni M, Mousavi Z, Soleymani M, Dehdashtian S (2020) Removal of aspirin from aqueous solutions using graphitic carbon nitride nanosheet: theoretical and experimental studies. Diam Relat Mater 101:107621

Chen C, Xie M, Kong L, Lu W, Feng Z, Zhan J (2020a) Mn_3O_4 nanodots loaded g-C_3N_4 nanosheets for catalytic membrane degradation of organic contaminants. J Hazard Mater 390:12214.

Chen M, Guo C, Hou S, Lv J, Zhang Y, Zhang H, Xu J (2020b) A novel Z-scheme AgBr/P-g-C_3N_4 heterojunction photocatalyst: excellent photocatalytic performance and photocatalytic mechanism for ephedrine degradation. Appl Catal B Environ 266:118614

Chuaicham C, Pawar RR, Karthikeyan S, Ohtani B, Sasaki K (2020) Fabrication and characterization of ternary sepiolite/g-C_3N_4/Pd composites for improvement of photocatalytic degradation of ciprofloxacin under visible light irradiation. J Colloid Interface Sci 577:397–405

Dadigala R, Bandi R, Gangapuram BR, Guttena V (2017) Carbon dots and Ag nanoparticles decorated g-C_3N_4 nanosheets for enhanced organic pollutants degradation under sunlight irradiation. J Photochem Photobiol, A 342:42–52

Fang S, Xia Y, Lv K, Li Q, Sun J, Li M (2016) Effect of carbon-dots modification on the structure and photocatalytic activity of g-C_3N_4. Appl Catal B 185:225–232

Fayad NM (1986) Identification of tar balls following the Nowruz oil spill. Mar Environ Res 18(3):155–163

Fernandes RA, Sampaio MJ, Dražić G, Faria JL, Silva CG (2020) Efficient removal of parabens from real water matrices by a metal-free carbon nitride photocatalyst. Sci Total Environ 716:135346

Fronczak M, Demby K, Strachowski P, Strawski M, Bystrzejewski M (2018) Graphitic carbon nitride doped with the s-block metals: adsorbent for the removal of methyl blue and Copper(II) ions. Langmuir 34(25):7272–7283

Fronczak M, Krajewska M, Demby K, Bystrzejewski M (2017) Extraordinary adsorption of methyl blue onto sodium-doped graphitic carbon nitride. J Phys Chem C 121(29):15756–15766

Gamshadzehi E, Nassiri M, Ershadifar H (2019) One-pot synthesis of microporous Fe_2O_3/g-C_3N_4 and its application for efficient removal of phosphate from sewage and polluted seawater. Colloids Surf A 567:7–15

Gan Q, Shi W, Xing Y, Hou Y (2018) A polyoxoniobate/g-C_3N_4 nanoporous material with high adsorption capacity of methylene blue from aqueous solution. Frontiers Chem 6(7)

Gao B, Wang J, Dou M, Huang X, Yu X (2020) Novel nitrogen-rich g-C_3N_4 with adjustable energy band by introducing triazole ring for cefotaxime removal. Sep Purif Technol 241:116576

Gawande S, Thakare SR (2012) Ternary polymer composite of graphene, carbon nitride, and Poly(3-hexylthiophene): an efficient photocatalyst. ChemCatChem 4(11):1759–1763

Ge L, Han C, Liu J (2011) Novel visible light-induced g-C_3N_4/Bi_2WO_6 composite photocatalysts for efficient degradation of methyl orange. Appl Catal B 108–109:100–107

Ge L, Han C, Liu J (2012) In situ synthesis and enhanced visible light photocatalytic activities of novel PANI–g-C_3N_4 composite photocatalysts. J Mater Chem 22(23):11843–11850

Ge L, Han C, Liu J, Li Y (2011) Enhanced visible light photocatalytic activity of novel polymeric g-C_3N_4 loaded with Ag nanoparticles. Appl Catal A 409–410:215–222

Gholami P, Khataee A, Vahid B, Karimi A, Golizadeh M, Ritala M (2020) Sonophotocatalytic degradation of sulfadiazine by integration of microfibrillated carboxymethyl cellulose with Zn-Cu-Mg mixed metal hydroxide/g-C_3N_4 composite. Sep Purif Technol 245:116866

Gu L, Dong G, Yu H, Qiao X, Zhang K, Lu X, Wen H.(2020) Graphitic carbon nitride–doped sewage sludge as a novel material for photodegradation of Eriochrome Black T. Environ Sci Pollut Res

Guo J, Chen T, Zhou X, Xia W, Zheng T, Zhong C, Liu Y (2020) Synthesis, Cr(VI) removal performance and mechanism of nanoscale zero-valent iron modified potassium-doped graphitic carbon nitride. Water Sci Technol

Guo S, Duan N, Dan Z, Chen G, Shi F, Gao W (2018) g-C_3N_4 modified magnetic Fe_3O_4 adsorbent: preparation, characterization, and performance of Zn(II), Pb(II) and Cd(II) removal from aqueous solution. J Mol Liq 258:225–234

Guo S, Duan N, Dan Z, Xu F, Zhang C, Shi F, Gao W (2018) Three-dimensional magnetic graphitic carbon nitride composites as high-performance adsorbent for removal Pb^{2+} from aqueous solution. J Taiwan Inst Chem Eng 89:169–182

Guo S, Wu K, Gao Y, Liu L, Zhu X, Li X, Zhang F (2018) Efficient removal of Zn(II), Pb(II), and Cd(II) in waste water based on magnetic graphitic carbon nitride materials with enhanced adsorption capacity. J Chem Eng Data 63(10):3902–3912

Hao R, Wang G, Tang H, Sun L, Xu C, Han D (2016) Template-free preparation of macro/mesoporous g-C_3N_4/TiO_2 heterojunction photocatalysts with enhanced visible light photocatalytic activity. Appl Catal B 187:47–58

Haque E, Jun JW, Talapaneni SN, Vinu A, Jhung SH (2010) Superior adsorption capacity of mesoporous carbon nitride with basic CN framework for phenol. J Mater Chem 20(48):10801–10803

He Y, Ma Z, Junior LB (2020) Distinctive binary g-C_3N_4/MoS_2 heterojunctions with highly efficient ultrasonic catalytic degradation for levofloxacin and methylene blue. Ceram Int 46(8, Part B), 12364–12372

Hernández-Uresti DB, Vázquez A, Sanchez-Martinez D, Obregón S (2016) Performance of the polymeric g-C_3N_4 photocatalyst through the degradation of pharmaceutical pollutants under UV–vis irradiation. J Photochem Photobiol A 324:47–52

Hong Y, Jiang Y, Li C, Fan W, Yan X, Yan M, Shi W (2016) In-situ synthesis of direct solid-state Z-scheme V_2O_5/g-C_3N_4 heterojunctions with enhanced visible light efficiency in photocatalytic degradation of pollutants. Appl Catal B 180:663–673

Horn SA, Neal CP (1981) (1981) The Atlantic empress sinking—a large spill without environmental disaster. Int Oil Spill Conf Proc 1:429–435

Hu C, Hung W-Z, Wang M-S, Lu P-J (2018) Phosphorus and sulfur codoped g-C_3N_4 as an efficient metal-free photocatalyst. Carbon 127:374–383

Hu R, Wang X, Dai S, Shao D, Hayat T, Alsaedi A (2015) Application of graphitic carbon nitride for the removal of Pb(II) and aniline from aqueous solutions. Chem Eng J 260:469–477

Hu S, Ma L, You J, Li F, Fan Z, Lu G, Liu D, Gui J (2014) Enhanced visible light photocatalytic performance of g-C_3N_4 photocatalysts co-doped with iron and phosphorus. Appl Surf Sci 311:164–171

Huang C, Zhang W, Yan Z, Gao J, Liu W, Tong P, Zhang L (2015) Protonated mesoporous graphitic carbon nitride for rapid and highly efficient removal of microcystins. RSC Adv 5(56):45368–45375

Huang J, Li D, Liu Y, Li R, Chen P, Liu H, Lv W, Liu G, Feng Y (2020) Ultrathin Ag_2WO_4-coated P-doped g-C_3N_4 nanosheets with remarkable photocatalytic performance for indomethacin degradation. J Hazard Mater 392:122355

Huy BT, Thao CTB, Dao V-D, Phuong NTK, Lee Y-I (2017) A mixed-metal oxides/graphitic carbon nitride: high visible light photocatalytic activity for efficient mineralization of rhodamine B. Adv Mater Interfaces 4(12):1700128

Islam MN, Taki G, Jung Y-J, Jung S-K, Park J-H (2018) Remediation of gulf war oil spill contaminated soil by a subcritical water extraction process: oil removal, recovery, and degradation. Soil Sedim Contam Int J 27(2):120–130

Jiménez-Salcedo M, Monge M, Tena MT (2019) Photocatalytic degradation of ibuprofen in water using TiO_2/UV and g-C_3N_4/visible light: study of intermediate degradation products by liquid chromatography coupled to high-resolution mass spectrometry. Chemosphere 215:605–618

Jin C, Kang J, Li Z, Wang M, Wu Z, Xie Y (2020) Enhanced visible light photocatalytic degradation of tetracycline by MoS_2/Ag/g-C_3N_4 Z-scheme composites with peroxymonosulfate. Appl Surface Sci 514:146076

Kumar A, Khan M, He J, Lo IMC (2020) Visible–light–driven magnetically recyclable terephthalic acid functionalized g–C_3N_4/TiO_2 heterojunction nanophotocatalyst for enhanced degradation of PPCPs. Appl Catal B Environ 270:118898

Kumar A, Kumar A, Sharma G, Al-Muhtaseb AAH, Naushad M, Ghfar AA, Stadler FJ (2018) Quaternary magnetic BiOCl/g-C_3N_4/Cu_2O/Fe_3O_4 nano-junction for visible light and solar powered degradation of sulfamethoxazole from aqueous environment. Chem Eng J 334:462–478

Kumar A, Sharma G, Naushad M, Al-Muhtaseb AAH, Kumar A, Hira I, Ahamad T, Ghfar AA, Stadler FJ (2019) Visible photodegradation of ibuprofen and 2,4-D in simulated waste water using sustainable metal free-hybrids based on carbon nitride and biochar. J Environ Manag 231:1164–1175

Kumar R, Barakat MA, Alseroury FA (2017) Oxidized g-C_3N_4/polyaniline nanofiber composite for the selective removal of hexavalent chromium. Sci Rep 7(1):12850

Lamkhao S, Rujijanagul G, Randorn C (2018) Fabrication of g-C_3N_4 and a promising charcoal property towards enhanced chromium(VI) reduction and wastewater treatment under visible light. Chemosphere 193:237–243

Lei J, Liu F, Wang L, Liu Y, Zhang J (2017) A binary polymer composite of graphitic carbon nitride and poly(diphenylbutadiyne) with enhanced visible light photocatalytic activity. RSC Adv 7(44):27377–27383
Li C, Yu S, Dong H, Liu C, Wu H, Che H, Chen G (2018) Z-scheme mesoporous photocatalyst constructed by modification of Sn_3O_4 nanoclusters on g-C_3N_4 nanosheets with improved photocatalytic performance and mechanism insight. Appl Catal B 238:284–293
Li F, Tang M, Li T, Zhang L, Hu C (2020a) Two-dimensional graphene/g-C_3N_4 in-plane hybrid heterostructure for enhanced photocatalytic activity with surface-adsorbed pollutants assistant. Appl Catal B Environ 268:118397
Li G, Nie X, Gao Y, An T (2016) Can environmental pharmaceuticals be photocatalytically degraded and completely mineralized in water using g-C_3N_4/TiO_2 under visible light irradiation?—Implications of persistent toxic intermediates. Appl Catal B 180:726–732
Li G, Wang R, Wang B, Zhang J (2020b) Sm-doped mesoporous g-C_3N_4 as efficient catalyst for degradation of tylosin: influencing factors and toxicity assessment. Appl Surf Sci 517:146212
Li H, Liu J, Hou W, Du N, Zhang R, Tao X (2014) Synthesis and characterization of g-C_3N_4/Bi_2MoO_6 heterojunctions with enhanced visible light photocatalytic activity. Appl Catal B 160–161:89–97
Li M, Wang B, Yang M, Li Q, Calatayud DG, Zhang S, Wang H, Wang L, Mao B (2020c) Promoting mercury removal from desulfurization slurry via S-doped carbon nitride/graphene oxide 3D hierarchical framework. Sep Purif Technol 239:116515
Li T, Zhao L, He Y, Cai J, Luo M, Lin J (2013) Synthesis of g-C_3N_4/$SmVO_4$ composite photocatalyst with improved visible light photocatalytic activities in RhB degradation. Appl Catal B 129:255–263
Li X (2019) Preparation and adsorption properties of Biochar/g-C_3N_4 composites for methylene blue in aqueous solution. J Nanomater 2019:2394184
Liao Q, Zou D, Pan W, Linghu W, Shen R, Jin Y, Feng G, Li X, Ye F, Asiri AM, Marwani HM, Zhu Y, Wu X, Dong W (2018) Highly efficient scavenging of P(V), Cr(VI), Re(VII) anions onto g-C_3N_4 nanosheets from aqueous solutions as impacted via water chemistry. J Mol Liq 258:275–284
Liao Q, Zou D, Pan W, Linghu W, Shen R, Li X, Asiri AM, Alamry KA, Sheng G, Zhan L, Wu X (2018) Highly efficient capture of Eu(III), La(III), Nd(III), Th(IV) from aqueous solutions using g-C_3N_4 nanosheets. J Mol Liq 252:351–361
Liebig JV (1834) About some nitrogen compounds. Ann Pharm 10(10)
Liu G, Liao M, Zhang Z, Wang H, Chen D, Feng Y (2020a) Enhanced photodegradation performance of Rhodamine B with g-C_3N_4 modified by carbon nanotubes. Sep Purif Technol 244:116618
Liu J, Chen Z, Yu K, Liu Y, Ge Y, Xie S (2019) Polyaniline/oxidation etching graphitic carbon nitride composites for U(VI) removal from aqueous solutions. J Radioanal Nucl Chem 321(3):1005–1017
Liu Q, Guo Y, Chen Z, Zhang Z, Fang X (2016) Constructing a novel ternary Fe(III)/graphene/g-C_3N_4 composite photocatalyst with enhanced visible-light driven photocatalytic activity via interfacial charge transfer effect. Appl Catal B 183:231–241
Liu Q, Zhou L, Gao J, Wang S, Liu L, Liu S (2020) Surface chemistry-dependent activity and comparative investigation on the enhanced photocatalytic performance of graphitic carbon nitride modified with various nanocarbons. J Colloid Interface Sci 569:12–21
Liu W, Li Y, Liu F, Jiang W, Zhang D, Liang J (2019) Visible-light-driven photocatalytic degradation of diclofenac by carbon quantum dots modified porous g-C(3)N(4): mechanisms, degradation pathway and DFT calculation. Water Res 150:431–441
Liu X, Li W, Hu R, Wei Y, Yun W, Nian P, Feng J, Zhang A (2020c) Synergistic degradation of acid orange 7 dye by using non-thermal plasma and g-C_3N_4/TiO_2: performance, degradation pathways and catalytic mechanism. Chemosphere 249:126093
Lu P, Hu X, Li Y, Zhang M, Liu X, He Y, Dong F, Fu M, Zhang Z (2018) One-step preparation of a novel $SrCO_3$/g-C_3N_4 nano-composite and its application in selective adsorption of crystal violet. RSC Adv 8(12):6315–6325

Miao X, Ji Z, Wu J, Shen X, Wang J, Kong L, Liu M, Song C (2017) g-C_3N_4/AgBr nanocomposite decorated with carbon dots as a highly efficient visible-light-driven photocatalyst. J Colloid Interface Sci 502:24–32

Moreira NFF, Sampaio MJ, Ribeiro AR, Silva CG, Faria JL, Silva AMT (2019) Metal-free g-C_3N_4 photocatalysis of organic micropollutants in urban wastewater under visible light. Appl Catal B 248:184–192

Mortazavi B, Cuniberti G, Rabczuk T (2015) Mechanical properties and thermal conductivity of graphitic carbon nitride: a molecular dynamics study. Comput Mater Sci 99:285–289

Ong W-J, Tan L-L, Ng YH, Yong S-T, Chai S-P (2016) Graphitic Carbon Nitride (g-C_3N_4)-based photocatalysts for artificial photosynthesis and environmental remediation: are we a step closer to achieving sustainability? Chem Rev 116(12):7159–7329

Ou Q, Xu S, Long Y, Zhang X (2020) Porous visible light-responsive Fe(3+)-doped carbon nitride for efficient degradation of sulfadiazine. Environ Sci Pollut Res Int

Pan T, Chen D, Fang J, Wu K, Feng W, Zhu X, Fang Z (2020a) Facile synthesis of iron and cerium co-doped g-C_3N_4 with synergistic effect to enhance visible-light photocatalytic performance. Mater Res Bull 125:110812

Pan T, Chen D, Xu W, Fang J, Wu S, Liu Z, Wu K, Fang Z (2020b) Anionic polyacrylamide-assisted construction of thin 2D-2D WO_3/g-C_3N_4 Step-scheme heterojunction for enhanced tetracycline degradation under visible light irradiation. J Hazard Mater 393:122366

Patton JSR, Boehm MW, Fiest PD, David L (1981) Ixtoc 1 oil spill: flaking of surface mousse in the Gulf of Mexico. Nature 290:235–238

Prabavathi SL, Govindan K, Saravanakumar K, Jang A, Muthuraj V (2019) Construction of heterostructure $CoWO_4$/g-C_3N_4 nanocomposite as an efficient visible-light photocatalyst for norfloxacin degradation. J Ind Eng Chem 80:558–567

Qian X, Wu Y, Kan M, Fang M, Yue D, Zeng J, Zhao Y (2018) FeOOH quantum dots coupled g-C_3N_4 for visible light driving photo-Fenton degradation of organic pollutants. Appl Catal B 237:513–520

Qu J, Du Y, Feng Y, Wang J, He B, Du M, Liu Y, Jiang N (2020) Visible-light-responsive K-doped g-C_3N_4/BiOBr hybrid photocatalyst with highly efficient degradation of Rhodamine B and tetracycline. Mater Sci Semicond Process 112:105023

Raha S, Ahmaruzzaman M (2020) Facile fabrication of g-C_3N_4 supported Fe_3O_4 nanoparticles/ZnO nanorods: a superlative visible light responsive architecture for express degradation of pantoprazole. Chem Eng J 387:123766

Rathinam K, Jayaram P, Sankaran M (2019) Synthesis and characterization of magnetic chitin composite and its application towards the uptake of Pb(II) and Cd(II) ions from aqueous solution. Environ Prog Sustain Energ 38(s1):S288–S297

Rathinam K, Singh SP (2020) In: Gupta T, Singh SP, Rajput P, Agarwal AK (eds) Measurement, analysis and remediation of environmental pollutants. Springer Singapore, Singapore, pp 445–474

Rathinam K, Singh SP, Arnusch CJ, Kasher R (2018) An environmentally-friendly chitosan-lysozyme biocomposite for the effective removal of dyes and heavy metals from aqueous solutions. Carbohyd Polym 199:506–515

Ren B, Xu Y, Zhang L, Liu Z (2018) Carbon-doped graphitic carbon nitride as environment-benign adsorbent for methylene blue adsorption: Kinetics, isotherm and thermodynamics study. J Taiwan Inst Chem Eng 88:114–120

Rong X, Qiu F, Jiang Z, Rong J, Pan J, Zhang T, Yang D (2016) Preparation of ternary combined ZnO-Ag_2O/porous g-C_3N_4 composite photocatalyst and enhanced visible-light photocatalytic activity for degradation of ciprofloxacin. Chem Eng Res Des 111:253–261

Sahoo SK, Padhiari S, Biswal SK, Panda BB, Hota G (2020) Fe_3O_4 nanoparticles functionalized GO/g-C_3N_4 nanocomposite: an efficient magnetic nanoadsorbent for adsorptive removal of organic pollutants. Mater Chem Phys 244:122710

Santha Kumar K, Vellaichamy B, Paulmony T (2019) Visible light active metal-free photocatalysis: N-doped graphene covalently grafted with g-C_3N_4 for highly robust degradation of methyl orange. Solid State Sci 94:99–105

Saravanakumar K, Karthik R, Chen S-M, Vinoth Kumar J, Prakash K, Muthuraj V (2017) Construction of novel Pd/CeO_2/g-C_3N_4 nanocomposites as efficient visible-light photocatalysts for hexavalent chromium detoxification. J Colloid Interface Sci 504:514–526
Sharma G, Kumar A, Sharma S, Naushad M, Ghfar AA, Al-Muhtaseb AAH, Ahamad T, Sharma N, Stadler FJ (2020) Carboxymethyl cellulose structured nano-adsorbent for removal of methyl violet from aqueous solution: isotherm and kinetic analyses. Cellulose 27(7):3677–3691
Shen C, Chen C, Wen T, Zhao Z, Wang X, Xu A (2015) Superior adsorption capacity of g-C_3N_4 for heavy metal ions from aqueous solutions. J Colloid Interface Sci 456:7–14
Shen W, An Q-D, Xiao Z-Y, Zhai S-R, Hao J-A, Tong Y (2020) Alginate modified graphitic carbon nitride composite hydrogels for efficient removal of Pb(II), Ni(II) and Cu(II) from water. Int J Biol Macromol 148:1298–1306
Song Y, Huang L, Zhang X, Zhang H, Wang L, Zhang H, Liu Y (2020) Synergistic effect of persulfate and g-C_3N_4 under simulated solar light irradiation: Implication for the degradation of sulfamethoxazole. J Hazard Mater 393:122379
Sui G, Li J, Du L, Zhuang Y, Zhang Y, Zou Y, Li B (2020) Preparation and characterization of g-C_3N_4/Ag–TiO_2 ternary hollowsphere nanoheterojunction catalyst with high visible light photocatalytic performance. J Alloys Comp 823:153851
Tahir MB, Sagir M, Shahzad K (2019) Removal of acetylsalicylate and methyl-theobromine from aqueous environment using nano-photocatalyst WO_3-TiO_2 @g-C_3N_4 composite. J Hazard Mater 363:205–213
Talukdar M, Behera SK, Bhattacharya K, Deb P (2019) Surface modified mesoporous g-C_3N_4@$FeNi_3$ as prompt and proficient magnetic adsorbent for crude oil recovery. Appl Surf Sci 473:275–281
Tong Z, Yang D, Xiao T, Tian Y, Jiang Z (2015) Biomimetic fabrication of g-C_3N_4/TiO_2 nanosheets with enhanced photocatalytic activity toward organic pollutant degradation. Chem Eng J 260:117–125
Vellaichamy B, Periakaruppan P (2018) Synergistic combination of a novel metal-free mesoporous band-gap-modified carbon nitride grafted polyaniline nanocomposite for decontamination of refractory pollutant. Ind Eng Chem Res 57(19):6684–6695
Vigneshwaran S, Preethi J, Meenakshi S (2019) Removal of chlorpyrifos, an insecticide using metal free heterogeneous graphitic carbon nitride (g-C_3N_4) incorporated chitosan as catalyst: photocatalytic and adsorption studies. Int J Biol Macromol 132:289–299
Wan X, Khan MA, Wang F, Xia M, Lei W, Zhu S, Fu C, Ding Y (2019) Facile synthesis of protonated g-C_3N_4 and acid-activated montmorillonite composite with efficient adsorption capacity for PO_4^{3-} and Pb(II). Chem Eng Res Des 152:95–105
Wang F, Chen P, Feng Y, Xie Z, Liu Y, Su Y, Zhang Q, Wang Y, Yao K, Lv W, Liu G (2017) Facile synthesis of N-doped carbon dots/g-C_3N_4 photocatalyst with enhanced visible-light photocatalytic activity for the degradation of indomethacin. Appl Catal B 207:103–113
Wang F, Zeng Q, Tang J, Peng L, Shao J, Luo S (2020) Synthesis of g-C_3N_4/CuS heterojunction with enhanced photocatalytic activity under visible-light. J Nanosci Nanotechnol 20(9):5896–5905
Wang H, Li J, Ma C, Guan Q, Lu Z, Huo P, Yan Y (2015) Melamine modified P25 with heating method and enhanced the photocatalytic activity on degradation of ciprofloxacin. Appl Surf Sci 329:17–22
Wang H, Li Q, Zhang S, Chen Z, Wang W, Zhao G, Zhuang L, Hu B, Wang X (2019) Visible-light-driven N_2-g-C_3N_4 as a highly stable and efficient photocatalyst for bisphenol A and Cr(VI) removal in binary systems. Catal Today 335:110–116
Wang H, Zhang J, Yuan X, Jiang L, Xia Q, Chen H (2020b) Photocatalytic removal of antibiotics from natural water matrices and swine wastewater via Cu(I) coordinately polymeric carbon nitride framework. Chem Eng J 392:123638
Wang J, Gao B, Dou M, Huang X, Ma Z (2020c) A porous g-C_3N_4 nanosheets containing nitrogen defects for enhanced photocatalytic removal meropenem: mechanism, degradation pathway and DFT calculation. Environ Res 184:109339

Wang S, Li D, Sun C, Yang S, Guan Y, He H (2014) Synthesis and characterization of g-C_3N_4/Ag_3VO_4 composites with significantly enhanced visible-light photocatalytic activity for triphenylmethane dye degradation. Appl Catal B 144:885–892

Wang W, Fang J, Shao S, Lai M, Lu C (2017) Compact and uniform TiO_2@g-C_3N_4 core-shell quantum heterojunction for photocatalytic degradation of tetracycline antibiotics. Appl Catal B 217:57–64

Wang Y, Jing B, Wang F, Wang S, Liu X, Ao Z, Li C (2020d) Mechanism Insight into enhanced photodegradation of pharmaceuticals and personal care products in natural water matrix over crystalline graphitic carbon nitrides. Water Res 180:115925

Wang YX, Chen MN, Tao H (2020) Preparation of nested g-C_3N_4 fiber surrounding graphene oxide and its photocatalytic degradation of rhodamine B. J Nanosci Nanotechnol 20(9):5445–5451

Wang Z, Muruganathan M, Zhang Y (2019) Graphitic carbon nitride based photocatalysis for redox conversion of arsenic(III) and chromium(VI) in acid aqueous solution. Appl Catal B 248:349–356

Wei K, Li K, Zeng Z, Dai Y, Yan L, Guo H, Luo X (2017) Synergistic photocatalytic effect of porous g-C_3N_4 in a Cr(VI)/4-chlorophenol composite pollution system. Chin J Catal 38(11):1804–1811

Wen J, Xie J, Chen X, Li X (2017) A review on g-C_3N_4-based photocatalysts. Appl Surf Sci 391:72–123

Wu H-H, Chang C-W, Lu D, Maeda K, Hu C (2019) Synergistic effect of hydrochloric acid and phytic acid doping on polyaniline-coupled g-C_3N_4 nanosheets for photocatalytic Cr(VI) reduction and dye degradation. ACS Appl Mater Interfaces 11(39):35702–35712

Wu J-H, Shao F-Q, Luo X-Q, Xu H-J, Wang A-J (2019) Pd nanocones supported on g-C_3N_4: an efficient photocatalyst for boosting catalytic reduction of hexavalent chromium under visible-light irradiation. Appl Surf Sci 471:935–942

Xiao G, Wang Y, Xu S, Li P, Yang C, Jin Y, Sun Q, Su H (2019) Superior adsorption performance of graphitic carbon nitride nanosheets for both cationic and anionic heavy metals from wastewater. Chin J Chem Eng 27(2):305–313

Xie H, Zhang J, Wang D, Liu J, Wang L, Xiao H (2020a) Construction of three-dimensional g-C_3N_4/attapulgite hybrids for Cd(II) adsorption and the reutilization of waste adsorbent. Appl Surface Sci 504:144456

Xie Y, Wu J, Sun C, Ling Y, Li S, Li X, Zhao J, Yang K (2020b) La_2O_3-modified graphite carbon nitride achieving the enhanced photocatalytic degradation of different organic pollutants under visible light irradiation. Mater Chem Phys 246:122846

Xie Z, Feng Y, Wang F, Chen D, Zhang Q, Zeng Y, Lv W, Liu G (2018) Construction of carbon dots modified MoO_3/g-C_3N_4 Z-scheme photocatalyst with enhanced visible-light photocatalytic activity for the degradation of tetracycline. Appl Catal B 229:96–104

Xin G, Xia Y, Lv Y, Liu L, Yu B (2016) Investigation of mesoporous graphitic carbon nitride as the adsorbent to remove Ni(II) ions. Water Environ Res 88(4):318–324

Xu T, Zou R, Lei X, Qi X, Wu Q, Yao W, Xu Q (2019) New and stable g-C_3N_4/HAp composites as highly efficient photocatalysts for tetracycline fast degradation. Appl Catal B 245:662–671

Xue J, Ma S, Zhou Y, Zhang Z, Liu X (2015) Fabrication of porous g-$C_3N_4/Ag/Fe_2O_3$ composites with enhanced visible light photocatalysis performance. RSC Adv 5(72):58738–58745

Yagub MT, Sen TK, Afroze S, Ang HM (2014) Dye and its removal from aqueous solution by adsorption: a review. Adv Coll Interface Sci 209:172–184

Yan SC, Li ZS, Zou ZG (2010) Photodegradation of Rhodamine B and methyl orange over boron-doped g-C_3N_4 under visible light irradiation. Langmuir 26(6):3894–3901

Yan T, Chen H, Jiang F, Wang X (2014) Adsorption of perfluorooctane sulfonate and perfluorooctanoic acid on magnetic mesoporous carbon nitride. J Chem Eng Data 59(2):508–515

Yan T, Chen H, Wang X, Jiang F (2013) Adsorption of perfluorooctane sulfonate (PFOS) on mesoporous carbon nitride. RSC Adv 3(44):22480–22489

Yazdankish E, Foroughi M, Azqhandi MHA (2020) Capture of I131 from medical-based wastewater using the highly effective and recyclable adsorbent of g-C_3N_4 assembled with Mg-Co-Al-layered double hydroxide. J Hazardous Mater 389:122151

Ye L, Liu J, Jiang Z, Peng T, Zan L (2013) Facets coupling of $BiOBr\text{-}g\text{-}C_3N_4$ composite photocatalyst for enhanced visible-light-driven photocatalytic activity. Appl Catal B 142–143:1–7

Yin Z, Han M, Hu Z, Feng L, Liu Y, Du Z, Zhang L (2020) Peroxymonosulfate enhancing visible light photocatalytic degradation of bezafibrate by $Pd/g\text{-}C_3N_4$ catalysts: the role of sulfate radicals and hydroxyl radicals. Chem Eng J 390:124532

Yousefi M, Villar-Rodil S, Paredes JI, Moshfegh AZ (2019) Oxidized graphitic carbon nitride nanosheets as an effective adsorbent for organic dyes and tetracycline for water remediation. J Alloys Comp 809:151783

Yu B, Meng F, Khan MW, Qin R, Liu X (2020) Synthesis of hollow $TiO_2@g\text{-}C_3N_4/Co_3O_4$ core-shell microspheres for effective photooxidation degradation of tetracycline and MO. Ceram Int 46(9):13133–13143

Yu Y, Wang J (2016) Direct microwave synthesis of graphitic C_3N_4 with improved visible-light photocatalytic activity. Ceram Int 42(3):4063–4071

Zhang H, Zhao L, Geng F, Guo L-H, Wan B, Yang Y (2016) Carbon dots decorated graphitic carbon nitride as an efficient metal-free photocatalyst for phenol degradation. Appl Catal B 180:656–662

Zhang J, Gao N, Chen F, Zhang T, Zhang G, Wang D, Xie X, Cai D, Ma X, Wu L, Wu Z (2019) Improvement of Cr (VI) photoreduction under visible-light by $g\text{-}C_3N_4$ modified by nano-network structured palygorskite. Chem Eng J 358:398–407

Zhao C, Liao Z, Liu W, Liu F, Ye J, Liang J, Li Y (2020a) Carbon quantum dots modified tubular g-C(3)N(4) with enhanced photocatalytic activity for carbamazepine elimination: mechanisms, degradation pathway and DFT calculation. J Hazard Mater 381:120957

Zhao C, Yan Q, Wang S, Dong P, Zhang L (2018) Regenerable $g\text{-}C_3N_4$–chitosan beads with enhanced photocatalytic activity and stability. RSC Adv 8(48):27516–27524

Zhao H, Tian C, Mei J, Yang S, Wong PK (2020b) Synergistic effect and mechanism of catalytic degradation toward antibiotic contaminants by amorphous goethite nanoparticles decorated graphitic carbon nitride. Chem Eng J 390:124551

Zhao L, Lv W, Hou J, Li Y, Duan J, Ai S (2020c) Synthesis of magnetically recyclable $g\text{-}C_3N_4/Fe_3O_4$/ZIF-8 nanocomposites for excellent adsorption of malachite green. Microchem J 152:104425

Zhong Q, Lan H, Zhang M, Zhu H, Bu M (2020) Preparation of heterostructure $g\text{-}C_3N_4/ZnO$ nanorods for high photocatalytic activity on different pollutants (MB, RhB, Cr(VI) and eosin). Ceram Int 46(8, Part B):12192–12199

Zhou J, Ji X, Zhou X, Guo J, Sun J, Liu Y (2019) Three-dimensional $g\text{-}C_3N_4/MgO$ composites as a high-performance adsorbent for removal of Pb(II) from aqueous solution. Sep Sci Technol 54(17):2817–2829

Zhou X, Jin B, Chen R, Peng F, Fang Y (2013) Synthesis of porous $Fe_3O_4/g\text{-}C_3N_4$ nanospheres as highly efficient and recyclable photocatalysts. Mater Res Bull 48(4):1447–1452

Zhu J, Xiao P, Li H, Carabineiro SAC (2014) Graphitic carbon nitride: synthesis, properties, and applications in catalysis. ACS Appl Mater Interfaces 6(19):16449–16465

Zhu Z, Lu Z, Wang D, Tang X, Yan Y, Shi W, Wang Y, Gao N, Yao X, Dong H (2016) Construction of high-dispersed $Ag/Fe_3O_4/g\text{-}C_3N_4$ photocatalyst by selective photo-deposition and improved photocatalytic activity. Appl Catal B 182:115–122

Zou Y, Wang X, Ai Y, Liu Y, Ji Y, Wang H, Hayat T, Alsaedi A, Hu W, Wang X (2016) β-Cyclodextrin modified graphitic carbon nitride for the removal of pollutants from aqueous solution: experimental and theoretical calculation study. J Mater Chem A 4(37):14170–14179

Chapter 6
Application of Metal and Metal Oxide Nanoparticles as Potential Antibacterial Agents

Chinmoy Mandal and Manoranjan Sahu

Abstract With the widespread use of antibiotics, the bacteria have evolved to build up mechanisms to resist the activity of antibiotics. The antibiotics target cell wall, protein synthesis and DNA replication, whereas microbes resist it through genetic basis or mechanistic basis of environmental resistance. The nanoparticles have shown a great potential as an alternative to antibiotics for treatment of microbial infections. Nanoparticles cause microbial inactivation via oxidative stress, dissolved metal ions or non-oxidative mechanisms and target cell barrier, bacterial protein, enzymes and DNA synthesis and its metabolism. The multiple mechanisms which nanoparticle employ simultaneously against microbes can have the potential to overcome the microbial resistance by providing the bacteria insufficient time to mutate and develop resistance. This book chapter focuses on mechanisms of antibiotic action and resistance developed by bacteria. Various metal nanoparticles (silver, copper, gold, aluminium) and metal oxides nanoparticles (copper oxide, titanium dioxide nanoparticles, zinc oxide and magnesium oxide nanoparticles) that are used for antibacterial action and impact of physicochemical properties are discussed. The chapter also elucidates the potential mode of action by various nanoparticles and impact of various important factors such as size, composition, shape, morphology, zeta potential and environmental conditions on antibacterial effectiveness.

Keywords Antibiotic action · Antibiotic resistance · Nanoparticle · Antibacterial action · Oxidative stress

6.1 Introduction

The field of nanotechnology has emerged as one of the most extensively researched fields involving multidisciplinary branches of science. It has been one of the major growing sectors in scientific field of research and also has large scale implication

C. Mandal · M. Sahu (✉)
Environmental Science and Engineering Department, Indian Institute of Technology, Bombay 400076, India
e-mail: mrsahu@iitb.ac.in

S. P. Singh et al. (eds.), *Nanomaterials and Nanocomposites for Environmental Remediation*, Energy, Environment, and Sustainability,
https://doi.org/10.1007/978-981-16-3256-3_6

in industry, hence often called as next generation revolution. The implications of nanotechnology extend to the environmental field, which has benefits that can be categorized into treatment and environmental remediation, pollution sensing and detection and prevention of pollution. Nanoparticles (NPs) can have applications in soil, wastewater treatment, groundwater treatment and also large part in air remediation (Yadav et al. 2017).

Ever since the discovery of antibiotics as antimicrobial agents, they have been used extensively to treat bacterial and fungal diseases. Use of antibiotics comes with two advantages: first it has been cost-effective and another it has given powerful outcomes (Wang et al. 2017). But with the increasing use of antibiotics, the bacteria are artificially selected and acquiring resistance against these substances. In fact, there has been evidences of bacteria, which have developed resistance to multiple drugs via various mechanisms of gene transfer (Hajipour et al. 2012).

With the emergence of these drug-resistant bacteria, it can prove fatal in future, since the resistant properties are evolutionary based or can be developed spontaneously (Wang et al. 2017). In many cases, the last generation of drugs is being used for treatment. If anyhow, there is a development of resistance in those bacteria, we would be left with no option and epidemic may break out. Even with the immense progress in the field of medicine, the rate of mortality and morbidity that has been caused by bacterial diseases is large in numbers (Rai and Bai 2011). Therefore, there is an urgent need for an alternative for treatment of microbes. So, that the last generation of antibiotics can be preserved for future urgency.

If we go through the ancient literature, we can find that they have been using several inorganic elements such as copper, silver, gold and other materials to combat the effects of microbes. The idea has given origin to the use of inorganic nanoparticles as a potential antimicrobial agent, which can extend in various fields such as agriculture, food industry, soil and wastewater remediation, textile industries and other numerous applications (Rai and Bai 2011).

Nanoparticles are small particles, which have a range of scale 1–100 nm and have at least one of its unit in this scale. The inorganic and organic nanoparticles have been found as having potential antimicrobial properties. Metallic nanoparticles of inorganic origin have been profoundly used and experimented with the microbes and are successful in either killing or stopping its metabolism or reproduction hence restricting the growth. These nanoparticles have shown to have effect on a broad spectrum of both gram-positive as well as gram-negative bacteria (Wang et al. 2017). This success of inorganic nanoparticles can be attributed to its property to endure the adverse conditions and the unique property of nanoparticles, which it possesses due to its small size and large surface area. These factors provide advantage to kill bacteria, as nanoparticles can show surface phenomenon of killing instead of going inside the cell to disrupt it. This property can also be hypothesized as it can give the bacteria a very little time to develop resistance against either antibiotic coated with nanoparticles or nanoparticles alone (Rai and Bai 2011). Hence, we can think of a probable solution to the antibiotic resistance complication. Nanoparticles can also be made as composites such as to tackle the bacterial resistance with more than one mechanism hence rendering it viable for destruction (Wang et al. 2017). This book chapter discusses in details various aspects of antibiotics action and mechanism of antibiotic resistance development. Additionally, also the chapter discusses various types of nanoparticles that can be used for antimicrobial activities and their mode of inaction potential for various biological agents.

6.2 Mechanism of Antibiotic Action

Antibiotics act as powerful medicines to fights infectious diseases. Their action is such that either they stop the bacteria from reproducing or they kill or destroy the bacteria. In humans, the immune system consists of white blood cells (WBCs) that is responsible for attacking infectious bacteria and fighting the infectious diseases but sometimes these harmful bacteria become too large in number, in this case the immune system fails to cope up with this number, antibiotics comes in to aid in this condition (Munita and Arius 2016).

There are two types of bacterial found in environmental systems: gram positive and gram negative (Fig. 6.1). The figure illustrates bacterial cells are of two types,

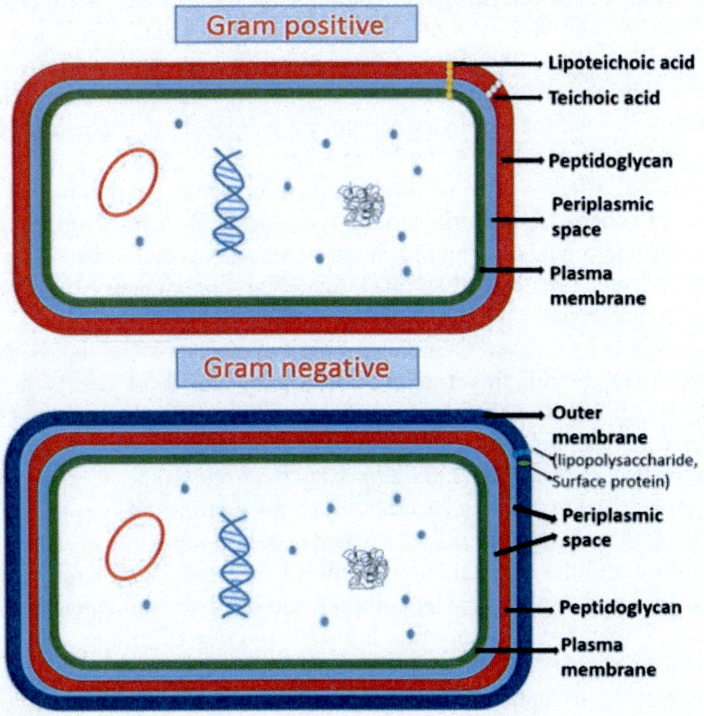

Fig. 6.1 Structure of bacterial cell envelope. Adopted from Kapoor et al. (2017)

gram positive and gram negative. Gram-positive bacteria consist of thick rigid peptidoglycan cell wall around the cell membrane that is attached to teichoic acids found only in gram-positive bacteria (Scott and Barnett 2006), whereas gram-negative bacteria have a thin peptidoglycan cell wall which is surrounded by second lipid membrane, the outer membrane (OM). The space between the outer membrane and the cell membrane is called periplasmic space consists of periplasm. Outer membrane provides additional protective layer against foreign substances and prevent them to enter to into bacterium (Munita and Arius 2016; Kapoor et al. 2017).

Outer membrane however also contains porins that allow entry of various molecules such as drug. Cell wall is a tough layer which provides the bacterium its characteristic shape and also prevent it from mechanical and osmotic stresses. The function of cell membrane is to prevent ions from flowing out or into the cell and also maintains cytoplasmic or cellular bacterial components in a defined space (Munita and Arius 2016).

The mechanism of an antibiotic action takes place in a biochemical way such as the drug works effectively. Antibiotics can be target-specific like a drug that binds like an enzyme (e.g. antibiotic or a receptor). Mechanism of action of antibiotics can be described as the biochemical process especially at a molecular level. (Munita and Arius 2016; Mazel and Davies 1999).

Classification of antibacterial agent is done on the basis of their activity spectrum, impact on bacteria and their way of action. Activity spectrum of antibiotics can be further divided into broad spectrum antibiotics which are effective against both the gram-negative and the gram-positive bacteria and narrow spectrum antibiotics which have restricted activity, their primary activities are against microorganism of only a particular species (Munita and Arius 2016; Kapoor et al. 2017), e.g. glycopeptides and bacitracin are effective against gram-positive bacteria, polymixin that is effective against gram-negative bacteria, aminoglycosides and sulphonamides are effective against aerobic organisms and nitroimadazoles against anaerobes (Kapoor et al. 2017). However, this spectrum of antibacterial activity may change if the bacterium acquires resistance genes.

Effect of antibiotics on bacteria involves different mechanism of individual antibiotics on respective bacteria therefore antibacterial has different effects on bacterial agents and it is either bacteriostatic or bactericidal. The latter consists of drugs that can kill bacteria (e.g. cephalosporins, aminoglycosides, quinolones, penicillins) whereas the former consists of drugs that inhibit or delay the bacterial activity of growth and their replication (e.g. tetracyclins, macrolides, sulphonamides). Some antibiotics can act as both bactericidal and bacteriostatic which depend upon the dose given, the duration of exposure and the original state of invading bacteria, e.g. fluoroquinolones, aminoglycosides and metronidazole, and they can kill bacteria depending upon its concentration, higher the concentration higher is the rate of killing. (Hajipur et al. 2012; Kapoor et al. 2017). Antibiotics can also show antagonistic and synergistic effect on each other like action of aminoglycosides is enhanced with antibiotics that inhibit cell wall synthesis (Hancock 2005).

Antibacterial action generally involves three mechanisms which involve inhibition and regulation of enzymes involved in the pathway of cell wall biosynthesis,

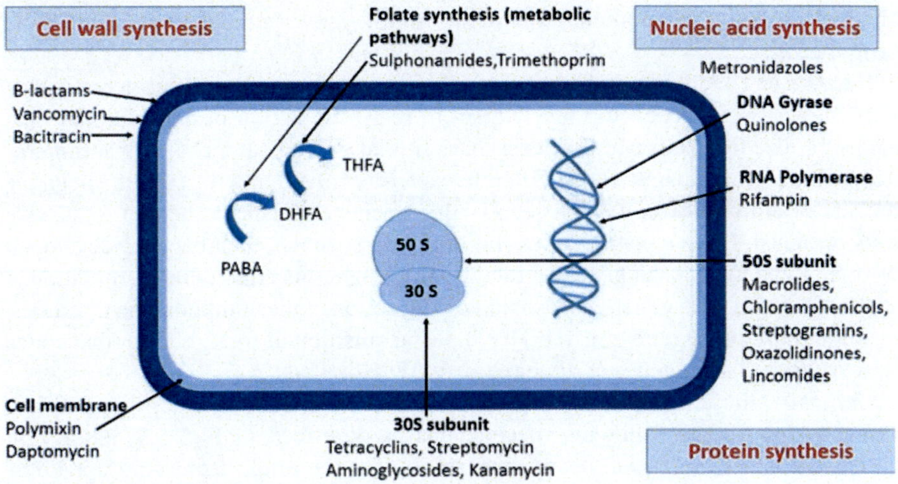

Fig. 6.2 Mechanism of antibiotics action. Adopted from Kapoor et al. (2017). *Note* PABA: Para-aminobenzoic acid, DHFA: dihydrofolic acid, THFA: tetrahydrofolic acid

metabolism of nucleic acid as well as deoxyribonucleic acid (DNA) repair or pathway of protein synthesis and disturbance in bacterial cell membrane structure. The detailed mechanisms are shown in Fig. 6.2. (Kapoor et al. 2017). Multiplying cells are the main target for antibiotics as these cells have the most of these cellular function. Sometimes there is overlapping in these functions that occur between prokaryotic bacteria cells and eukaryotic mammalian cells. Some antibiotics have also been function as anticancer agents (Munita and Arius 2016).

6.2.1 *Antibiotic Resistance in Bacteria*

Antimicrobial resistance has developed over a period with interaction with many organisms and their environment. Mostly antimicrobial compounds were naturally produced molecules and as such the bacteria that were co-resident evolved mechanisms to resist and overcome the action of antimicrobials to survive this gave rise to the organisms that often considered as "intrinsically resistant" to more than one antimicrobials (Munita and Arius 2016). In general, we refer to the expression of the "acquired resistance" in particular bacteria, which was previously susceptible to antimicrobial compound. The antimicrobial resistant microorganisms can be either attributed to either mutation in their genetic material or horizontal transfer of the gene from some intrinsically resistant microorganism present in the environment (Mazel and Davies 1999; Munita and Arius 2016). Another important factor in antimicrobial resistance or susceptibility is the effectiveness of antimicrobial agents on microbes which is a relative phenomenon involving multilayers of complexity (Munita and Arius 2016).

Bacteria can adapt to a wide range of environmental conditions; this genetic plasticity allows them to develop mechanism to resist the action of antimicrobial that can jeopardize their survival. Bacteria share similar ecological niche with antimicrobial producing organisms. Hence, in time bacteria evolved to withstand the negative impacts of antibiotics and subsequently this intrinsic resistance helped to survive in its presence. If we consider an evolutionary perspective, then bacteria developed two major mechanisms in genetic strategies to combat this effect. First is mutation at gene(s) level which is generally associated with action of the compound on microbes. Second is horizontal gene transfer (HGT), i.e. acquisition of foreign DNA that codes for resistance coding genes (Kapoor et al. 2017; Hopwood 2007).

Generally, these mutations effect the antibiotic resistance by altering antibiotic action by one of the following mechanisms as described in Fig. 6.3, modifying the antimicrobial target, i.e. decreasing affinity for the drug, decreasing uptake of drug, extrusion of harmful molecules by activating efflux mechanisms, modulation of regulatory networks and changing important global metabolic pathway. Therefore, resistance arising due to acquired mutational changes in diverse groups and varies in complexity (Singh et al. 2014; Wright 2011).

As the bacteria share the same ecological niche with the antimicrobial producing agents, they develop resistance in themselves, and this "environmental resistome" gives a robust source for other bacteria for acquisition of resistance genes. HGT can occur through three ways, transformation, which occurs through incorporation

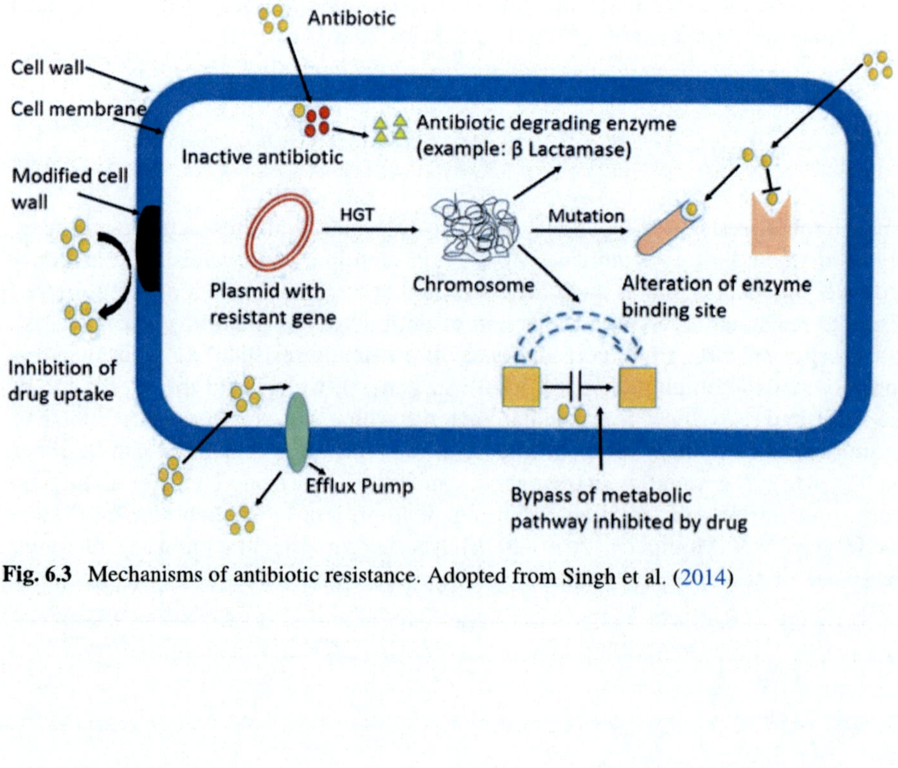

Fig. 6.3 Mechanisms of antibiotic resistance. Adopted from Singh et al. (2014)

of naked DNA, transduction which is phage mediated and conjugation which is considered as bacterial sex.

Among the three transformations is simplest way of acquisition of genes, but only few species of pathogenic bacteria are naturally able to incorporate naked DNA and develop resistance (Munita and Arius 2016). Bacteria have evolved over millions of years of evolution and have developed sophisticated mechanisms towards antimicrobial molecules. The resistant microbial class usually has attained this characteristic property via multiple biochemical pathways whereas a bacterial cell can achieve this resistant property by using more than one mechanism. The biochemical route for the attaining antibiotic resistance can be categorized as, modifying the antimicrobial molecule, preventing the antibiotic from reaching its target (either by decreasing its penetration or by active efflux of antimicrobial compound), changing and/or bypassing of the target sites and resistance through global cell adaptive processes. Every category mentioned can have multiple specific biochemical pathways (Kapoor et al. 2017; Munita and Arius 2016).

6.3 Nanoparticles Antibacterial Application

The detail mechanistic path of NPs interaction with bacteria and various types of NPs used for antibacterial are discussed along with different physicochemical properties that are critical for effective antibacterial action.

6.3.1 Interaction Mechanism of Nanoparticles with Microbes

There are different types of nanoparticles such as metallic, metal oxides, doped oxides, composites with different size, morphology and composition. Depending on the nanomaterial's properties like size and composition different NPs like copper, gold, silver, aluminium, zinc, magnesium and titanium, nanoparticles will show different effectiveness. These nanoparticles have different mechanism of action as bactericidal, antiviral and antifungal agents. Cell wall and cell membranes act as essential defensive barriers and protect the bacterial cell from osmotic and physical pressures, therefore plays a crucial role in the maintenance of the shape of bacterium. The gram-positive and gram-negative bacteria differ in their cell membrane's components hence have different adsorption pathways for NPs. Gram-negative bacteria have a unique structure lipopolysaccharide (LPS) which provides a negative charge, and it is responsible for attraction of NPs, whereas in gram-positive bacterial cell wall consists of teichoic acid, so NPs get dispersed along the molecular chain of phosphate, and this prevents their aggregation (Hajipour et al. 2012; Wang et al. 2017). The mechanism of NPs bactericidal action also depends on structure and components of bacterial cell. The mechanism of NPs penetration was through diffusion and adsorption (Melander et al. 2018; Slavin et al. 2017).

There are several mechanisms that were proposed as the emergence of NPs as antimicrobial agent. For example, bacterial cell metabolism can be changed by using metal NPs. The NPs have a promising potential as a cure for bacterial diseases. The NPs have shown their ability to enter biofilms and inhibit its formation shown by Ag-inhibited expression of genes (Wang et al. 2017). NPs to carry out their antibacterial action must come in contact with the bacteria. This contact can be in the form of electrostatic attraction, interaction between receptor–ligand, van der Waal force and hydrophobic interactions (Hajipour et al. 2012). NPs after crossing bacterial membrane can affect the metabolism of bacterial cell, influencing a change in its shape and the cell membrane's function. The natural bacterial metabolism cycle plays an important role in sustaining bacterial growth and development and in addition can also be a cause of disease. Bacterial metabolism disruption damages the membrane of bacterial cell and induces oxidative stress, eventually leading to death of bacterial cells (Wang et al. 2017). Further, NPs can also react with bacteria's genetic components like DNA, ribosomes, lysosomes and interfere with enzymes which can lead to oxidative stress, change in permeability of the cell membrane, heterogeneous alteration, disorder in electrolyte balance, inhibition of an enzyme, deactivation of a protein and interfere with the expression of genes as shown in Fig. 6.4 (Singh et al. 2014). CuO NPs influence was studied on bacterial denitrification. It was seen that there was significant alteration in the expression of the key proteins. CuO NPs lead to interference with the proteins that were involved in the nitrogen metabolism, transfer of electrons and transfer of substance (Pelgrift and Friedman 2013). But the following three mechanisms have been mostly proposed and followed in research. They are oxidative stress, release of metal ions and non-oxidative mechanisms.

Among NPs antimicrobial mechanisms, one of the important ways is production of reactive oxygen species (ROS)-induced oxidative stress. ROS are those molecules

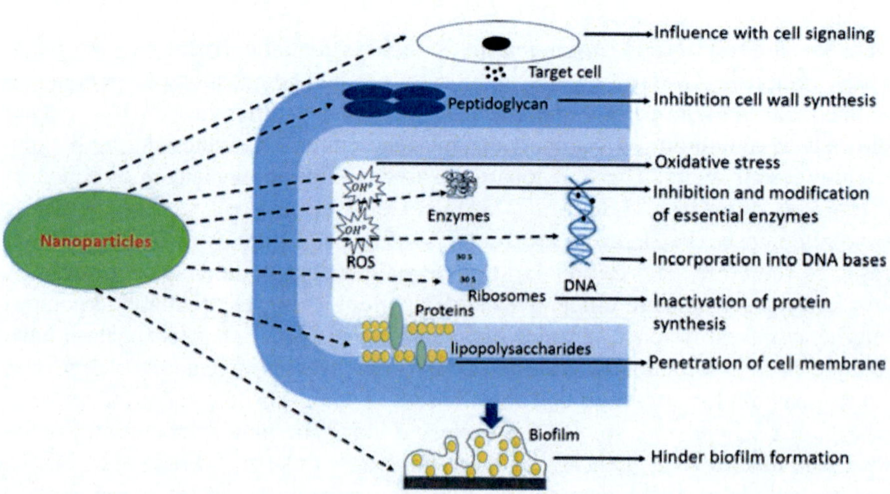

Fig. 6.4 Mechanisms for antimicrobial activity of nanoparticles. Adopted from: Singh et al. (2014)

with reactive intermediate having a strong positive redox potential, and various other types of ROS were produced by reduction of oxygen molecules using different types of NPs. There are generally four types of ROS exhibiting different dynamics level and activity, they are superoxide radical (O^{2-}), hydrogen peroxide (H_2O_2), hydroxyl radical (OH^o) and singlet oxygen (O_2) (Dwyer et al. 2009). Studies have showed that (OH^o) and O_2 have a more potent bactericidal activity as compared to O^{2-} and H_2O_2 as they may be neutralized by various endogenous antioxidants like catalase and superoxide enzymes (Wang et al. 2017). TiO_2 nanoparticles let the bacterial DNA's compression, degeneration and fragmentation, thus reducing physiological activity of the genes. The affinity and mode of binding of TiO_2 NPs with DNA were predicted using molecular docking, that showed TiO_2 NPs targets G:C-rich DNA. Further, analysis of the whole genome was used to identify the molecular mechanism of the bacterial apoptosis (Wang et al. 2017). In normal conditions, the ROS production and its clearance inside the bacterial cell are in balanced condition, but with the excessive ROS production oxidation is favoured by the redox balance in cell. The oxidative stress is thus produced by this unbalanced redox balance, that leads to the damage of individual components of the cell (Nel et al. 2006).

Metal oxides release their ions in the environment slowly that are absorbed by the bacterial cell membranes, and these metal ions can directly interact with the proteins and nucleic acid's functional groups like mercapto (–SH), carboxyl (–COOH) and amino (–NH) groups, causing damage to enzyme activity, disruption of cell structure, interfering with normal physiological metabolic processes and finally inhibiting the microorganisms (Wang et al. 2017). Metabolic pathway of bacteria is not isolated, but is incorporated into living cell's complex activity. Therefore, deliberate alteration of the bacterial metabolism can be utilized to regulate the pathogenicity of bacterial cell. Various mechanisms for the effects of nanoparticles on bacterial metabolism have been suggested, including a mechanism for reactive oxygen and a mechanism for metal ion dissolution (Melander et al. 2018).

6.3.2 Different Types of NPs and Their Action

There are different types of metallic nanoparticles like copper, gold, silver, aluminium, zinc, magnesium and titanium nanoparticles. These nanoparticles have different mechanism of action as bactericidal, antiviral and antifungal agents. Some of the metal and metal oxide nanoparticles and their activity are briefly discussed as follows:

Silver Nanoparticles

Silver nanoparticles have a wide range of applications in medical field like in treatment of wounds, burns and infections. The salts derived from silver and in its various forms are found to be antimicrobial in nature. It is also reported that Ag nanoparticles can be used as medium for delivering the antibiotic to target sites, and there is a possibility of antibiotic coating with Ag to enhance antibiotic activity (Wang et al. 2017). Multiple studies have been conducted to explain the mechanism of Ag nanoparticles as microbial growth inhibitors. According to one of the mechanisms, Ag nanoparticles show high affinity for sulphur moiety and phosphorous moiety of proteins and genetic elements in bacterial cell, Ag nanoparticles react with these moieties of protein and effects viability of cell of the bacterial or it can also interfere with enzyme action. Ag nanoparticles also react with the phosphorous of genetic material (DNA mostly) and inactivate its replication, hence stopping its growth and reproduction (Pelgrift and Friedman 2013; Qu et al. 2013). Silver NPs also have the ability of targeting bacterial membrane ultimately that causes dissipation of proton motive force that leads to blockage of oxidative phosphorylation (Singh et al. 2014). It was reported that Ag nanoparticles with a characteristic size which has less than 20 nm diameter are able to attach itself to the sulphur moiety of proteins on cell membranes which ultimately makes it more permeable and finally bacterial death (Rai and Bai 2011). Silver NPs have ability to modulate bacterial signal transduction. Ag^+ ions have the ability to interact with thiol groups of important enzymes causing their inactivity leading to the disruption of cellular functions, and this can collapse the membrane potential and inhibit ribosomal binding to the ribosomal ribonucleic acid (rRNA) (Singh et al. 2014). NPs that are within the size of 10–15 nm have been found to show concentration dependent effects against both the groups of bacteria (gram positive and negative). At microlevel concentration, it has showed effect such as in oxidative phosphorylation, uncoupling of respiratory electron transport chain, interfering with the permeability of bacterial cell membrane with respect to the exchange of phosphate and protons or inhibition of enzymes of respiratory chain. Further increase in concentration of Ag^+ ions has shown effects on cytoplasmic components and genetic materials (Rai et al. 2009). Ag NPs smaller than 10 nm were found to be toxic for *E.coli* and *Pseudomonas aeruginosa* (Li et al. 2008).

A study of silver NPs effect on bacterial cell morphology was done which on *E.coli*. The results of Ag^+ ions showed similar effects on morphology of bacteria. It detached the cell wall from the cell membrane (Choi et al. 2008). It was found out that the gram-negative bacteria had more inhibitory effect as compared to the gram-positive bacteria. This might be due to the composition of gram-positive bacteria that has a thick peptidoglycan layer in cell wall which might prevent it from the inhibitory action of silver ions (Rai and Bai 2011). The silver and sulphur ions form dense granules of electrons inside the cytoplasm which suggest that Ag NPs may have interacted with the nucleic acids and somehow impaired the DNA replication that lead to the cell viability loss and finally death of cell (Rai et al. 2009). Ag NPs have also proved to inhibit biofilms (Sheng and Liu 2011).

Ag NPs of size 2–5 nm using integrated with green fluorescent protein (GFP) in E. coli was studied. It was found that Ag NPs below 100 nm caused perforation in cell wall via getting attached to it and lead to bacterial death. Studies showed the activity of Ag NPs on bacteria were size- and shape-dependent. An experiment on E. coli was done using truncated triangular-, spherical- and rod-shaped Ag NPs, and the maximum effect was seen in case of triangular nanoplates. The smaller the particle higher was its antimicrobial activity (Wang et al. 2017). Silver NPs produced from biosynthetically from fungus, plants extract and bacteria have shown strong antimicrobial efficacy against multidrug-resistant (MDR) mycobacterium tuberculosis (Singh et al. 2014). Agarwal et al. (2013) studied the effect of biosynthesized silver nanoparticle against standard *Mycobacterium tuberculosis* along with 26 other chemical isolates that induced "multidrug resistance (MDR), drug sensitive (DS), extensive drug resistant (XDR) and mycobacteria other than tuberculosis (MOTT) strains" and found out they have effective bactericidal activity. Ag-ACF/CNF (silver nanoparticle composite with activated carbon nanofibres (ACF) or carbon nanofibres (CNF)) was found to have lethal inhibitory effect on bacteria *E.coli* and *S.aureus* that were completely inhibited in duration of 72 h (Singh et al. 2013; Singh et al. 2014). Silver nanoparticles synthesized in the presence of sesame oil cake were found to have good inhibitory effect on gram-negative bacteria like *P. aeruginosa, K. pneumoniae* and *E.coli* (Alfuraydi et al. 2019).

Copper Oxide and Copper Nanoparticles

CuO has a monoclinic structure and semiconducting in nature. Among the copper compound family, it is considered as the simplest with potentially advantageous properties with respect to superconductivity, high temperature effects on electron correlation and with spin dynamics which makes it valuable for large array of applications (Rai and Bai 2011). The crystal structure of CuO possesses photovoltaic and photocatalytic properties as well as photoconductive functionalities (Santo et al. 2008). CuO is used as antimicrobial agent as it is cheaper compared to that of silver, readily mixes with the polymers and comparably stable if we consider its physical and chemical properties which make it useful in wide applications. CuO as ionic nanoparticle can act as potent antimicrobial owing to its unique crystal morphology and high surface area such as its active action against a range of hospital acquired infection but is concentration dependent (Nel et al. 2006). It has been hypothesized that if the amount of peptidoglycans (negatively charged) is reduced, gram-negative bacteria (*Pseudomonas aeruginosa* and *Proteus spp.*) become relatively less vulnerable to those positively charged NPs. However, combined NPs of CuO and Ag showed greater activity towards gram-negative strains in time-kill experiments (Haipour et al. 2012). Studies suggested that for the optimal activity as antimicrobials, local environments need the presence of ions in it. *B.subtilis* have been found to susceptible towards Cu NPs action as the affinity of Cu is higher towards the amines and carboxyl groups that are abundant on its cell surface (Huh and Kwon 2011). Copper ions can intercalate with the nucleic acid strands and interact with DNA molecules. It can also disrupt the biochemical processes by getting inside the cells of bacteria (Melander et al. 2018).

The mechanism of action of Cu and Cu-based nanoparticles is reported to be associated primarily to the large surface area-to-volume ratio of the nanoparticles (Pramanik et al. 2012). Cu nanoparticle-specific mechanism of action on bacterial cells is mainly threefold, as shown in Fig. 6.5. Firstly accumulation and diffusion of the nanoparticles through the bacterial cell membrane, subsequently altering the permeability of the cell membrane by blocking or inducing excess release of one or several lipopolysaccharides, membrane proteins and transmembrane proteins which may constitute one mechanism of cell rupture. Secondly, the nanoparticles may

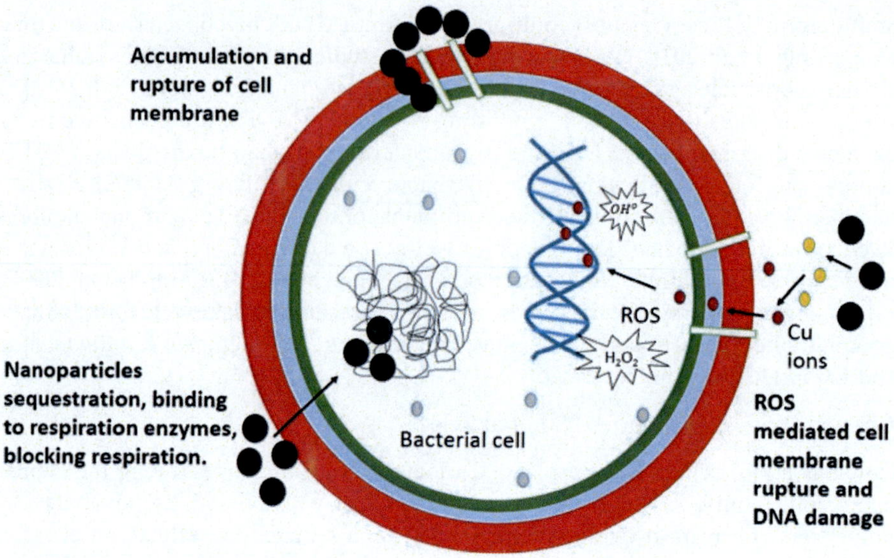

Fig. 6.5 Known mechanisms of action of Cu and CuO nanoparticles on bacteria. Adopted from Chatterjee et al. (2014)

induce release of their constituent metal ions, which may trigger a host of reactive oxygen species (ROS) generation, leading to oxidative damage of the cell membrane and DNA damage. The third reported mechanism of action of Cu-based nanoparticles involves the sequestering of the nanoparticles inside the cell, following which the nanoparticles bind to specific enzymes, blocking respiration (Chatterjee et al. 2014).

The nature and antimicrobial efficacy in a designed nanoparticle system may include one or a combination of these mechanisms. Additionally, the mechanism of action is also dependent on the size, shape, charge, coating and crystallinity of the nanoparticles apart from the nature of the microbes (Chatterjee et al. 2014; Wang et al. 2017). Some studies also indicated the broad spectrum activity of Cu-based nanoparticles, by demonstrating their mode of action by interaction and subsequent distortion and damage of phosphorus and sulphur containing biomolecules, such as DNA and proteins, respectively (Ruparelia et al. 2008). The hydroxyl radicals produced by copper ions and metals were found to damage the DNA and necessary proteins of *E.coli* (Baek and An 2011). Cu-ACF/CNF was found to have lethal inhibitory effect on bacteria *E.coli* and *S.aureus* (Singh et al. 2013; Singh et al. 2014).

Titanium Dioxide Nanoparticles

The most widely used photocatalyst is TiO_2 because of its low cost, high photostability and efficiency in generating reactive oxygen species (Singh et al. 2014). The semiconductor TiO_2 is utilized as a photocatalyst to induce a series of reductive and oxidative reactions on its surface (Musee et al. 2011). TiO_2 is shown to have antimicrobial activity hence it is utilized in various applications, such as water disinfection, paints, protection of marbles from microbial corrosion, coating on wood, fabrics, food packing films and as a surface disinfectant (Foster et al. 2011). There are mainly three crystal phases of TiO_2 referred as anatase, rutile and brookite. Composition of these crystal structures is shown to have an impact on its photocatalytic properties (Mahmoud et al. 2018; Mcwan et al. 2011).

The bandgap of TiO_2 is reported as 3.2 eV. TiO_2 shows photocatalytic activity when it is illuminated with ultraviolet light at wavelength less than 385 nm, this gives it strong oxidizing power and is the basis of its inhibitory activity and bactericidal action. When photoexcited by UV light TiO_2 particle generates Reactive oxygen species (ROS) which are responsible for the killing of bacteria as shown in Fig. 6.6 (Sahu et al. 2011). The effect of photocatalysis has been studied on *E.coli* present in water. TiO_2 NPs show photocatalysis which can provide an alternative means to self-disinfect surfaces that are contaminated and its further application can be extended to find out potential disinfecting solutions to prevent formation of biofilm, which can be used in food processing industries (King et al. 2018). TiO_2 has been found to have effective bactericidal action on *E.coli* which has been the basis of development of the photocatalytic methods for bactericidal action against bacteria in aqueous environment (Sunanda et al. 1998). There has been studies on the effect of TiO_2 NPs which

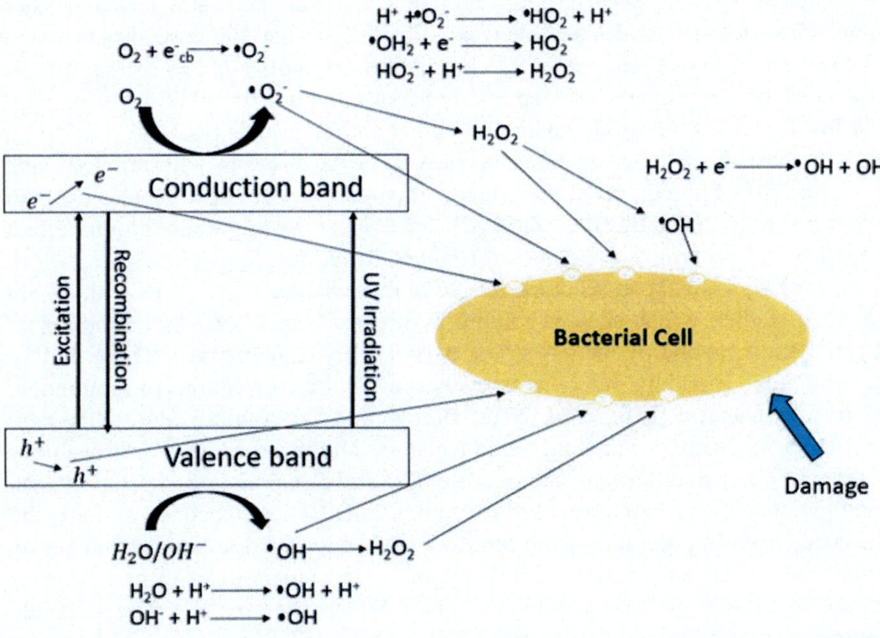

Fig. 6.6 Mechanism of disinfection when TiO_2 is used as photocatalyst under UV irradiation. Adopted from Wang et al. (2013)

showed that on UV irradiation it can be used effectively to reduce the time of disinfection, eliminating microorganisms from food that are pathogenic and to increase the food safety (Baruah and Dutta 2009). Though using TiO_2 has a major disadvantage, to activate the photocatalysis and initiate bactericidal action it requires UV irradiation. Nowadays, $Ag/AgBrTiO_2$ that are visible light absorbing photocatalyst have been proved effective in killing *E.coli* and *S.aureus* (Hu et al. 2006). Stoyanova et al. (2013) prepared TiO_2-ZnO nanocomposites by nonhydrolytic sol–gel method. Synthesized TiO_2-ZnO nanocomposites were found to be highly effective against *E. coli* in the presence of UV irradiation. In the case of visible irradiation only one log reduction achieved under 2 h. Wu et al. (2010) demonstrated the effect of Cu doped TiO_2 on *Mycobacterium smegmatis* and found that the growth rate of this bacteria reduced by three folds probably due to the release of Cu^{2+} ions from parent NPs. In *Salmonella typhimurium* TiO_2 was found to induce weak frameshift mutations (Pan et al. 2010) and also the NP was found toxic to *Psuedomonas aeruginosa* (Maness et al. 1999). TiO_2 nanoparticles can be used in dental applications because of its antibacterial properties against bacteria like *S. mutans* and *S. sanguinis* (Magraner et al. 2020).

Zinc Oxide Nanoparticles

Zinc oxide NPs among various other metal oxides that were studied have been found to be remarkably toxic. ZnO consisting of a band gap of ~3.2 eV similar to TiO_2 was treated as an alternative for photocatalytic activity against pathogens and other pollutants, such as pesticides and pharmaceuticals. ZnO had the same disadvantage as TiO_2, i.e. it is active only under UV irradiation. Its application as antimicrobials is favoured by its property of stability is harsh condition in addition to comparably low toxicity when conjugated with potent antimicrobials (Rai and Bai 2011). ZnO NPs have been found to be selectively toxic on bacteria with minimal side effects on cells of human body, these factors favour its recommendation to use it in food and agricultural industries. ZnO NPs have shown strong antimicrobial effect on food borne bacteria like *Salmonella typhimurium* and *Staphylococcus aureus*. Where ZnO NPs could cause complete lyse of these bacteria. ZnO NPs with 12 nm have been studied which showed inhibition of growth in *E. coli* by disruption of the cell membrane and by increasing the permeability of membrane. These finding support that the ZnO NPs can be effectively used in the applications in food industries to treat bacteria (King et al. 2018; Baruah and Dutta 2009). ZnO NPs were found to cause death by inducing oxidative stress and increasing cell permeability in methicillin-resistant *Streptococcus agalactiae* and *S.aureus* (Huang et al. 2008). Among various mechanism proposed to explain ZnO NPs antibacterial activity, the hydrogen peroxide generation from the ZnO surface is found to have potent action on inhibition of growth in bacteria. It is hypothesized, smaller size of particles and increase in number per unit volume increases the surface area and is responsible for generation of hydrogen peroxide. ZnO may also release Zn^{2+} ions which can be possibly another mechanism by which it can damage the cell membrane and interact with the intracellular contents (Brayner et al. 2006).

Studies also showed the effect of ZnO NP on methicillin-resistant *Staphylococcus aureus (*MRSA) and it was demonstrated that these NPs can get internalized inside cell that can cause cell membrane damage and disorganization of the cell wall. It is also known to increase oxidative stress inside the cell causing damage to the lipid, protein and DNA (Singh et al. 2014). Luo et al. (2013) showed the synergistic effect of ZnO with 25 different antibiotics against *S.aureus* and *E.coli.* They found that ZnO somehow enhanced the antimicrobial activity of penicillin, aminoglycosides, clarithromycin and tetracycline. ZnO NPs coated over glass surfaces were found to interfere with biofilm formation of *E.coli* and *S.aureus* (Applerot et al. 2012). Biocompatible nano-ZnO-bacterial cellulose (BC) has been found to show significant antibacterial activity against bacteria *B. subtilis* and *E. coli* (Dinca et al. 2020). Biogenic zinc oxide nanoparticles developed from aqueous *Pandanus odorifer* leaf extract (POLE) were found to show excellent antimicrobial activity against gram-positive bacteria *Bacillus subtilis* and gram-negative bacteria *E.coli* (Hussain et al. 2019).

Gold Nanoparticles

The most important property of gold NPs is its biocompatibility which makes it an extensively used material in organisms. The biologically inert nature of gold NPs can be altered to have chemical functionality and also photothermal functionality. Gold nanoparticles, cages, nanorods and spheres on exposure to near-infrared radiation (NIR) showed destruction of cancer cells and killing of bacterial cells through photothermal heating. Combination of photodynamic antimicrobial chemotherapy with NIR photothermal radiation of Au nanorods that was conjugated with the photosensitizers killed methicillin-resistant *Staphylococcus aureus* (MRSA) (Rai and Bai 2011; Brown et al. 2012). Au NPs with light absorbing capacity combined with specific antibodies have shown to destroy Staphylococcus aureus with the help of laser. In case of Vancomycin Resistant *Enterococci* (VRE) it was found out that the effect of antibiotic vancomycin showed synergistic effect with gold NP coating (Wang et al. 2017). Cefaclor which is a second-generation β-lactam antibiotic, when used with Au-NPs showed higher effect on both types of bacteria *S.aureus* (gram positive) and *E.coli* (gram negative) as compared to when they are used alone. Peptidoglycan layer of cell wall becomes porous by the action of cefaclor which acts as a cell wall synthesis inhibitor. This action gets enhanced by the action of Au NPs which generated holes on cell wall, leading to cell leakage and death of bacteria. Possibility is also that gold NPs inhibit uncoiling of DNA and also its transcription via binding to it (Hajipour et al. 2012). In solution Au NPs produced Au^{3+} ions along with decarboxylation of citrate produced free radicals in presence of light and were found to be responsible for photomutagenicity in *Salmonella typhimurium* (Wang et al. 2011).

Magnesium Oxide Nanoparticles

Nanoparticle metal oxides that are highly ionic can be prepared having a high surface area in addition to unusual morphologies of the crystal that have numerous edge/corners as well as reactive surface sites. Aerogel procedure (AP-MgO) is used for preparation of magnesium oxide (MgO) and MgO produced could be of varying shapes like polyhedral or square with size around 4 nm that is arranged within an extensive porous structure with ample pore volume (Rai and Bai 2011). AP-MgO NPs possess an interesting property for adsorbing and retaining elemental chlorine and bromine for a longer time (months), this combination of Ag-MgO/X_2 NPs can show potent killing activity against both types of bacteria and their spores. It can be used as a potent disinfectant. This property of Ag-MgO NPs to carry big amount of active halogens can be attributed to its high surface area and also enhanced surface reactivity (Pelgrift and Friedman 2013). The small size of NPs lets it cover around the cell of the bacteria to a large extent and this brings active halogens in high concentration in closeness to bacterial cell. These conditions in test against *Bacillus megaterium* and *E.coli* have shown good results and also against spores of Bacillus subtilis. AP-MgO/X_2 has a positive charge in water suspension, which is opposite to the charges on bacteria and spore cells increasing the effect of NPs and responsible for its bioactivity. It was observed using confocal microscopy that when the bacteria and NPs are together in water suspension, their opposite charges tend to bring them together in the form of aggregates. It was found that halogenated magnesium oxide has an active influence on bacteria and in particular their cell membranes. This was done using atomic force microscopy and electron microscopy studies. Hence, it was seen that the NPs of MgO with X_2 (Cl, Br) showed synergistic effects (both strong and fast) on bactericidal action and also more effective against the spores (Melander et al. 2018). Biofilm formation of *E.coli* and *S.aureus* was found to be inhibited with the use of MgF_2 NPs (Musee et al. 2011).

Aluminium Nanoparticles

Alumina NPs showed effect of growth inhibition on *E.coli* over a concentration range of 10–1000 μg/mL but this effect was observed only when the concentrations were very high. This can be associated with the surface–charge interaction of the particles with the cells. The prevention of cell wall disruption and desperate antimicrobial action is possibly because of its property of free-radical scavenging. Alumina has a corundum-like structure which is thermodynamically stable over a wide range of temperature. The corundum structure consisted of oxygen atoms with hexagonal close packing and two third of octahedral sites filled by Al^{3+} ions. Near neutral pH alumina NPs surface carried a positive charge. This resulted in adhesion of NPs to the negatively charged surface of *E.coli* which increased the concentration of NPs around it and negatively influenced its growth. The bacterial adhesion may be due to the electrostatic interaction of particle surface with bacterial surface in addition to the hydrophobic interaction as well as polymer bridging. Reactive oxygen species (ROS) generation can be causable factor for the antimicrobial effect of the metal oxides which disrupts the cell wall, finally cell death. But alumina as free radical scavengers have also been reported (Rai and Bai 2011). Al_2O_3 NPs of size 50–70 nm was found to cause damage to bacterial cell wall and increase its membrane permeability by binding to its cell wall in *E.coli* (Jiang et al. 2009).

6.3.3 *Factors Affecting Antibacterial Properties of Nanoparticles*

There are several factors that affect the nanoparticles (NPs) bactericidal properties, like size, shape, charge and others. Further environmental condition also has significant impact on its properties. It also depends on bacterial strain and the exposure time for its activity. The following factors are described briefly that affect the physicochemical properties of nanoparticles. Size of nanoparticle, smaller size provides it with large surface area that increases the probability of contact between bacteria and nanoparticles and help in passing through the cell membrane of bacteria compared to larger NPs. Shape was found that activity of NPs changed significantly with change in its shape with similar NPs. On increasing the roughness of NPs adsorption of bacterial proteins increases with the reduction in bacterial adhesion (Wang et al. 2017). Zeta potential has a significant influence on bacterial adhesion. The charges on NPs and bacterial cell membrane are important factors for the electrostatic force between them. Oppositely charged NPs tend to gather selectively at sites of bacterial infection and increase the vascular permeability. Doping modification was found to be one of the most adequate method by which we can regulate and have control over the NPs and bacterial interaction. Sahu and Biswas (2011) synthesized Cu-doped TiO_2 and found that with increasing Cu doping the crystal phase changed from anatase to rutile and the growth of particles were restricted and also the band gap came to visible region. Environmental conditions, the antimicrobial activities of NPs, were found to vary with different environmental conditions. The medium characteristics like pH and osmotic pressure influence the surface charge, aggregation and solubility of NPs which in turn affect its activity (Wang et al. 2017; Hajipour et al. 2012).

6.4 Conclusion and Future Perspectives

The bacterial strains have developed resistance against many generations of drugs, leaving very little options for treatment of bacterial infections. The gram-positive and gram negative-bacteria respond differently to the antibiotics due to the difference in their cell wall composition, and gram-negative bacteria tend to be more effected as compared to the gram-positive bacteria due to the difference in thickness and composition of cell wall. The mechanism of action of antibiotics on bacterial cell is mainly through three mechanisms, first by targeting its cell wall, second by inhibiting the protein synthesis or third by inhibiting the DNA replication of bacteria. The bacteria have also evolved various mechanisms to negate the effect of antibiotics on them, like increase in function of efflux pumps or changing antibiotic binding receptor sites. This change is brought about in bacteria either by genetical basis which includes mutational resistance and horizontal gene transfer both of which are inheritable to the next generation. Various kinds of NPs like Ag, Au, MgO, aluminium, TiO_2 and ZnO have antimicrobial action on different types of bacteria either individually or in doped condition and generally follows mechanism of oxidative stress, dissolved metallic ions or non-oxidative stress conditions. Some NPs involve multiple mechanism simultaneously to kill bacteria. The antimicrobial action of NPs on bacteria are mainly through one of the following, firstly interaction with its cell barrier, either by creating pores in it or preventing it from formation or disrupting the cell barriers. The NPs penetrate the cell barrier either by diffusion or get adsorbed on the cell membrane. Secondly, by inhibiting the synthesis of bacterial protein and DNA. Third by regulating the expression of bacterial metabolic genes and also by inhibiting the formation of bacterial biofilm. The NPs advantage as antimicrobials can be summarized as it can be helpful in overcoming existing antibiotic resistance organisms, or it can combat microbes via multiple mechanism simultaneously or it can act as a good carrier of antibiotic. Several characteristics of NPs such as size, charge, shape, zeta potential, surface morphology and crystal structure and environmental conditions such as pH, osmotic pressure impact the antibacterial potential. With the desired characteristics, NPs can be utilized as an antimicrobial agent. Furthermore, NPs can employ several mechanisms simultaneously on bacteria and let it unlikely to develop resistance against it, thus having a potential of solving MDR in bacteria.

References

Agarwal P, Mehta A, Kachhwaha S, Kothari SL (2013) Green synthesis of silver nanoparticles and their activity against Mycobacterium tuberculosis. Adv Sci Eng Med 5(7):709–714

Alfuraydi AA, Devanesan S, Al-Ansari M, AlSalhi MS, Ranjitsingh AJ (2019) Eco-friendly green synthesis of silver nanoparticles from the sesame oil cake and its potential anticancer and antimicrobial activities. J Photochem Photobiol B 192:83–89

Applerot G, Lellouche J, Perkas N, Nitzan Y, Gedanken A, Banin E (2012) ZnO nanoparticle-coated surfaces inhibit bacterial biofilm formation and increase antibiotic susceptibility. Royal Soc Chem Adv 2(6):2314–2321

Baek YW, An YJ (2011) Microbial toxicity of metal oxide nanoparticles (CuO, NiO, ZnO, and Sb_2O_3) to Escherichia coli, *Bacillus subtilis*, and *Streptococcus aureus*. Sci Total Environ 409(8):1603–1608

Baruah S, Dutta J (2009) Nanotechnology applications in pollution sensing and degradation in agriculture: a review. Environ Chem Lett 7(3):191–204

Brayner R, Ferrari-Iliou R, Brivois N, Djediat S, Benedetti MF, Fiévet F (2006) Toxicological impact studies based on Escherichia coli bacteria in ultrafine ZnO nanoparticles colloidal medium. Nano Lett 6(4):866–870

Brown AN, Smith K, Samuels TA, Lu J, Obare SO, Scott ME (2012) Nanoparticles functionalized with ampicillin destroy multiple-antibiotic-resistant isolates of Pseudomonas aeruginosa and Enterobacter aerogenes and methicillin-resistant Staphylococcus aureus. Appl Environ Microbiol 78(8):2768–2774

Chatterjee AK, Chakraborty R, Basu T (2014) Mechanism of antibacterial activity of copper nanoparticles. Nanotechnology 25(13):101–135

Choi O, Deng KK, Kim NJ, Ross L Jr, Surampalli RY, Hu Z (2008) The inhibitory effects of silver nanoparticles, silver ions, and silver chloride colloids on microbial growth. Water Res 42(12):3066–3074

Dincă V, Mocanu A, Isopencu G, Busuioc C, Brajnicov S, Vlad A, Icriverzi M, Roseanu A, Dinescu M, Stroescu M, Stoica-Guzun A (2020) Biocompatible pure ZnO nanoparticles-3D bacterial cellulose biointerfaces with antibacterial properties. Arab J Chem 13(1):3521–3533

Dwyer DJ, Kohanski MA, Collins JJ (2009) Role of reactive oxygen species in antibiotic action and resistance. Curr Opin Microbiol 12(5):482–489

Ferrando-Magraner E, Bellot-Arcís C, Paredes-Gallardo V, Almerich-Silla JM, García-Sanz V, Fernández-Alonso M (2020) Antibacterial properties of nanoparticles in dental restorative materials. A systematic review and meta-analysis. Medicina 56(2):55–77

Foster HA, Ditta IB, Varghese S, Steele A (2011) Photocatalytic disinfection using titanium dioxide: spectrum and mechanism of antimicrobial activity. Appl Microbiol Biotechnol 90(6):1847–1868

Hajipour MJ, Fromm KM, Ashkarran AA, de Aberasturi DJ, de Larramendi IR, Rojo T, Mahmoudi M (2012) Antibacterial properties of nanoparticles. Trends Biotechnol 30(10):499–511

Hancock RE (2005) Mechanisms of action of newer antibiotics for Gram-positive pathogens. Lancet Infect Dis 5(4):209–218

Hopwood DA (2007) How do antibiotic-producing bacteria ensure their self-resistance before antibiotic biosynthesis incapacitates them. Mol Microbiol 63(4):937–940

Hu C, Lan Y, Qu J, Hu X, Wang A (2006) Ag/AgBr/TiO_2 visible light photocatalyst for destruction of azodyes and bacteria. J Phys Chem B 110(9):4066–4072

Huang Z, Zheng X, Yan D, Yin G, Liao X, Kang Y, Yao Y, Huang D, Hao B (2008) Toxicological effect of ZnO nanoparticles based on bacteria. Langmuir 24(8):4140–4144

Huh AJ, Kwon YJ (2011) "Nanoantibiotics": a new paradigm for treating infectious diseases using nanomaterials in the antibiotics resistant era. J Control Release 156(2):128–145

Hussain A, Oves M, Alajmi MF, Hussain I, Amir S, Ahmed J, Rehman MT, El-Seedif HR, Ali I (2019) Biogenesis of ZnO nanoparticles using Pandanus odorifer leaf extract: anticancer and antimicrobial activities. Royal Soc Chem Adv 9(27):15357–15369

Jiang W, Mashayekhi H, Xing B (2009) Bacterial toxicity comparison between nano-and micro-scaled oxide particles. Environ Pollut 157(5):1619–1625

Kapoor G, Saigal S, Elongavan A (2017) Action and resistance mechanisms of antibiotics: a guide for clinicians. J Anaesthesiol Clin Pharmacol 33(3):300–305

King T, Osmond-McLeod MJ, Duffy LL (2018) Nanotechnology in the food sector and potential applications for the poultry industry. Trends Food Sci Technol 72:62–73

Li Q, Mahendra S, Lyon DY, Brunet L, Liga MV, Li D, Alvarez PJ (2008) Antimicrobial nanomaterials for water disinfection and microbial control: potential applications and implications. Water Res 42(18):4591–4602

Luo Z, Wu Q, Xue J, Ding Y (2013) Selectively enhanced antibacterial effects and ultraviolet activation of antibiotics with ZnO nanorods against Escherichia coli. J Biomed Nanotechnol 9(1):69–76

Macwan DP, Dave PN, Chaturvedi S (2011) A review on nano-TiO_2 sol–gel type syntheses and its applications. J Mater Sci 46(11):3669–3686

Mahmoud HA, Narasimharao K, Ali TT, Khalil KM (2018) Acidic peptizing agent effect on anatase-rutile ratio and photocatalytic performance of TiO_2 nanoparticles. Nanoscale Res Lett 13(1):48–65

Maness PC, Smolinski S, Blake DM, Huang Z, Wolfrum EJ, Jacoby WA (1999) Bactericidal activity of photocatalytic TiO_2 reaction: toward an understanding of its killing mechanism. Appl Environ Microbiol 65(9):4094–4098

Mazel D, Davies J (1999) Antibiotic resistance in microbes. Cellular Molecular Life Sci CMLS 56(9–10):742–754

Melander RJ, Zurawski DV, Melander C (2018) Narrow-spectrum antibacterial agents. RSC Med Chem 9(1):12–21

Munita JM, Arias CA (2016) Mechanisms of antibiotic resistance. Microbiol Spectrum 4(2):742–754

Musee N, Thwala M, Nota N (2011) The antibacterial effects of engineered nanomaterials: implications for wastewater treatment plants. J Environ Monit 13(5):1164–1183

Nel A, Xia T, Mädler L, Li N (2006) Toxic potential of materials at the nanolevel. Science 311(5761):622–627

Pan X, Redding JE, Wiley PA, Wen L, McConnell JS, Zhang B (2010) Mutagenicity evaluation of metal oxide nanoparticles by the bacterial reverse mutation assay. Chemosphere 79(1):113–116

Pelgrift RY, Friedman AJ (2013) Nanotechnology as a therapeutic tool to combat microbial resistance. Adv Drug Deliv Rev 65(13–14):1803–1815

Pramanik A, Laha D, Bhattacharya D, Pramanik P, Karmakar P (2012) A novel study of antibacterial activity of copper iodide nanoparticle mediated by DNA and membrane damage. Colloids Surf B 96:50–55

Qu X, Alvarez PJ, Li Q (2013) Applications of nanotechnology in water and wastewater treatment. Water Res 47(12):3931–3946

Rai M, Yadav A, Gade A (2009) Silver nanoparticles as a new generation of antimicrobials. Biotechnol Adv 27(1):76–83

Rai RV, Bai JA (2011) Nanoparticles and their potential application as antimicrobials. In: Méndez-Vilas A (ed). Formatex, Mysore

Ruparelia JP, Chatterjee AK, Duttagupta SP, Mukherji S (2008) Strain specificity in antimicrobial activity of silver and copper nanoparticles. Acta Biomater 4(3):707–716

Sahu M, Biswas P (2011) Single-step processing of copper-doped titania nanomaterials in a flame aerosol reactor. Nanoscale Res Lett 6(1):441

Sahu M, Wu B, Zhu L, Jacobson C, Wang WN, Jones K, Biswas P (2011) Role of dopant concentration, crystal phase and particle size on microbial inactivation of Cu-doped TiO_2 nanoparticles. Nanotechnology 22(41):415–704

Santo CE, Taudte N, Nies DH, Grass G (2008) Contribution of copper ion resistance to survival of Escherichia coli on metallic copper surfaces. Appl Environ Microbiol 74(4):977–986

Scott JR, Barnett TC (2006) Surface proteins of gram-positive bacteria and how they get there. Annual Rev Microbiol 60:397–423

Sheng Z, Liu Y (2011) Effects of silver nanoparticles on wastewater biofilms. Water Res 45(18):6039–6050

Singh R, Smitha MS, Singh SP (2014) The role of nanotechnology in combating multi-drug resistant bacteria. J Nanosci Nanotechnol 14(7):4745–4756

Singh S, Ashfaq M, Singh RK, Joshi HC, Srivastava A, Sharma A, Verma N (2013) Preparation of surfactant-mediated silver and copper nanoparticles dispersed in hierarchical carbon micro-nanofibers for antibacterial applications. New Biotechnol 30(6):656–665

Singh S, Joshi HC, Srivastava A, Sharma A, Verma N (2014) An efficient antibacterial multi-scale web of carbon fibers with asymmetrically dispersed Ag–Cu bimetal nanoparticles. Colloids Surf A 443:311–319

Slavin YN, Asnis J, Häfeli UO, Bach H (2017) Metal nanoparticles: understanding the mechanisms behind antibacterial activity. J Nanobiotechnol 15(1):65–85

Stoyanova A, Hitkova H, Bachvarova-Nedelcheva A, Iordanova R, Ivanova N, Sredkova M (2013) Synthesis and antibacterial activity of TiO_2/ZnO nanocomposites prepared via nonhydrolytic route. J Chem Technol Metall 48(2):154–161

Sunada K, Kikuchi Y, Hashimoto K, Fujishima A (1998) Bactericidal and detoxification effects of TiO_2 thin film photocatalysts. Environ Sci Technol 32(5):726–728

Wang L, Hu C, Shao L (2017) The antimicrobial activity of nanoparticles: present situation and prospects for the future. Int J Nanomed 12:12–27

Wang S, Lawson R, Ray PC, Yu H (2011) Toxic effects of gold nanoparticles on Salmonella typhimurium bacteria. Toxicol Ind Health 27(6):547–554

Wright GD (2011) Molecular mechanisms of antibiotic resistance. Chem Commun 47(14):4055–4061

Wu B, Huang R, Sahu M, Feng X, Biswas P, Tang YJ (2010) Bacterial responses to Cu-doped TiO_2 nanoparticles. Sci Total Environ 408(7):1755–1758

Yadav KK, Singh JK, Gupta N, Kumar V (2017) A review of nanobioremediation technologies for environmental cleanup: a novel biological approach. J Mater Environ Sci 8(2):740–757

Chapter 7
Perovskite $BiFeO_3$ Nanostructure Photocatalysts for Degradation of Organic Pollutants

Simant Kumar Srivastav, Swatantra P. Singh, and Kamlesh Kumar

Abstract Environmental remediation is a need of time, due to rapid industrialization, urbanization, and changing lifestyles of people. Nanomaterials offer great potential for efficient detection and removal of pollutants due to their vast array of useful properties. Recently, the magnetic property and catalytic activity of Bi-based oxide nanostructures have been exploited for environmental applications such as catalytic oxidation of recalcitrant pollutants and simultaneous magnetic recovery. The loss of the catalyst over cycles can be prenevted and makes these catalyst cost-effective. Thereby, bismuth (Bi)-based oxide nanostructures can serve as the right candidate for sustainable pollutant remediation. Among bismuth (Bi) -based oxide, perovskite type-$BiFeO_3$ nanostructures are unique potential photocatalyst due to its multiferroic behaviour, narrow band gap and chemical stability. In this chapter, crystal structure, the state-of-the-art in the synthesis, charaterizations and photocatalytic degradation mechanism of $BiFeO_3$ are discussed. The strategies to improve the photocatalytic performances of $BiFeO_3$ for organic pollutants are discussed.

Keywords Nanostructures · Nanoparticles · Nanocomposite · Water remediation · Bismuth ferrite · Graphene · Photocatalysis · Organic pollutants

7.1 Introduction

The spread of pollutants in the environment has become a global menace, due to rapidly growing industrialization, urbanization, and changing lifestyles, which contaminates environment through distribution in the water, air, and soil and causing

S. K. Srivastav (✉)
University Department of Chemistry, L. N. Mithila University, Darbhanga, Bihar 846008, India

S. P. Singh
Department of Environmental Science and Engineering (ESED), Indian Institute of Technology Bombay, Bombay, Maharashtra 400076, India

K. Kumar
Department of Chemistry, Institute of Science, Banaras Hindu University, Varanasi 221005, India

S. P. Singh et al. (eds.), *Nanomaterials and Nanocomposites for Environmental Remediation*, Energy, Environment, and Sustainability,
https://doi.org/10.1007/978-981-16-3256-3_7

great harm to human health (Peng et al. 2004; Singh et al. 2010; Fu and Wang 2011). In view of this, easy and efficient methods for environmental decontamination are necessary. The advent of nanotechnology has offered significant scope and opportunities to revolutionize approaches towards the environmental applications of nanomaterials by enhancing the multifunctionality and versatility (Yunus et al. 2012; Khin et al. 2012; Chang et al. 2018). Nanomaterials shows unique functionalities due to large surface-to-volume ratios i.e. high specific surface area which is difficult to obtain using traditional chemicals or bulk materials. The high specific surface area and hence high reactivity at nanoscale makes them excellent adsorbents, catalysts and sensors (Yunus et al. 2012; Khin et al. 2012; Chang et al. 2018; Lu et al. 2016; Yaqoob et al. 2020). These unique properties of nanomaterials have been utilized to degrade and scavenge pollutants in water and air (Fu and Wang 2011; Yunus et al. 2012; Khin et al. 2012; Yaqoob et al. 2020). Therefore, nanomaterials play an important role in environmental remediation.

In the past decades, nanostructured semiconducting materials such as, ZnS, CdS, ZnO, and TiO_2 have shown significant photosensitive properties which have been utilized as photocatalysts for remediation of water and air from pollutants after photocatalytic oxidation (Maeda 2011; Shaheen et al. 2020; Lam et al. 2012; Sin et al. 2012). Nevertheless, the application of most of these semiconductor photocatalysts are practically limited due to two main factors; (1) wide band-gap that absorbs light only in UV-region, (2) the fast recombination of charge carriers, which lowers the efficiency of photocatalyst (Herrmann 1999). To circumvent these two limitations, a new group of materials having perovskite structure is getting importance for photocatalysis (Shi and Guo 2012; Kanhere and Chen 2014; Wang et al. 2015a; Chen et al. 2015). The perovskite structure has a general formula, ABO_3, where 'O' is oxygen, 'A' represents a cation (divalent or trivalent) with larger ionic radius, and 'B' represents a cation (tetravalent or trivalent) with smaller ionic radius. ABO_3 materials have shown better photocatalytic performances than other semiconductor photocatalyst due to narrow band-gap and photo-physical properties of A and B cations (Shi and Guo 2012; Kanhere and Chen 2014). Several perovskite oxides have been studied as photocatalyst such as ferrites, tantalates, and titanates (Shi and Guo 2012; Kanhere and Chen 2014; Wang et al. 2015a).

Among these perovskite oxides, nanostructure $BiFeO_3$ (BFO) have shown great potential as a promising visible light-responsive photocatalyst for the degradation of organic pollutants due to the following properties: (i) room temperature multiferroic behaviour, which favours charge carrier separation efficiently, (ii) band-gap exists in visible-light region, and, (iii) excellent chemical stability (Bai et al. 2016; Lam et al. 2017; Irfan et al. 2019; Yin and Mi 2020). Recent increasing interest in developing methods to synthesize nanostructured $BiFeO_3$ due to its enhanced functionality and applications at nanoscale (Bai et al. 2016; Lam et al. 2017; Irfan et al. 2019; Yin and Mi 2020). For example, bulk $BiFeO_3$ is not magnetic material but it starts showing magnetic properties when its size reduced to nanoscale (Yin and Mi 2020). Its magnetic property along with catalytic activity, can be utilized in the environmental applications (Bai et al. 2016; Lam et al. 2017; Irfan et al. 2019; Yin and Mi 2020). BFO nanostructures have shown excellent catalytic properties with different dyes such as methyl orange (MO) (Wang et al. 2016a; Gao et al. 2015), Rhodamine B (RhB) (Zhang et al. 2012), and methylene blue (MB) (Li, Y.a., Li, Chen et al. 2019). A number of factors such as morphologies, doping, formation of nanocomposites with other materials can affect the photocatalysis behaviour of $BiFeO_3$ nanostructures.

This chapter discusses crystal structure, state of the art in synthesis, characterization, and photocatalytic performances of $BiFeO_3$ nanostructures. The sol–gel and hydrothermal methods for the $BiFeO_3$ nanostructure are discussed. This chapter provides a comprehensive understanding for perovskite $BiFeO_3$ nanostructure in terms of its synthesis process, possible ways of improving its photocatalytic performance and its application in for the degradation of organic pollutants.

7.2 Crystal Structure of $BiFeO_3$

$BiFeO_3$ is, perhaps, the only single phase magnetoelectric multiferroic material that exhibits ferroelectric ($T_C = 1103$ K) and antiferromagnetic ($T_N = 640$ K) properties at or above room temperature (Catalan and Scott 2009). At room temperature $BiFeO_3$ has a rhombohedrally distorted perovskite structure with a space group *R3c (*Catalan and Scott 2009*)*. The rhombohedral unit cell has lattice parameter of $a_{\mathrm{rh}} = 5.63$ Å, distorted angle of $\alpha_{\mathrm{rh}} = 59.4°$ at room temperature. In the cubic $BiFeO_3$ perovskite type unit cell, where **Bi**-ion sits at the corners of the cube and **Fe**-ion sits at the body center position of the cube and oxygen at the center of each face positions as shown in Fig. 7.1a (Catalan and Scott 2009; Chu et al. 2007). Another view of cubic structure is the tree dimensional network formed by the corner sharing of the FeO6 octahedra.

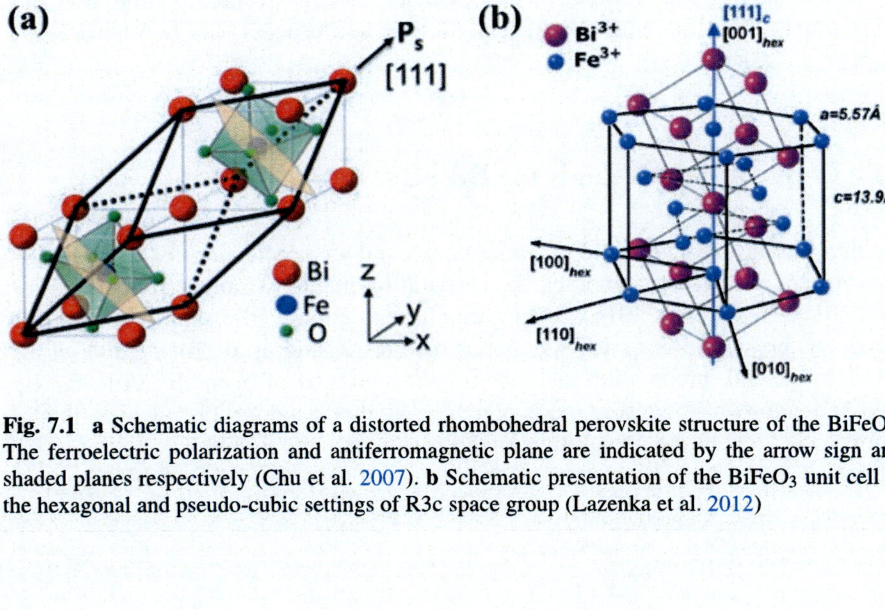

Fig. 7.1 **a** Schematic diagrams of a distorted rhombohedral perovskite structure of the $BiFeO_3$. The ferroelectric polarization and antiferromagnetic plane are indicated by the arrow sign and shaded planes respectively (Chu et al. 2007). **b** Schematic presentation of the $BiFeO_3$ unit cell in the hexagonal and pseudo-cubic settings of R3c space group (Lazenka et al. 2012)

The structure of BFO can also be represented as pseudo-cubic structure. In pseudo-cubic representation two distorted perovskite unit cells (lattice parameter $a_c = 3.96$ Å and pseudo cubic angle $\alpha_c = 89.3–89.4°$) are connected along their body diagonal $[111]_{\text{pseudocubic (c)}}$ to form a rhombohedral unit cell as shown in Fig. 7.1a (Chu et al. 2007). A rhombohedral unit cell representation can be transformed to an equivalent hexagonal representation as shown in Fig. 7.1b, in which the pseudo-cubic direction $[111]_c$ corresponding to the hexagonal $[001]_{hex}$. The hexagonal lattice parameters are $a_h = 5.57$ Å and $c_h = 13.90$ Å at room temperature (Catalan and Scott 2009; Lazenka et al. 2012).

$BiFeO_3$ is a room temperature ferroelectric material. The ferroelectric polarization of $BiFeO_3$ is due to the dislocation of Bi^{3+} ions relative to the FeO6 octahedral and directed along the diagonals of perovskite unit cell i.e. $[111]_c$ or $[011]_{hex}$ as shown in Fig. 7.1 (Catalan and Scott 2009; Chu et al. 2007). $BiFeO_3$ is G-type antiferromagnetic where Fe magnetic moments couples ferromagnetically coupled within planes and antiferromagnetically between adjacent planes (Catalan and Scott 2009; Chu et al. 2007; Lazenka et al. 2012).

In the perovskite structures, the rotation angle of oxygen octahedra is a critical structural parameter. For a cubic perovskite, the rotation angle of oxygen octahedra is 0° (Catalan and Scott 2009). The stability and distortion of a perovskite unit cell is measured by Goldschmid tolerance factor, t, which is defined as $(r_{\text{Bi}} + r_O)/\sqrt{2}(r_{\text{Fe}} + r_O)$, where r is the ionic radius of the respective ions (Catalan and Scott 2009). For $BiFeO_3$, t is 0.88 (Catalan and Scott 2009). The 't' smaller than one results in the buckling of the oxygen octahedra in order to fit into a perovskite unit cell. The doping and/or co-doping for Bi^{3+} and Fe^{3+} ions changes the value of 't' and also effect the crystallographic symmetry in numerous perovskites (Catalan and Scott 2009). The oxygen octahedra (FeO6) rotation have a strong effect on crystal field that significantly changes the dipole moments, electronic band structures, thus affects the behaviours of photo-generated charge carriers such as excitation, migration and redox processes in the entire photocatalytic reaction process (Shi and Guo 2012; Kanhere and Chen 2014).

7.3 Preparation Methods for $BiFeO_3$

The morphology, structure and potential photocatalytic applications of BFO nanostructures are strongly governed by the preparation methods(Lam et al. 2017; Irfan et al. 2019; Zhang et al. 2016). A number of preparation methods have been introduced for the synthesis of $BiFeO_3$ nanostructures(Zhang et al. 2016; Silva et al. 2011). In general, preparation methods for the synthesis of bismuth ferrite can be classified in to two categories: oxide precursor methods and wet chemical methods. A bird's eye view of the preparation methods is mentioned in Fig. 7.2.

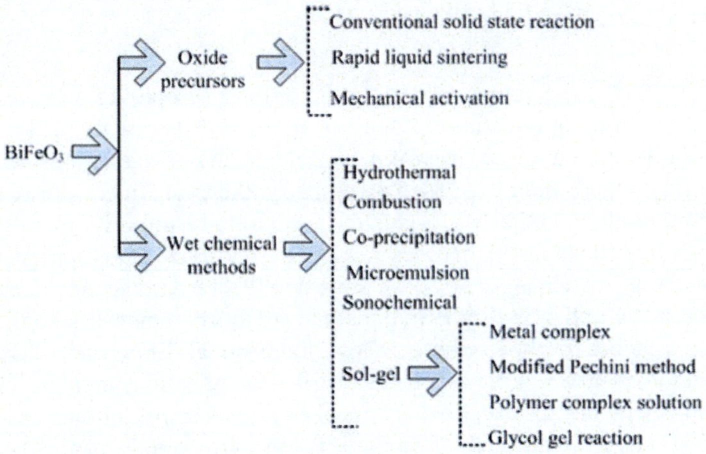

Fig. 7.2 Diagram showing the preparation techniques used for $BiFeO_3$

7.3.1 *Oxide Precursor Methods*

Bismuth oxide (Bi_2O_3) and iron oxide (Fe_2O_3) are used as precursors in this method. The phase diagram of Bi_2O_3 and Fe_2O_3 system suggest that synthesis of phase pure $BiFeO_3$ is a challenging task due to very narrow temperature stability range, which causes the formation of impurity phases along with pure phase (Catalan and Scott 2009). In this method, bismuth and iron oxides are mixed and calcined at high temperatures which leads to formation of Bi-poor phase $Bi_2Fe_4O_9$ due to high volatility of Bi_2O_3. To compensate for the loss of Bi_2O_3 during calcination, small excess amount of Bi_2O_3 is used which again leads to formation of Bi-rich phase $Bi_{25}FeO_{39}$. $BiFeO_3$ decomposes above 675 °C into $Bi_2Fe_4O_9$ and rapidly into $Bi_2Fe_4O_9$ above 830 °C (Mukherjee and Wang 1971; Tabares-Munoz et al. 1984). To prevent the formation of impurity phase $Bi_2Fe_4O_9$, the initial attempts were made for the synthesis of $BiFeO_3$ using solid state reaction method by adding large excess of Bi_2O_3 (100%). The disadvantage of this method is to form a large quantity of impurity $Bi_{25}FeO_{39}$ phase that cannot be separated from pure $BiFeO_3$ phase with a good yield. The impurity phases can be leached out by using dilute nitric acid but this process can change the stoichiometry of $BiFeO_3$ by selective dissolution some elements. This method involves high crystallization temperature of $BiFeO_3$ is very high (Achenbach et al. 1967; Kumar et al. 2000). However, pure phase $BiFeO_3$ can be synthesized by rapid liquid-phase sintering technique but involvement of high calcination temperature of nearly 880 °C for the $BiFeO_3$ phase formation leads to the high energy consumption and cost (Wang et al. 2004).

7.3.2 *Wet Chemical Methods*

Perovskite $BiFeO_3$ have shown intriguing the physical properties at nanoscale and offers great potential applications (Zhang et al. 2016; Silva et al. 2011). To avoid bismuth volatilization and necessity of nanoscale $BiFeO_3$, a variety of low temperature synthesis methods have been used to for the synthesis of $BiFeO_3$ nanostructures. The low-temperature synthesis prevents the formation of impurity phases. Various low temperature wet chemical methods are used to synthesize $BiFeO_3$ nanostructures as outlined in Fig. 7.2 (Zhang et al. 2016; Silva et al. 2011). Among these methods, the sol–gel method as well as hydrothermal method are most efficient methods and most common approaches for the synthesis of low-dimensional $BiFeO_3$ nanomaterials. In sol–gel method, proper mixing as well as distribution of cations in the solutions can be achieved which ensures excellent chemical as well as compositional as compared to other methods. In hydrothermal method, Phase-pure $BiFeO_3$ nanostructures are synthesized at much lower temperature but high pressure. Therefore, the salient features of sol–gel and hydrothermal synthesis methods are discussed here for the fabrication of $BiFeO_3$ nanostructures.

7.3.2.1 Sol–gel Synthesis

Metal source, a solvent and chelation agent act as precursors for this method. The synthesis process starts from the mixing of these precursors to form precursor solution. The precursor solutions are mixed under continuous stirring and appropriate heating conditions which transform this solution into a polymeric state, called as "gel". Since the gel is liquid in nature, the final-product material can be fabricated in different morphologies by using different deposition approaches (Zhang et al. 2016). For example, heat treatment is used to obtain nanoparticles (0 D), pouring into template is used to fabricate nanowires or nanotubes (1D), electrospinning is used to fabricate nanofibers, and spin-coating is used for thin films (2D). This is graphically shown in Fig. 7.3.

For the synthesis of $BiFeO_3$ nanostructures, metal nitrates (i.e.$Bi(NO_3)_3.5H_2O$ and $Fe(NO_3)_3.9H_2O$) are used as universal metal source. For solvent generally ethylene glycol (EG) or 2-methoxyethanol (2-MOE) are used due to the good solubility of various starting reagents. Chelating agents such as acetic anhydride, acetic acid, citric acid, tartaric acid, EDTA, succinic acid, malonic acid, polyvinyl alcohol or propylene glycol are used for the synthesis of $BiFeO_3$ nanostructures. The reaction between chelating agent and the metal source, known as chelation reactions, facilitate the gelation process. The choice of a chelating agent is important for the fabrication of $BiFeO_3$ with specific morphologies because the phases and morphologies of the final products are controlled by the molecular structures of the chelating agents (Zhang et al. 2016). For example, in a typical synthesis process of $BiFeO_3$ nanoparticles, equimolar amount of bismuth and iron nitrates are dissolved in solvent separately and mixed followed by the addition of appropriate chelating agent under constant

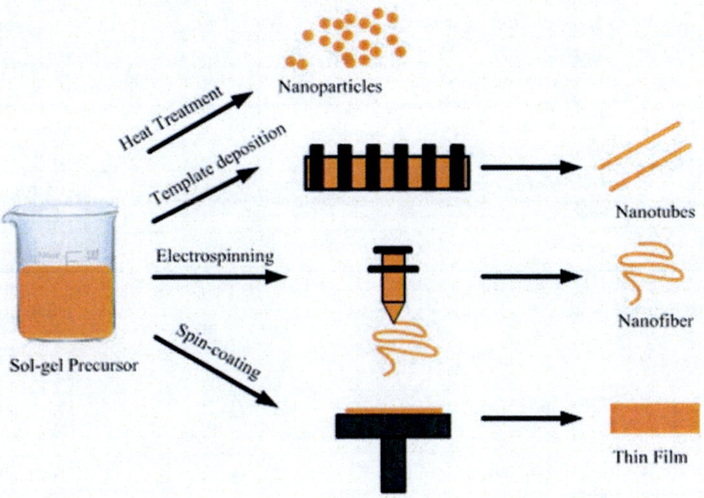

Fig. 7.3 Schematic for the fabrication of various nanostructures via sol–gel method

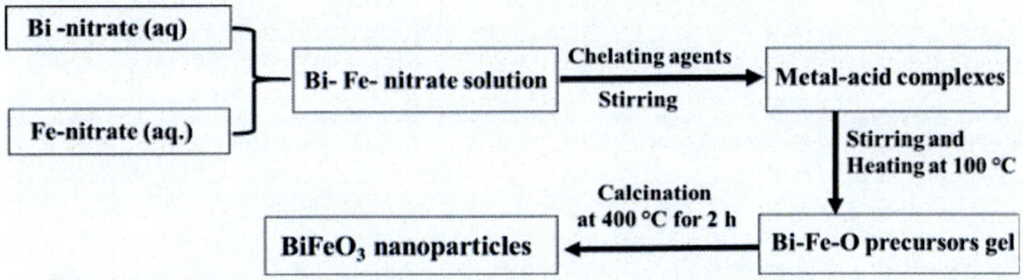

Fig. 7.4 Flow diagram for the sol–gel synthesis of $BiFeO_3$ nanoparticles

stirring and heating at about 100 °C until liquid changes to fluffy gel. Then this fluffy gel is calcined at 400–650 °C for 2–6 h to obtain $BiFeO_3$ nanoparticles with different sizes. A flow digram for the sol–gel synthesis of $BiFeO_3$ nanoparticles is presented in Fig. 7.4. The synthesis of BFO have been reported via sol–gel method by using variety of chelating agents and the results are listed in Table 7.1.

7.3.2.2 Hydrothermal Synthesis

The hydrothermal method has been used widely for the synthesis of $BiFeO_3$ due to its low-temperature synthesis process which prevent the production of unwanted impurities (Zhang et al. 2016). In a typical hydrothermal method, precursors which includes, metal source, solvent (deionized (DI) water), and mineralizer are mixed and heated in a sealed vessel (autoclave, bomb, etc.) so that the autogenous pressure far exceeds the ambient pressure. This automatically raises the effective boiling point

Table 7.1 Summary of chelating agents and results in sol–gel sythsis of $BiFeO_3$ nanoparticles

Chelating	Solvent	Synthesis Temp. (°C)	Particle size (nm)	References
Tartaric acid	DI water, Nitric acid	600	16	Ghosh et al. 2005)
Tartaric acid	EG/ Nitric Acid	500	60–90	Wang et al. 2010)
Citric acid	EG	600	11	Popa et al. 2723)
Citric acid/Tartaric acid	DI water, Nitric acid	300	4 (C.A.)	Yongming et al. 2011)
			12 (T.A.)	Yongming et al. 2011)
Succinic acid	DI water, Nitric acid	600	60–70	Selbach et al. 2007)
Malic aacid	DI water, Nitric acid	600	60–70	Selbach et al. 2007)
Malonic acid	DI water, Nitric acid	600	60–70	Selbach et al. 2007)
EDTA	DI water, Nitric acid	600	110	Wei and Xue 2008)
Glycerol	DI water, Nitric acid	400	100	Liu et al. 2011a)
Propylene glycol	DI water, Nitric acid	450	35	Srivastav and Gajbhiye 2012)

of the solvent. When the nonaqueous solvent is used instead of water, the technique is often called "solvothermal"(Zhang et al. 2016). The temperature, pH value and mineralizer are the critical parameters in hydrothermal process which controls the morphologies and properties of the desired materials. The temperature plays a key role in stabilization of the material phase, pH value controls the ionization equilibrium conditions, the mineralizer act as a "catalyst" to aid formation of the seed crystals for the growth of desired nanomaterials (Zhang et al. 2016). The size, shape, and morphology of desired nanomaterials can be controlled by the optimization of these parameters. Therefore, the size, shape and morphology of $BiFeO_3$ can be easily controlled by hydrothermal process (Zhang et al. 2016). It is well-known that the shape, size and morphology can significantly affect the magnetic, electrical and optical properties (Zhang et al. 2016; Mao et al. 2005).

For the hydrothermal synthesis of $BiFeO_3$, the precursor solution is prepared by dissolving bismuth and iron nitrates (metal sources), in deionized (DI) water or nitric acid (solvent) and KOH or NaOH, often used as the mineralizer, is added. After obtaining the precursor, the solution is put in a Teflon linear autoclave for hydrothermal processing at 150–220 °C for 5–72 hours. Then the autoclave is cooled naturally to room temperature. The obtained precipitate is filtered and washed using deionized water and absolute ethanol to remove all the soluble salts. Finally, the wet

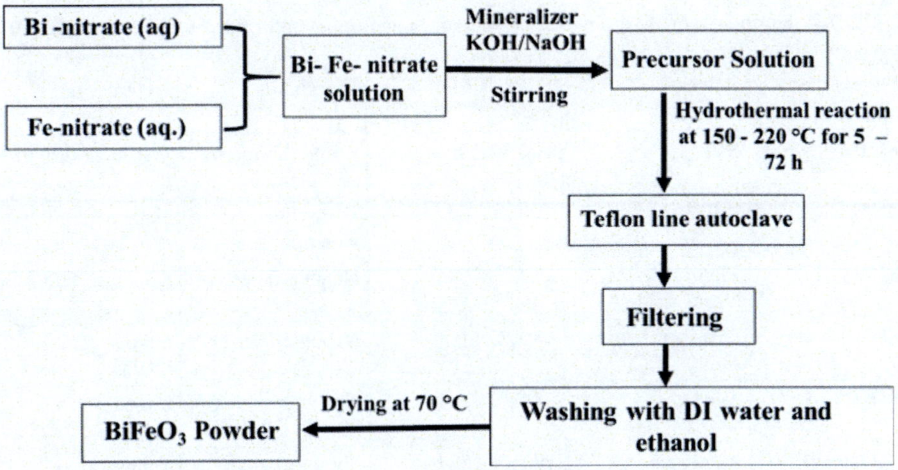

Fig. 7.5 Flow diagram for the hydrothermal synthesis of $BiFeO_3$

sample is dried at 80–150 °C for few hours to obtain the $BiFeO_3$ nanomaterials. A flow digram for the hydrothermal synthesis of $BiFeO_3$ is presented in Fig. 7.5. In this process, the bismuth and iron nitrates first changes into $Bi(OH)_3$ and $Fe(OH)_3$ precipitates in the precursor and further addition of the hydroxide mineralizer dissolves the $Bi(OH)_3$ and $Fe(OH)_3$ in hydrothermal conditions. The $BiFeO_3$ phase formation starts when the ion concentration in the alkaline solution surpasses the saturation point and precipitate from the supersaturated hydrothermal fluid (Zhang et al. 2016). A distinct advantage of the hydrothermal synthesis process over sol–gel methods is that nanocrystalline materials can be synthesized at much lower temperatures.

The phase and morphologies of the final $BiFeO_3$ phase shows dependence on mineralizer and reaction conditions. A summary of this is presented in Table 7.2.

7.4 Structural Characterization of $BiFeO_3$ Nanostructures

The following instrumental methods are used for structural and morphological characterization of the $BiFeO_3$ nanostructures.

7.4.1 X-ray Diffraction (XRD)

The XRD technique is used to identify the phase, crystal structure, and lattice parameters of a material. In this technique the diffracted X-ray intensity is measured as a function of diffraction angle 2θ. This X-ray diffraction pattern is used to identify the crystalline phases and to measure its structural properties. The crystallite size,

Table 7.2 Summary of hydrothermal synthetic conditions and morphologies of $BiFeO_3$ nanostructures

Mineralizer	Solvent	Synthesis temp /time	Morphology	References
KOH	PVP	200 °C/5 h	Cubic submicron particles	Gao et al. (2015)
KOH	PEG200	200 °C /6 h	Wafer like structure	Jiang et al. (2011)
KOH	TEA/DI water Nitric Acid	130 °C /24 h	Nanoparticles	Cho et al. (2008)
KOH	DI water, Nitric acid	180 °C 1 h	Nanoparticles	Glenda et al. (2011)
KOH	Acetone	180 °C /72 h	Nanowire	Liu et al. (2011b)
NaOH/PVP	Acetone	180 °C /72 h	Spindle, Cube. Plate- like structure	Yang et al. (2011)
NaOH	Acetone, Ammonia	180 °C /72 h	Nanocubes	Wang et al. (2015b)

D, of the materials can be calculated from the diffraction peak broadening using the Scherrer's formula as follows

$$D = \frac{K\lambda}{B\cos\theta_B} = \frac{0.9\lambda}{B\cos\theta_B}$$

where λ is the X-ray wavelength, *B* is the full width of half maximum (FWHM) of diffraction peak in "radians" at diffraction angle θ_B. K (= 0.9) is the Scherrer's constant.

The XRD pattern of rhombohedral $BiFeO_3$ with space group *R3c* is presented in Fig. 7.6. The diffraction peaks appear for the reflections from (012), (104), (113), (110), (006), (202), (024), (116), (122), (018), and (214) planes for diffraction angle 2θ between 20° and 60°. From the analysis of the XRD pattern, one can find the different lattice parameters for $BiFeO_3$ as $a = b = 5.57$, $c = 13.86$, and $V = 373$ Å^3 (Srivastav et al. 2013).

7.4.1.1 Scanning Electron Microscopy (SEM)

$BiFeO_3$ nanostructures can be synthesized in various morphologies using different synthesis methods for desired applications. SEM is one of the widely used technique for the characterization of shape, size, and surface morphology for $BiFeO_3$. As an example, the SEM images of rod, cube, and pill shaped $BiFeO_3$ nanostructures prepared by hydrothermal method are presented in Fig. 7.7 (Fei et al. 2011).

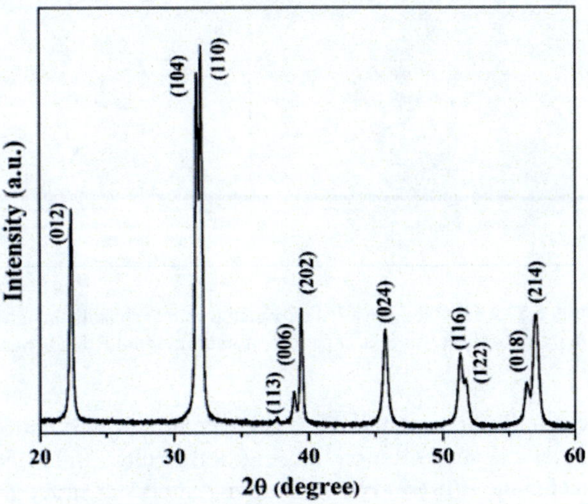

Fig. 7.6 XRD patterns of perovskite $BiFeO_3$ nanoparticles

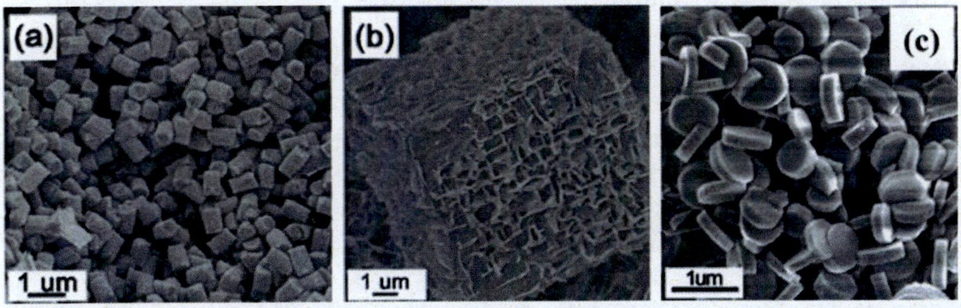

Fig. 7.7 SEM micrographs of **a** rods, **b** cubes, and, **c** pills of $BiFeO_3$ (Fei et al. 2011)

Figure 7.7a shows a well-defined $BiFeO_3$ rods having an average length of 1–2 μm and an average diameter of 0.5–1 μm. Figure 7.7b shows $BiFeO_3$ cubes having rough surfaces and an edge length ranges from 5–10 μm. Figure 7.7c shows $BiFeO_3$ pills having an average diameter ranges from 0.5–1 μm and an average thickness ranges 100–300 nm. Therefore, SEM characterization is efficient technique for understanding the surface morphologies of the prepared photocatalyst.

7.4.1.2 Transmission Electron Microscopy (TEM)

TEM technique is used to characterize the shape, size, morphology, crystallinity, and crystal structure of nanostructured $BiFeO_3$. Selected area electron diffraction (SAED) is a TEM technique in which electron beam is diffracted from the selected area of the sample to obtain diffraction patterns based on Bragg's law which offers unique

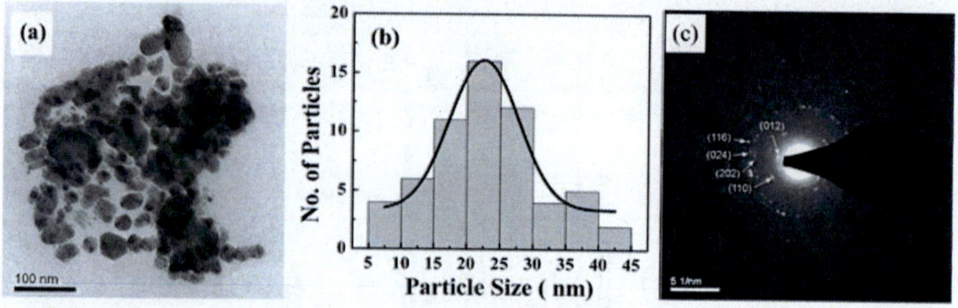

Fig. 7.8 **a** TEM image of $BiFeO_3$ nanoparticles, **b** statistic particle size distribution obtained from (**a**), and (**c**) Indexed SAED pattern of an individual $BiFeO_3$ nanoparticle

capability to determine the Bravais lattices and lattice parameters of the nanomaterials. High-resolution transmission electron microscopy (HRTEM) mode of TEM technique is used to direct imaging of crystal structure of a sample at atomic scale. HRTEM images are used to extract information about grain boundaries, interface formation, defects, stacking faults, and precipitates. As an example, TEM images in different condition are presented in Fig. 7.8 for perovskite $BiFeO_3$ nanoparticles synthesized by sol–gel method (Srivastav and Gajbhiye 2012; Srivastav et al. 2013).

The TEM image shown in Fig. 7.6a for perovskite-type $BiFeO_3$ nanoparticles (Fig. 7.8a) has spherical shaped morphology with particle sizes of 10–45 nm. The average particle size of the $BiFeO_3$ nanoparticles are obtained from statistic particle size distribution after lognormal fitting (Fig. 7.8b) and in this case, the average particle size is ~ 27 nm. The SAED pattern of $BiFeO_3$ nanoparticles is shown in Fig. 7.8c. The SAED pattern is consists of ring pattern which is characteristic of polycrystalline samples. These diffraction rings are indexed as (012), (110), (202), (024), (116), and (214) planes of rhombohedral distorted perovskite phase of $BiFeO_3$ nanoparticle which is in well agreement with the XRD results (Fig. 7.6) (Srivastav and Gajbhiye 2012). Therefore, TEM is very invaluable characterization technique for the nanostructured materials.

7.5 Photocatalytic Mechanism for the Degradation of Organic Pollutants

Photocatalysis mechanism of semiconductor photocatalyst involves a series of photochemical reactions and starts from the production of electron and hole ($e^- - h^+$) pairs in semiconductor material. The general model for photocatalysis process is illustrated in Fig. 7.9. When the light with energy greater or equal to that the band gap of the semiconducting material falls on it then light is absorbed and e^- –h^+ pairs are produced. These e^- –h^+ pairs migrate towards the surface of photocatalyst and redox reactions occur. The surface defects prevent recombination of these charge carriers by acting as trapping sites. The redox reactions take place with the compounds bounded on the

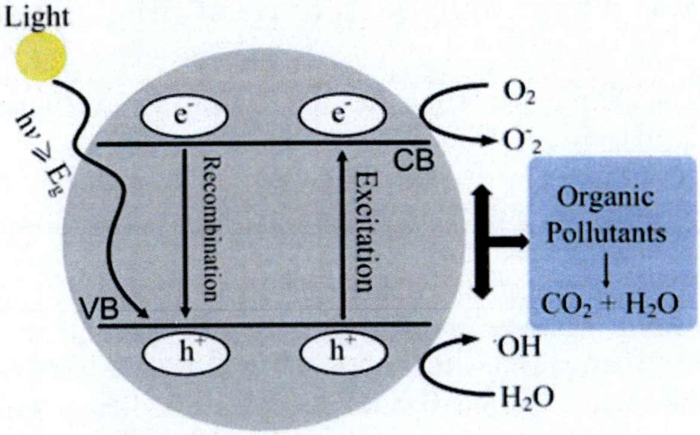

Fig. 7.9 The general model for photocatalysis process

surface of that catalyst. The hole oxidizes H_2O molecules to yield hydroxyl radicals ($^{\bullet}OH$) and electron reduces the dissolved oxygen in water to produce super oxide anion radicals (O_2^-). These $^{\bullet}OH$ and O_2^- species causes the redox reactions with organic molecules and degrades various organic contaminants. The O_2^- reacts with H^+ ions to produce more $^{\bullet}OH$ radicals (Yaqoob et al. 2020; Irfan et al. 2019).

The steps involved in the photocatalytic redox reactions of dye molecule are described as:

$$\text{Photocatalyst} + h\nu \rightarrow \underset{(CB)}{e^-} + \underset{(VB)}{h^+}$$
$$H_2O + h^+ \rightarrow {}^{\bullet}OH + H^+$$
$$O_2 + e^- \rightarrow O_2^-$$
$$OH + \text{dye} \rightarrow \text{oxidation process}$$
$$e^- + \text{dye} \rightarrow \text{reduction process}$$

Bismuth ferrite has a narrow optical band gap at room temperature ranging from ~2.2–2.8 eV(Lam et al. 2017; Irfan et al. 2019; Gao et al. 2015). This narrow band gap extends the absorption up to 750 nm i.e. visible region and also favours the separation of charge carriers. Due to ferroelectric nature of $BiFeO_3$, the photocurrent is generated by the depolarization of electric filed. This photocurrent facilitates the separation of the photogenerated charged carriers and consequently restrict the charge carrier recombination loss as in conventionally known semiconductors. The band gap of $BiFeO_3$ can be tuned by changing the processing temperature and this can be reduce from 2.5 eV at ambient temperature to about 1.5 eV at 550 °C (Palai et al. 2008). It has been also found that the position of CB, oxidation potential and VB, reduction potential are at about +0.44 and +2.60 V respectively (Fan et al. 2015; Niu et al. 2015). These uniqueness makes $BiFeO_3$ more favourable candidate for photocatalysis. Therefore, $BiFeO_3$ with narrow band gap and suitable band position meets the basic requirements for solar light induced photocatalytic process.

7.6 Photocatalytic Activity of $BiFeO_3$ Nanostructures

The $BiFeO_3$ nanostructures have shown an exceptional photocatalytic activity for degradation of various organic dyes as model organic pollutants. The photocatalytic performance of pure $BiFeO_3$ photocatalyst mainly depends on nano-structuring, surface area, photocatalyst loading, initial concentration of pollutant, pollutant type and light source (Lam et al. 2017; Irfan et al. 2019). The photocatalytic degradation of MO, for example, as a model organic pollutant for treating the wastewater from dyeing industries is discussed here for examining the photocatalytic performances of $BiFeO_3$ nanoparticles. The UV–vis diffuse reflectance spectrum of the $BiFeO_3$ nanoparticles synthesized by sol–gel method shows that the absorption cut-off wavelength is about 565 nm, suggests that $BiFeO_3$ nanoparticles can absorb visible light. The energy bandgap for $BiFeO_3$ nanoparticles is 2.18 eV. Taking methyl orange as organic pollutant, the photocatalytic degradation studies shows that the degradation rate of MO is less than 3% after 15 h without $BiFeO_3$ nanoparticles and more than 90% after 8 h with $BiFeO_3$ nanoparticles under UV–vis irradiation. This suggests that MO is stable under UV–vis irradiation without $BiFeO_3$ photocatalyst. More than 90% of MO decolorizes after 16 h under visible light irradiation with $BiFeO_3$ nanoparticles photocatalyst. Therefore, $BiFeO_3$ nanoparticles shows advantage over the normal photocatalyst TiO_2 because it can make use of visible light. The degradation rate of MO is about 70% after 16 h with bulk $BiFeO_3$ under UV–vis irradiation which implies that bulk $BiFeO_3$ is significantly less efficient than nanoparticles in similar condition (Gao et al. 2007). This clearly indicates that nano-structuring improves the photocatalytic performances of $BiFeO_3$.

7.6.1 Strategy to Improve the Photocatalytic Efficiency $BiFeO_3$

The $BiFeO_3$ material is not commercialized yet in photocatalytic technology due to its low photocatalytic performance relative to other commercially available materials (Lam et al. 2017; Irfan et al. 2019). The e^-–h^+ pairs formed after band-gap excitation are very short separated. Therefore, they recombine quickly and this results in a decrease of the quantum yield of the process. The efficiency and photocatalytic activity of a good photocatalyst are directly affected by many factors such as e^-–h^+ pairs separation and optical absorption properties i.e. electronic band structure, morphology, particle size, porosity, and surface area. Different strategy has been taken to enhance the efficiency as well as the activity of photocatalysis of $BiFeO_3$ such as;

(a) Effect of surface morphology
(b) Effect of doping and co-doping
(c) Effect of O_2 vacancies
(d) Formation of Heterojunction

(e) Formation nanocomposites with carbon materials

7.6.1.1 Effect of Surface Morphology

The photocatalytic activity of $BiFeO_3$ nanostructures has been improved by changing thc morphology (Wang et al. 2016a; Liu et al. 2015). $BiFeO_3$ nanostructured catalysts with ball like, honeycomb-like and flower-like morphologies have shown different photocatalytic activity toward Rhodamine B (RhB) pollutant. The band gaps for ball like, honeycomb-like and flower-like $BiFeO_3$ catalysts are 2.08, 2.06, and 1.93 eV, respectively. The specific surface area for ball like, honeycomb-like and flower-like $BiFeO_3$ nanostructures are 7.48, 3.58, and 12.38 m^2/g, respectively. The rate of degradation for RhB after 4 h is 49, 61, and 87% by ball like, honeycomb-like and flower-like $BiFeO_3$ catalysts, respectively. This increase in the rate of degradation of RhB is due to difference morphology and increased specific surface area of $BiFeO_3$ catalysts. Therefore, surface area of the photocatalyst is one of the key factors for improving the photocatalytic performances.

The cube-like, spindles-like, and plate-like morphologies of $BiFeO_3$ catalysts have shown different photocatalytic behaviour for MO (Wang et al. 2016a). The specific surface area of cubes, spindles, and plates is 0.526, 0.874, and 0.766 m^2/g, respectively and the rate of degradation of MO by cubes, spindles, and plates like $BiFeO_3$ catalyst in 3 h is found to be 38.7, 49.8, and 69.1%. The photocatalytic activity of plates like nanostructure is the highest among them although its specific surface area is not the maximum because in the plates like nanostructure, the incoming light faces more surface area. Therefore, more photons will be absorbed as a result large number of photo-generated charge carriers will be generated which significantly improves the photocatalytic activity of plat-like $BiFeO_3$ structure. This result suggests that the optical properties of the $BiFeO_3$ nanostructures are strongly related to their shape and exposed facet. The degradation rate of MB dye by mesh like $BiFeO_3$ nanostructure photocatalyst is ~98% within 4 h under sunlight irradiation. This is attributed to the band-bending which reduces recombination rate of charge carriers and improves photocatalytic performance(Bharathkumar et al. 2019). These results clearly suggest that morphology of $BeFeO_3$ photocatalyst has strong effect on efficiency and the photoactivity and this can be tuned by changing the morphology.

7.6.1.2 Effect of Doping

The doping is an efficient way to improve the photocatalytic performances by controlled addition of small impurity into photocatalysts. Addition of small quantity of dopant can reduce the recombination rate of charge carriers thereby enhances the photocatalytic activity and if the amount of dopant exceeds to certain critical limit then it may act as recombination centres for charge carriers, which can reduce the photocatalytic performance of photocatalyst. The dopant act like traps for photo-generated charge carriers as a result there is a decrease in the recombination rate and increase in lifetime of these charge carriers. Therefore, by proper selection of dopants, the photocatalytic activity can be improved by tuning the charge carrier recombination rates. The doping and co-doping of the Bi^{3+} and Fe^{3+} sites with transition metal, alkaline earth metal and rare earth metals have been investigated in order to improve the photocatalytic performances of $BiFeO_3$.

The Gd^{3+} doping on Bi^{3+} Site in $BiFeO_3$ shows enhancement for the photocatalytic degradation rate of RhB. The Gd^{3+} doping up to 10% increases the photocatalytic activity and further increase of Gd^{3+} concentration decreases the photocatalytic activity (Guo et al. 2010). The Dy-doped $BiFeO_3$ nanofibers shows enhanced photocatalytic efficiency under visible light compared to the pure phase for the degradation of methylene blue (Sakar et al. 2015). The Dy-doping modifies the band-gap of pure $BiFeO_3$ which helps in decrease of recombination rate of charge carriers. The pure $BiFeO_3$ degrades only 69% of MB dye but on doping with Sc^{3+}, it degrades MB dye completely within 3 h sunlight irradiation (Sakar et al. 2016). The co-doping of Nd and Ni for Bi^{3+} and Fe^{3+}-sites, respectively, into $BiFeO_3$ enhances the photocatalytic activity under visible-light by facilitating the charge transfer and decrease in the recombination rate of charge carriers (Vanga et al. 2015). The co-doping of La and Mn as well as co-doping of La and Sc into Bi^{3+} and Fe^{3+} site of $BiFeO_3$, respectively, enhances the photocatalytic activity by decreasing the recombination time due to significantly enhanced surface area with the large reduction of band-gap (Irfan et al. 2017a). These results clearly suggest that the photocatalytic activity of $BiFeO_3$ is very sensitive to the doping as well as co-doping.

7.6.1.3 Effect of Oxygen (O_2) Vacancies

It has been observed that light absorption capability of $BiFeO_3$ can be improved by the creation of oxygen vacancies into $BiFeO_3$ structure. The creation of proper amount of O_2 vacancies reduces the band gap by inducing O2 vacancy levels in the energy gap and also increases the charge mobility and charge separation effectively (Samadi et al. 2013; Verma et al. 2015; Tan et al. 2014). The oxygen vacancies can be created in the $BiFeO_3$ nanoparticles synthesized by high pressure hydrogenation process(Wang et al. 2016b). The amount of O_2 vacancies can be controlled by the hydrogenation temperature and the concentration of O_2 vacancies increases with increase of hydrogenation temperature. The hydrogenated $BiFeO_3$ samples shows enhanced absorption capability especially in the visible light region compared to the pristine $BiFeO_3$. The band gap decreases gradually on increasing hydrogenation temperature. The hydrogenated $BiFeO_3$ shows higher photocatalytic activity in comparison to the pristine $BiFeO_3$ for the degradation of MO under visible light irradiation. A similar behaviour is observed by 2D laminated cylinder-like nanostructured $BiFeO_3$ photocatalysts. This degrades RhB efficiently under visible light due to its cylinder-like shape and large numbers of oxygen vacancies (Gao et al. 2016). These studies revealed that the photocatalytic activity of $BiFeO_3$ photocatalyst can greatly improved by creation of oxygen vacancies.

7.6.1.4 Formation of Heterojunction

In the photocatalytic reaction, heterojunction can be formed by the overlapping of the band gaps of different semiconducting materials. The photocatalyst with heterojunction such as, $TiO_2/BiFeO_3$, $Fe_2O_3/BiFeO_3$, g-$C_3N_4/BiFeO_3$, and $SrTiO_3/BiFeO_3$ shows improved photocatalytic performance using visible light (Wang et al. 2015b; Liu et al. 2015; Humayun et al. 2016; Luo and Maggard 2006). This is due to the formation of Schottky barrier at the heterojunction which inhibits the movement of $e^- - h^+$ pairs from semiconductor to semiconductor and reduces the charge carrier recombination rate and thus increases the lifetime of e^- –h^+ pairs. Therefore, formation of heterojunction can significantly improve the photoactivity of pristine $BiFeO_3$.

Pt-$BiFeO_3$ heterostructure photocatalyst shows a superior photocatalytic activity for the degrading of MO and it degrades MO five times faster than pure $BiFeO_3$ in visible light. This shows that the formation of heterojunction between the Pt and $BiFeO_3$ is favourable for production of charge carriers and electronic interaction at the photocatalyst interface (Niu et al. 2015). The nanocomposites of Ag and Au with $BiFeO_3$ shows enhanced photocatalytic activity for degradation of RhB under visible light irradiation (Zhang et al. 2015). These results inferred that the formation of heterojunction is efficient method for improving the photocatalytic performance of $BiFeO_3$ nanostructures.

7.6.1.5 Formation Nanocomposites with Carbon Materials

The formation of nanocomposite of $BiFeO_3$ with carbon based materials can also improve the photocatalytic performances of $BiFeO_3$ because of the formation of the junctions between these two materials result into the modification of band gap which hinders the e^- –h^+ pair recombination process as well as enhances the pollutants adsorption capacity on the composite thereby leads to the improvement of photoactivity of $BiFeO_3$ (Wang et al. 2015b; Fan et al. 2015; Li et al. 2019, 2013). $BiFeO_3$ shows extraordinary photocatalytic activity toward organic pollutants when coupled with graphitic carbon nitride (g-C_3N_4) and graphene (GR) (Wang et al. 2015b; Fan et al. 2015; Li et al. 2019, 2013). The coupling between $BiFeO_3$ and graphene takes place through formation of Fe–O–C bonds which facilitates the adsorption of –OH groups on the graphene surface. In addition to this the improvement in photocatalytic performances of graphene– $BiFeO_3$ nanocomposites is due to π–π stacking process on the surface of graphene provided by the large p-conjugation plane of graphene which facilitates large absorption of organic pollutants.

7.7 Reusability and Stability of $BiFeO_3$ Photocatalyst

The reusability and stability are the important factors which determines practical applicability of any photocatalyst. Pure $BiFeO_3$ nanostructure shows magnetic properties due to this $BiFeO_3$ photocatalyst can be easily collected after photocatalytic process from the solution. For example, the structure of $BiFeO_3$ microsphere is nearly same after five cyclic runs and can be removed easily from the solution (Yuning et al. 2010). 3D mesoporous $BiFeO_3$ photocatalyst can be recycled three times without changing its structure (Papadas et al. 2015). Sm–and Mn–co-doped $BiFeO_3$ photocatalysts can be recycled after four cyclic runs (Irfan et al. 2017b). Similarly, Gd and Sn- co-doped $BiFeO_3$ nanoparticles photocatalyst shows excellent stability in visible, UV, and NIR regions of lights even after four cyclic runs (Irfan et al. 2017c). These studies clearly show the reusability and stability of $BiFeO_3$ nanostructure as an efficient photocatalyst.

7.8 Conclusions and Future Prospects

This chapter summerized the $BiFeO_3$ crytal sturucture, sysntheis, photocatalysts activity, degradation mechanism, and how to improve the photocatalytic activites. $BiFeO_3$ has a narrow optical band gap at room temperature ranging from $\sim$2.2 – 2.8 eV(Lam et al. 2017; Irfan et al. 2019; Gao et al. 2015), extends the absorption up to visible region and also favours the separation of charge carriers. The ferroelectric nature of $BiFeO_3$ restrict the charge carrier recombination loss as seen in conventionally known semiconductors. The band gap of $BiFeO_3$ can be tuned from 2.5 at to 1.5 eV by changing temerpature. It has been also found that the position of CB, oxidation potential and VB, reduction potential are at about + 0.44 V and + 2.60 V respectively. These uniqueness makes $BiFeO_3$ more favourable candidate for photocatalysis. Therefore, $BiFeO_3$ narrow band gap and suitable band position meets the basic requirements for visible light photocatalytic process.

Doping and formation of nanocomposite of $BiFeO_3$ with metals and carbon based materials can also improve the photocatalytic performances of $BiFeO_3$. The formation of the junctions between these two atoms and materials result into the modification of band gap which hinders the e^- –h^+ pair recombination process as well as enhances the pollutants adsorption capacity on the composite thereby leads to the improvement of photoactivity of $BiFeO_3$. $BiFeO_3$ shows extraordinary photocatalytic activity toward organic pollutants when coupled with graphitic carbon nitride (g-C_3N_4), graphene (GR), TiO_2/$BiFeO_3$, and Fe_2O_3/$BiFeO_3$. $BiFeO_3$ has shown huge potential for contaminenets removel form contaminated wastewater. It has shown extraordinary results with modification for degradation of organic pollutant can compete with exisiting commercial catalysist.

References

Achenbach GD, James WJ, Gerson R (1967) Preparation of single-phase polycrystalline $BiFeO_3$. J Am Ceram Soc 50:437

Bai X, Wei J, Tian B, Liu Y, Reiss T, Guiblin N, Gemeiner P, Dkhil B, Infante IC (2016) Size effect on optical and photocatalytic properties in $BiFeO_3$ nanoparticles. J Phys Chem C 120:3595–3601

Bharathkumar S, Sakar M, Balakumar S (2019) Fabrication of $BiFeO_3$ nanostructures and their visible light photocatalytic degradation and water splitting properties. AIP Conf Proceed 2115:030167

Catalan G, Scott JF (2009) Physics and applications of bismuth ferrite. Adv Mater 21:2463

Chang J, Zhang L, Wang P (2018) Intelligent environmental nanomaterials. Environ Sci NANO 5:811–836

Chen D, Wang Q, Wang R, Shen G (2015) Ternary oxide nanostructured materials for supercapacitors: a review. J Mater Chem A 3:10158–10173

Cho CM, Noh JH, Cho I-S, An J-S, Hong KS, Kim JY (2008) Low-temperature hydrothermal synthesis of pure $BiFeO_3$ nanopowders using triethanolamine and their applications as visible-light photocatalysts. J Am Ceram Soc 91:3753–3755

Chu Y, Martin L, Holcomb M, Ramesh R (2007) Controlling magnetism with multiferroics. Mater Today 10:16

Fan T, Chen C, Tang Z, Ni Y, Lu C (2015) Synthesis and characterization of g-C3N4/$BiFeO_3$ composites with an enhanced visible light photocatalytic activity. Mater Sci Semicond Process 40:439–445

Fei L, Yuan J, Hu Y, Wu C, Wang J, Wang Y (2011) Visible light responsive perovskite $BiFeO_3$ pills and rods with dominant facets. Cryst Growth Des 11:1049–1053

Fu F, Wang Q (2011) Removal of heavy metal ions from wastewaters: a review. J Environ Manage 92:407–418

Gao F, Chen XY, Yin KB, Dong S, Ren ZF, Yuan F, Yu T, Zou ZG, Liu JM (2007) Visible-light photocatalytic properties of weak magnetic $BiFeO_3$ nanoparticles. Adv Mater 19:2889

Gao T, Chen Z, Niu F, Zhou D, Huang Q, Zhu Y, Qin L, Sun X, Huang Y (2015) Shape-controlled preparation of bismuth ferrite by hydrothermal method and their visible-light degradation properties. J Alloy Compd 648:564–570

Gao X, Dai Y, Fu F, Hua X (2016) 2D laminated cylinder-like $BiFeO_3$ composites: hydrothermal preparation, formation mechanism, and photocatalytic properties. Solid State Sci 62:6–12

Ghosh S, Dasgupta S, Sen A, Maiti HS (2005) Low-temperature synthesis of nanosized bismuth ferrite by soft chemical route. J Am Ceram Soc 88:1349

Glenda B, Simões AZ, Foschini CR, Antônio SG, Zaghete MA, A., Varela J (2011) A novel synthesis of perovskite bismuth ferrite nanoparticles. Process Appl Ceramics 5

Guo R, Fang L, Dong W, Zheng F, Shen M (2010) Enhanced photocatalytic activity and ferromagnetism in Gd doped $BiFeO_3$ nanoparticles. J Phys Chem C 114:21390

Herrmann J-M (1999) Heterogeneous photocatalysis: fundamentals and applications to the removal of various types of aqueous pollutants. Catal Today 53:115–129

Humayun M, Zada A, Li Z, Xie M, Zhang X, Qu Y, Raziq F, Jing L (2016) Enhanced visible-light activities of porous $BiFeO_3$ by coupling with nanocrystalline TiO_2 and mechanism. Appl Catal B 180:219–226

Irfan S, Shen Y, Rizwan S, Wang H-C, Khan SB, Nan C-W (2017b) Band-gap engineering and enhanced photocatalytic activity of Sm and Mn doped BiFeO3 nanoparticles. J Am Ceram Soc 100:31–40

Irfan S, Rizwan S, Shen Y, Li L, Asfandiyar, Butt S, Nan C-W (2017) The Gadolinium (Gd3+) and Tin (Sn4+) Co-doped BiFeO3 nanoparticles as new solar light active photocatalyst. Scientific Reports 7:42493

Irfan S, Li L, Saleemi AS, Nan C-W (2017a) Enhanced photocatalytic activity of La3+ and Se4+ co-doped bismuth ferrite nanostructures. J Mater Chem A 5:11143–11151

Irfan S, Zhuanghao Z, Li F, Chen Y-X, Liang G-X, Luo J-T, Ping F (2019) Critical review: bismuth ferrite as an emerging visible light active nanostructured photocatalyst. J Market Res 8:6375–6389
Jiang J, Zou J, Anjum MN, Yan J, Huang L, Zhang Y, Chen J (2011) Synthesis and characterization of wafer-like $BiFeO_3$ with efficient catalytic activity. Solid State Sci 13:1779–1785
Kanhere P, Chen Z (2014) A review on visible light active perovskite-based photocatalysts. Molecules 19
Khin MM, Nair AS, Babu VJ, Murugan R, Ramakrishna S (2012) A review on nanomaterials for environmental remediation. Energy Environ Sci 5:8075–8109
Kumar MM, Palkar VR, Srinivas K, Suryanarayana SV (2000) Ferroelectricity in a pure $BiFeO_3$ ceramic. Appl Phys Lett 76:2764
Lam S-M, Sin J-C, Abdullah AZ, Mohamed AR (2012) Degradation of wastewaters containing organic dyes photocatalysed by zinc oxide: a review. Desalin Water Treat 41:131–169
Lam S-M, Sin J-C, Mohamed AR (2017) A newly emerging visible light-responsive $BiFeO_3$ perovskite for photocatalytic applications: a mini review. Mater Res Bull 90:15–30
Lazenka VV, Zhang G, Vanacken J, Makoed II, Ravinski AF, Moshchalkov VV (2012) Structural transformation and magnetoelectric behaviour in Bi 1−x Gd x FeO_3 multiferroics. J Phys D: Appl Phys 45:125002
Liu T, Xu Y, Feng S, Zhao J (2011a) A Facile Route to the synthesis of $BiFeO_3$ at low temperature. J Am Ceram Soc 94:3060
Liu B, Hu B, Du Z (2011b) Hydrothermal synthesis and magnetic properties of single-crystalline $BiFeO_3$ nanowires. Chem Commun (camb) 47:8166–8168
Liu G, Wang T, Zhou W, Meng X, Zhang H, Liu H, Kako T, Ye J (2015) Crystal-facet-dependent hot-electron transfer in plasmonic-Au/semiconductor heterostructures for efficient solar photocatalysis. J Mater Chem C 3:7538–7542
Li Z, Shen Y, Yang C, Lei Y, Guan Y, Lin Y, Liu D, Nan C-W (2013) Significant enhancement in the visible light photocatalytic properties of BiFeO3–graphene nanohybrids. J Mater Chem A 1:823–829
Li YA, Li J, Chen L, Sun H, Zhang H, Guo H, Feng L (2019) In situ synthesis of au-induced hierarchical nanofibers/nanoflakes structured $BiFeO_3$ homojunction photocatalyst with enhanced photocatalytic activity. Front Chem 6
Li J, Wang Y, Ling H, Qiu Y, Lou J, Hou X, Bag SP, Wang J, Wu H, Chai G (2019) Significant enhancement of the visible light photocatalytic properties in 3D $BiFeO_3$/graphene composites. Nanomaterials 9:65
Luo J, Maggard PA (2006) Hydrothermal synthesis and photocatalytic activities of $SrTiO_3$-Coated Fe_2O_3 and $BiFeO_3$. Adv Mater 18
Lu H, Wang J, Stoller M, Wang T, Bao Y, Hao H (2016) An overview of nanomaterials for water and wastewater treatment. Adv Mater Sci Eng 2016:4964828
Maeda K (2011) Photocatalytic water splitting using semiconductor particles: history and recent developments. J Photochem Photobiol C 12:237–268
Mao Y, Park T-J, Wong SS (2005) Synthesis of classes of ternary metal oxide nanostructures. Chem Commun 5721–5735
Mukherjee JL, Wang FY (1971) Kinetics of solid-state reaction of Bi_2O_3 and Fe_2O_3. J Am Ceram Soc 54:31
Niu F, Chen D, Qin L, Gao T, Zhang N, Wang S, Chen Z, Wang J, Sun X, Huang Y (2015) Synthesis of Pt/BiFeO3 heterostructured photocatalysts for highly efficient visible-light photocatalytic performances. Sol Energy Mater Sol Cells 143:386–396
Palai R, Katiyar RS, Schmid H, Tissot P, Clark S, Robertson J, Redfern S, Catalan G, Scott J (2008) β phase and γ-β metal-insulator transition in multiferroic $BiFeO_3$. Phys Rev B 77:014110
Papadas IT, Subrahmanyam KS, Kanatzidis MG, Armatas GS (2015) Templated assembly of BiFeO3 nanocrystals into 3D mesoporous networks for catalytic applications. Nanoscale 7:5737–5743
Peng S-H, Wang W-X, Li X, Yen Y-F (2004) Metal partitioning in river sediments measured by sequential extraction and biomimetic approaches. Chemosphere 57:839–851

Popa M, Crespo D, Calderon-Moreno JM, Preda S, Fruth V (2007) Synthesis and structural characterization of single-phase $BiFeO_3$ powders from a polymeric precursor. J American Ceramic Soc 90:2723

Sakar M, Balakumar S, Saravanan P, Bharathkumar S (2015) Compliments of confinements: substitution and dimension induced magnetic origin and band-bending mediated photocatalytic enhancements in $Bi1-xDyxFeO_3$ particulate and fiber nanostructures. Nanoscale 7:10667–10679

Sakar M, Balakumar S, Saravanan P, Bharathkumar S (2016) Particulates versus fibers: dimension featured magnetic and visible light driven photocatalytic properties of Sc modified multiferroic bismuth ferrite nanostructures. Nanoscale 8:1147–1160

Samadi M, Shivaee HA, Pourjavadi A, Moshfegh AZ (2013) Synergism of oxygen vacancy and carbonaceous species on enhanced photocatalytic activity of electrospun ZnO-carbon nanofibers: charge carrier scavengers mechanism. Appl Catal A 466:153–160

Selbach SM, Einarsrud M-A, Tybell T, Grande T (2007) Synthesis of $BiFeO_3$ by wet chemical methods. J Am Ceram Soc 90:3430

Shaheen K, Suo H, Arshad T, Shah Z, Khan SA, Khan SB, Khan MN, Liu M, Ma L, Cui J, Ji YT, Wang Y (2020) Metal oxides nanomaterials for the photocatalytic mineralization of toxic water wastes under solar light illumination. J Water Process Eng 34:101138

Shi J, Guo L (2012) ABO3-based photocatalysts for water splitting. Progress in Natural Sci: Mater Int 22:592–615

Silva J, Reyes A, Esparza H, Camacho H, Fuentes L (2011) $BiFeO_3$: a review on synthesis, doping and crystal structure. Integr Ferroelectr 126:47

Singh A, Sharma RK, Agrawal M, Marshall FM (2010) Health risk assessment of heavy metals via dietary intake of foodstuffs from the wastewater irrigated site of a dry tropical area of India. Food Chem Toxicol 48:611–619

Sin J-C, Lam S-M, Mohamed AR, Lee K-T (2012) Degrading endocrine disrupting chemicals from wastewater by TiO_2 photocatalysis: a review. Int J Photoenergy 185159

Srivastav SK, Gajbhiye NS (2012) Low temperature synthesis, structural, optical and magnetic properties of bismuth ferrite nanoparticles. J Am Ceram Soc 95:3678–3682

Srivastav SK, Gajbhiye NS, Banerjee A (2013) Structural transformation and enhancement in magnetic properties of single-phase $Bi1-xPrxFeO_3$ nanoparticles. J. Appl. Phys. 113:203917

Tabares-Munoz C, Rivera JP, Schmid H (1984) Ferroelectric domains, birefringence and absorption of single crystals of $BiFeO_3$. Ferroelectrics 55:235

Tan H, Zhao Z, Zhu W-B, Coker EN, Li B, Zheng M, Yu W, Fan H, Sun Z (2014) Oxygen vacancy enhanced photocatalytic activity of pervoskite $SrTiO_3$. ACS Appl Mater Interfaces 6:19184–19190

Vanga PR, Mangalaraja RV, Ashok M (2015) Effect of (Nd, Ni) co-doped on the multiferroic and photocatalytic properties of $BiFeO_3$. Mater Res Bull 72:299–305

Verma R, Samdarshi SK, Bojja S, Paul S, Choudhury B (2015) A novel thermophotocatalyst of mixed-phase cerium oxide (CeO_2/$Ce2O_3$) homocomposite nanostructure: Role of interface and oxygen vacancies. Sol Energy Mater Sol Cells 141:414–422

Wang YP, Zhou L, Zhang MF, Chen XY, Liu JM, Liu ZG (2004) Room-temperature saturated ferroelectric polarization in $BiFeO_3$ ceramics synthesized by rapid liquid phase sintering. Appl Phys Lett 84:1731

Wang X, Zhang YG, Wu Z (2010) Magnetic and optical properties of multiferroic bismuth ferrite nanoparticles by tartaric acid-assisted sol–gel strategy. Mater Lett 64:486–488

Wang X, Mao W, Zhang J, Han Y, Quan C, Zhang Q, Yang T, Yang J, Li XA, Huang W (2015) Facile fabrication of highly efficient g-C3N4/BiFeO3 nanocomposites with enhanced visible light photocatalytic activities. J Colloid Interface Sci 448:7–23

Wang W, Tadé MO, Shao Z (2015a) Research progress of perovskite materials in photocatalysis- and photovoltaics-related energy conversion and environmental treatment. Chem Soc Rev 44:5371–5408

Wang X, Mao W, Zhang Q, Wang Q, Zhu Y, Zhang J, Yang T, Yang J, Li XA, Huang W (2016) PVP assisted hydrothermal fabrication and morphology-controllable fabrication of $BiFeO_3$ uniform nanostructures with enhanced photocatalytic activities. J Alloys Compounds 677:288–293

Wang S, Chen D, Niu F, Zhang N, Qin L, Huang Y (2016b) Hydrogenation-induced surface oxygen vacancies in $BiFeO_3$ nanoparticles for enhanced visible light photocatalytic performance. J Alloy Compd 688:399–406

Wei J, Xue D (2008) Low-temperature synthesis of $BiFeO_3$ nanoparticles by ethylenediaminetetraacetic acid complexing sol–gel process. Mater Res Bull 43:3368

Yang J, Li X, Zhou J, Tang Y, Zhang Y, Li Y (2011) Factors controlling pure-phase magnetic $BiFeO_3$ powders synthesized by solution combustion synthesis. J Alloy Compd 509:9271–9277

Yaqoob AA, Parveen T, Umar K, Mohamad Ibrahim MN (2020) Role of nanomaterials in the treatment of wastewater: a review. Water 12:495

Yin L, Mi W (2020) Progress in BiFeO3-based heterostructures: materials properties and applications. Nanoscale 12:477–523

Yongming H, Linfeng F, Yiling Z, Jikang Y, Yu W, Haoshuang G (2011) Synthesis of bismuth ferrite nanoparticles via a wet chemical route at low temperature. J Nanomater

Yuning H, Yi J, Ya Z (2010) Citric acid assisted solvothermal synthesis of BiFeO3 microspheres with high visible-light photocatalytic activity. J Molecular Catalysis A: Chem 331

Yunus IS, Harwin, Kurniawan A, Adityawarman D, Indarto A (2012) Nanotechnologies in water and air pollution treatment. Environ Technol Rev 1:136–148

Zhang J, Gondal MA, Wei W, Zhang T, Xu Q, Shen K (2012) Preparation of room temperature ferromagnetic $BiFeO_3$ and its application as an highly efficient magnetic separable adsorbent for removal of Rhodamine B from aqueous solution. J Alloy Compd 530:107–110

Zhang X, Wang B, Wang X, Xiao X, Dai Z, Wu W, Zheng J, Ren F, Jiang C (2015) Preparation of $M@BiFeO_3$ nanocomposites ($M = Ag, Au$) bowl arrays with enhanced visible light photocatalytic activity. J Am Ceram Soc 98:2255–2263

Zhang Q, Sando D, Nagarajan V (2016) Chemical route derived bismuth ferrite thin films and nanomaterials. J Mater Chem C 4:4092–4124

Chapter 8
Photocatalysis for the Removal of Environmental Contaminants

Sukanya Krishnan, Ansaf V. Karim, and Amritanshu Shriwastav

Abstract The presence of toxic organic pollutants in environmental matrices is a serious concern to humans and other living organisms. The majority of these contaminants are recalcitrant in nature, because of which conventional treatment technologies are inefficient in their removal. Photocatalytic oxidation processes which produce highly reactive oxygen species are capable of complete mineralization of these organic compounds. The development of highly efficient visible light active and cost-effective photocatalysts remains the key challenge among the researchers. Therefore, currently, the researchers are emphasizing more on the development of various surface modification techniques of catalysts and processes necessary for improving the photocatalytic efficiency. This chapter gives a comprehensive review focusing on the basics of photocatalysis, its mechanism, and different surface modification techniques of catalyst for improving the process efficiency. It also discusses the photoreduction process for inorganic compounds removal and the factors affecting the photocatalysis process. Further, the chapter also reviews the various researches conducted in the field for photocatalytic oxidation of organic compounds.

Keywords Heterogeneous photocatalysis · Visible light activity · Surface modification · Degradation · Organic Pollutant

8.1 Introduction

The increasing occurrence of organic pollutants in aquatic bodies may pose a serious risk to the living organisms. Most of these contaminants are originated from human activities and being released knowingly or unknowingly into the aquatic environment. The biodegradation of these pollutants are relatively slow and the conventional wastewater treatment methods, such as coagulation-flocculation, adsorption,

S. Krishnan · A. V. Karim · A. Shriwastav (✉)
Environmental Science and Engineering Department, Indian Institute of Technology Bombay, Mumbai 400 076, India
e-mail: amritan@iitb.ac.in

S. P. Singh et al. (eds.), *Nanomaterials and Nanocomposites for Environmental Remediation*, Energy, Environment, and Sustainability,
https://doi.org/10.1007/978-981-16-3256-3_8

are insufficient in achieving complete mineralization of these pollutants, and only transfer them from one phase to another without destroying them.

A significant improvement in wastewater treatment was observed in the last few decades with Advanced Oxidation Processes (AOP) which overcome the potential limitations of available conventional treatment methods. The effectiveness of AOPs is proportional to the in-situ generation of unselective and powerful Hydroxyl radicals (·OH) having a higher oxidation potential. The way of choosing appropriate AOP depends on the characteristics of the wastewater, environmental regulations, and cost-effectiveness of the treatment. Among the different AOPs, photolytic oxidation processes are effective for the mineralization of refractory organic compounds. Photolysis is the process by which a chemical change occurs on a species due to the absorption of photons (Avisar et al. 2010). The efficiency of the process depends upon the ability of the pollutant to absorb photons, the intensity of light, presence of oxidants or catalysts in the solution, pH, etc. Heterogeneous photocatalysis using semiconductor materials such as TiO_2 and ZnO have been widely used by the researchers due to their superior photocatalytic activity under UV light irradiation. Semiconductors are characterized by a fully occupied valence band and a vacant conduction band. The degradation of an organic compound by photocatalyst involves the absorption of a photon of energy greater than the bandgap energy of the catalyst to form electron–hole pairs (Dong et al. 2015b; Yap et al. 2011). The valence band holes are also powerful oxidants (Bahnemann 2004).

The key advantage of the photocatalysis process is that it can lead to the complete mineralization of organic compounds under ambient conditions without involving any mass transfer (Khataee and Kasiri 2010). Photocatalytic processes can maximize the production of reactive oxygen species for the degradation of a variety of aliphatic or aromatic compounds under suitable conditions (Son et al. 2009). The recombination of electron–hole pairs and lack of visible light activity due to wide bandgap are the major limitations of semiconductor photocatalysis. It can be improved by doping with metal/non-metal ions, dye sensitization, coupling with semiconductors or hybrid materials, etc. (Pelaesz et al. 2012; Le et al. 2017). Even though the photocatalytic degradation process is efficient for the degradation of organic compounds, the separation of catalyst from aqueous solution after the treatment is a major hurdle. To overcome these limitations and to improve the photocatalytic performance, solid substrates can be used to support the photocatalyst. The selection of catalyst depends upon its ability to resist agglomeration, robustness, availability, and cost. This chapter provides an overview of the photocatalysis process and degradation mechanisms of organic compounds. The chapter is more focused on heterogenous photocatalysis based on semiconductor materials. Further, the chapter gives an idea about the different techniques used for improving the visible light activity of the catalyst and their photocatalytic application in organic compound removal.

8.2 Photolytic Oxidation Process

Photo-degradation of pollutants in an aqueous solution occurs by the absorption of light which introduces sufficient energy to break the bonding by overcoming the activation energy (Pelaez et al. 2012). In direct photolysis process, the compound itself absorbs photons resulting in the excitation and undergoes a chemical transformation. When the atoms or molecules absorb enough radiant energy from the light sources, they reach an excited (or activated) state where several photochemical processes occur. As a result of excitation, atoms may undergo photodissociation, or intramolecular rearrangement, or reaction with other molecules. The indirect photolysis process involves the generation of powerful radicals such as hydroxyl or singlet oxygen due to the absorption of light. The removal of compounds by photochemical reactions is further featured by criteria such as quantum yield, molar absorption, and radical reactions. The molar absorption coefficient is the ability of a chemical species to absorb a given wavelength of light, while quantum yield is the ratio of the number of molecules participating in each photochemical process upon the number of photons absorbed (Sanches et al. 2010; Wols and Hofman-Caris 2012). There are different factors which affect the photolytic oxidation of organic compounds.

8.2.1 Effect of pH

pH is a vital factor which determines the surface charge properties of the catalyst. The variation in the pH of the solution has an impact on the surface charge of the catalyst, thereby causing a change in adsorption and reaction rate (Rauf and Ashraf 2009). The degradation of a pollutant occurs in three ways; by the attack of hydroxyl radicals ($^{\bullet}OH$), oxidation by holes formed in the valence band, and reduction by the electron in the conduction band depending upon the pH of the aqueous solution and nature of the contaminant (Elmolla and Chaudhuri 2010). Therefore, to understand the pH effect on the photocatalytic degradation, the pH for a photocatalyst at which the surface charge density becomes zero (point of zero charges (PZC)) must be found out. If the pH of the solution is less than pH_{pzc}, then the catalyst surface charge will be positive. Whereas the catalyst surface charge will be negative if the pH of the solution is more than pH_{pzc} (Bora and Mewada 2017). Therefore, it affects the pollutant adsorption on the catalyst surface thereby affecting the photocatalytic treatment efficiency. For example, Chekir et al. (2016) reported that alkaline pH is favourable for the degradation of methylene blue by TiO_2 catalyst (point of zero charge around pH 6) under UV light irradiation due to the attraction of cationic methylene blue and negatively charged TiO_2 surface.

Besides, the charge variation of the catalyst with pH of the solution is mainly due to the protonation and deprotonation of the catalyst (eg. TiO_2), resulting in the formation of $TiOH_2^+$ (positive), TiOH (neutral), and TiO^- (negative) as shown in Eqs. 8.1 and 8.2 (Chiou et al. 2008).

$$\mathrm{TiOH} + \mathrm{H}^{+} \rightarrow \mathrm{TiOH}_2^{+} \tag{8.1}$$

$$\mathrm{TiOH} + \mathrm{OH}^{-} \rightarrow \mathrm{TiO}^{-} + \mathrm{H_2O} \tag{8.2}$$

Therefore, the holes (h^+) are the predominant oxidation species at acidic pH, Whereas the hydroxyl radical ($^{\bullet}OH$) predominates at alkaline or neutral pH since the hydroxide ions will get oxidized with holes (Ani et al. 2018). Also, the pH of aqueous solutions influences the molecular structure of compounds. Studies have demonstrated that the pH of the solution is related to the molar absorption coefficient and photolysis quantum yield. The change in pH of the solution can also influence the electron distribution in the target contaminant molecule, the absorbance of light, and hence the degradation potential (Avisar et al. 2010). Chiou et al. (2008) reported that the phenol having a pKa value of 9.95 can be in its non-ionic form under acidic to neutral pH whereas exists as phenolates anions at higher alkaline pH. The study by Nguyen et al. (2018) observed that maximum degradation of Methylene blue occurs at pH 10 due to enhanced adsorption of cationic MB on negatively charged Pd/TiO_2 surface ($pH_{pzc} = 6.7$). Another study by Nasr et al. (2019) showed maximum degradation of acetaminophen (AP) at acidic pH, because of the electrostatic attraction between AP (pKa = 9.5) and positively charged TiO_2 (pHpzc = 6.3). In another study, the presence of Cl^- inhibited the production of $^{\bullet}OH$ radicals by forming Fe(III)-chloride complex at acidic pH during the UV/H_2O_2/Fe(II) process for the degradation of pharmaceuticals wastewater (Monteagudo et al. 2014). In UV/H_2O_2 system, the alkaline pH is less preferred due to the formation of hydroperoxide anion (HO_2^-). Since the reaction of $^{\bullet}OH$ with HO_2^- is 100 times faster than reaction with H_2O_2, it can act as a scavenger for the radical. The stability of H_2O_2 at alkaline pH is comparatively less, which results in its self-decomposition and loses its oxidant characteristics (Deng et al. 2013).

8.2.2 *Effect of Catalyst Dose*

The catalyst dose is a major factor in photocatalysis treatment since it controls the formation of electrons-holes and other reactive oxygen species (Bechambi et al. 2015). It is observed that increasing the catalyst loading enhances the generation of electron–hole pairs and which ultimately leads to the formation of more hydroxyl and superoxide radicals. Many researchers reported that the initial rates of photo-degradation of pollutants were found to be directly proportional to catalyst concentration. Whereas after a certain limit of catalyst concentration, the rate of photocatalysis could even decrease (Akpan and Hameed 2009). This is because as the concentration of catalyst increases above the optimum value the degradation rate decreases due to the unfavourable light scattering and reduction of light penetration into the solution (Gaya and Abdullah 2008). For example, Bechambi et al. (2015) studied the effect of copper doped ZnO catalyst loading towards the degradation of Bisphenol A (BPA) by differing the amount of the catalyst from 0.5 to 2 g L^{-1}. The optimum loading is found to be 1 g L^{-1} beyond that the degradation of BPA decreases due to the screening effect and decrease in specific surface area with particle agglomeration. Similarly, another study by Nguyen et al. (2018) reported a substantial enhancement in the degradation of methylene blue dye with an increase in the catalyst Pd-TiO_2 concentration from 0.5 to 1 g L^{-1} and then lowering with further increase up to 1.5 g L^{-1}. Also, the apparent rate constant decreases after the optimum dosage of catalyst concentration.

8.2.3 *Effect of a Light Source*

The sources of light for photolysis studies include low pressure (LP) and medium pressure (MP) mercury lamps, Xenon arc lamp simulating sunlight, and LEDs (Lin et al. 2016). For the successful removal of pollutants using photolysis, the emission spectra of the light should overlap with the absorption spectra of the pollutant (Miller and Olejnik 2001). LP lamps have a higher energy conversion of 30–38% in the UV range of while the MP lamps have a comparatively lower conversion of 10–20% (Sichel et al. 2011). Low-pressure (LP) UV lamps (Monochromatic source -254 nm) are being used as a potential disinfectant of water whereas Medium-pressure (MP) lamps (polychromatic source 200–400 nm) had shown noticeable potential for degradation of organic pollutants (Pereira et al. 2007). The application of artificial light sources in photochemical processes has proven to be efficient for pollutant removal. Use of UV radiations for contaminant removal has the advantages such as no chemical addition, less sludge generation which are odourless, less pH and temperature dependence, no requirement of any adsorbent media, and mostly acceptability among the public (Shayeghi et al. 2012). However, UV lamps have certain limitations related to high-power consumption, cost, and lifetime. LEDs can be used as an alternative light source for photocatalytic applications and can be employed in any kind of reactor with ease and less cooling arrangement (Jagannath et al. 2009). They are efficient than conventional light sources with less power consumption and have opened new possibilities for photocatalytic degradation of pollutants (Jo and Tayade 2014). The basic element of an LED is the semiconductor chip comprising the junction of an n-type region and a p-type region where current flows when enough voltage is applied.

The light intensity of the source is an important factor in photocatalytic treatment. As the intensity of light increases, more and more photons get absorbed and which leads to an increase in the rate of excitation of the ground-state electron to an excited state. Therefore, it has been observed that the rate of reaction also increases with the intensity of light. When the intensity of light increases, a stage is reached at which the rate of electron–hole recombination becomes greater than the photocatalytic reaction rate (Nasr et al. 2019). Also, increasing the light intensity causes the formation of more electron–hole pairs thereby generating more hydroxyl radicals for the direct oxidation of the pollutant (Giraldo-Aguirre et al. 2015). For example, the study by

Chowdhury et al. (2012) carried out photocatalytic degradation of phenol with Eosin Y sensitized TiO_2/Pt catalyst under four different light intensities 25, 50, 80 and 100 mW cm^{-2}. It was observed that at higher light intensities, the degradation rate of phenol followed first order, with a maximum reaction rate at 100 mW cm^{-2}. Similarly, in another study, the degradation of methyl orange dye by chlorophyll sensitized Meso titania increased with light intensities of 1.03×10^2, 2.06×10^2 and 4.13×10^2 mW/cm^2. It was observed that when light intensity increased beyond 4.13×10^2 mW/cm^2, there was no significant change in the rate of the reaction. Therefore at the higher light intensity, the rate is independent of the light intensity (Joshi et al. 2009).

8.2.4 Effect of Oxidants

The addition of oxidants to photocatalytic reactions can improve the degradation efficiency by suppressing the electron recombination with holes, and produces more reactive oxygen species (Velegraki et al. 2015). The presence of oxidants such as H_2O_2, peroxymonosulfate (PMS), persulfate (PS), and ozone is also effective in increasing the efficiency of photochemical processes (Benitez et al. 2013; Ao et al. 2018; Deng et al. 2013; Sabaté et al. 2001). When UV treatment is combined with oxidants, such as H_2O_2 and ozone, the probability of final removal of contaminant increases (Tezcanli-Güyer et al. 2004). As the amount of oxidants increases, radical production also increases in the oxidation system resulting in higher degradation. Stronger oxidizing agents such as $^{\bullet}OH$ and $SO_4^{\bullet-}$ are generated due to the photon absorption as shown in Eqs. 8.3 and 8.4 (Ao and Liu 2017).

$$H_2O_2 + hv \rightarrow 2^{\bullet}OH \quad (8.3)$$

$$S_2O_8^{2-} + hv \rightarrow SO_4^{\bullet-} \quad (8.4)$$

The effectiveness of UV/H_2O_2 for the removal of pharmaceuticals in a secondary effluent was studied by Kim et al. (2009) and the removal efficiency was increased from less than 50% to 90% for 39 pharmaceuticals at a UV dosage of 923 mJ/cm^2 and H_2O_2 dosage of 7.8 mg L^{-1} indicating that H_2O_2 addition contributed significantly for the degradation. Photolysis of pollutants using chlorine as an oxidant mainly depends upon the functional group of the compound. Hydroxyl radicals are reactive with most organic compounds, while UV/chlorine is selective reacting only with phenols, anilines, olefins, and amines (Sichel et al. 2011). Degradation of antipyrine (AP) in water was compared by Tan et al. (2013) with PS and H_2O_2 as oxidants with LP UV lamps and concluded that the presence of oxidants significantly improved the degradation in the following order UV/ H_2O_2 > UV/PS > UV. Excess oxidant concentration can also act as scavengers for hydroxyl or sulphate radicals resulting in the inhibition of the degradation (Deng et al. 2013). A scavenging effect of H_2O_2 was observed beyond a concentration of 2 mM during the photolytic degradation of oxytetracycline (Liu et al. 2016c). If chloride ions are present in a system containing Fe^{2+}/UV/H_2O_2, the photocatalytic step that converts Fe^{3+} back to Fe^{2+} will be inhibited due to the scavenging effect of chloride ions (Monteagudo et al. 2014).

8.2.5 *Effect of Radical Scavengers and Inorganic Ions*

The presence of ions in aqueous solution can influence the photo-degradation in two ways; either accelerating the degradation by increasing the radical production or inhibiting the process due to competition among ions for solar radiation. Their presence can also inhibit the performance of the catalyst by causing fouling (Chong et al. 2010). The presence of alkalinity and organic matter have higher reactivity towards Reactive Oxygen species (ROS), which can inhibit the removal efficiency of pollutants. At lower concentrations, carbonate and bicarbonate ions react with $^{\bullet}OH$ and produce reactive carbonate radicals ($CO_3^{\bullet-}$), which are selective and react by electron or hydrogen transfer (Aleboyeh et al. 2012). The absorptivity of photons by carbonates and bicarbonates were at lower rates and generation of $CO_3^{\bullet-}$ slightly promoted the destruction of compounds (Duan et al. 2017). Higher organic matter content and dissolved organic materials decreased the rate constant due to the scavenging of the monochromatic light in the case of degradation of PAH due to their hydrophobic nature (Sanches et al. 2011). The presence of metal cations such as Mg^{2+} and Ca^{2+} hardly affected the removal of oxytetracycline in UV/H_2O_2, while Fe^{2+} and Cu^{2+} slightly enhanced the removal rate by participating in Fenton's like reaction (Liu et al. 2016c).

8.3 Photoreduction Processes

Photocatalytic reduction of inorganic compounds and harmful metals depends upon the redox potential of the compound relative to the conduction band of the photocatalyst (Djellabi and Ghorab 2015). The basis of photocatalytic reduction depends on the selection of the photocatalyst, surface properties of the catalyst, and the photogenerated electrons (Nu Hoai Nguyen et al. 2005; Wang et al. 2008). TiO_2 is most commonly used for the photocatalytic reduction process. Photocatalytic reduction of nitrates to molecular nitrogen using TiO_2 had been extensively studied in the presence of different hole scavengers. The process depends on many variables such as pH, the surface area of the catalyst, and the recombination rate (Doudrick et al. 2012). The presence of metals on TiO_2 particles can extend the role as an electron collector and decrease the recombination rate (Anderson 2011). An electron donor as a hole scavenger is usually required for the reduction of metals by photogenerated electrons (Djellabi and Ghorab 2015). Sowmya and Meenakshi (2015) studied the photocatalytic reduction of nitrate using Ag-TiO_2 with oxalic acid as the hole scavenger. They obtained 90% denitrification with 100% nitrogen selectivity at 0.012 N oxalic acid dose at an initial pH of 2.3. In another study, C-TiO_2 was used for the photocatalytic denitrification from seawater with formic acid as hole scavenger and the maximum nitrate removal was observed at an initial pH of 3, catalyst dose of $0.5gL^{-1}$ and 0.04 M formic acid (Shaban et al. 2016). The hole scavenger acted as the electron donor and inhibited the charge recombination.

Photoreduction of several metal ions such as Cr^{6+}, Cu^{2+}, Hg^{2+}, Pb^{2+}, U^{6+}, etc. in presence of different photocatalyst has been investigated by many research groups (Sun et al. 2005; Canterino et al. 2008; Li et al. 2013b; Feng et al. 2018). Cr^{6+} reduction using C doped TiO_2 was also studied by Shaban (2013) under sunlight irradiation and observed that 3 ppm of Cr^{6+} was removed in 10 min at a catalyst dosage of 2 g L^{-1} and pH 5. In another study, Bi_2O_3- ZrO_2 nanocomposite synthesized by Vignesh et al. (2013) was found more effective in the reduction of Cr^{6+} compared to Bi_2O_3, ZrO_2, TiO_2 and ZnO under visible light irradiation. Under irradiation, the electrons produced in the VB of Bi_2O_3 were transferred to the CB of ZrO_2, whereas the holes in the catalyst moved in the opposite direction and avoided the electron–hole pair recombination. The catalyst was highly stable during the photoreduction process and was effectively reused for four successive cycles. Simultaneous oxidation of organic compounds together with reduction processes can enhance the reduction capacity of the catalyst by acting as an electron donor (Chakrabarti et al. 2009). Simultaneous oxidation and reduction of Acid Orange 8 dye and Cr^{6+} was conducted using Au loaded TiO_2 particles using visible light treatment within the sight of formic acid (Dozzi et al. 2012). Improvement in photocatalytic reduction was observed in the presence of Au-TiO_2 which helped in oxidizing formic acid and it helped in doubling the reduction process when compared to P25 TiO_2. In another study, dye-sensitized TiO_2 film was used for the reduction of Cr^{6+} under visible light irradiation (Wu et al. 2013). The maximum removal efficiency was observed in acidic conditions and the presence of oxygen in the system enhanced the reduction process by acting as an electron scavenger.

In a study conducted by Murruni et al. (2008), Pb^{2+} was removed by N doped TiO_2 in presence of different electron donors such as formic acid, 2- propanol, methanol, ethanol, and citric acid. Among the different electron donors, formic acid at their lower concentrations was found to be best suited for the reduction production with less production of toxic products in the system. However, when the electron donor in the system got completely exhausted, the redissolution of Pb was observed. Apart from this study, TiO_2 embedded on PVA–alginate beads were effectively used for the removal of Pb^{2+} from an aqueous solution (Idris et al. 2016). They have observed a maximum removal of 99.1% of Pb^{2+} within 150 min of reaction at pH 7. Apart from these studies, U^{6+} reduction using a heterostructure of anatase–rutile TiO_2 particles was studied by Li et al. (2019) and the effect of crystal type on the reduction efficiency was investigated. They have observed that rutile was more efficient for the process and the maximum reduction of U^{6+} was observed for a mixture containing 70% rutile and 30% anatase phase. The physical interaction between the anatase and rutile phases prolonged the lifetime of charge carriers and led to more transfer of more electrons from anatase to rutile phase facilitating the reduction process. The use of visible light active catalysts for the reduction of metal ions can have become an imperative task. Recently, magnetic $ZnFe_2O_4$, a transition metal spinal ferrite with higher photostability and visible light activity was effectively used for the photoreduction of U^{6+} (Liang et al. 2020). The rod-shaped catalyst effectively reduced 98% of U^{6+} in 60 min at a catalyst dosage of 0.2 g L^{-1} when compared to microspheres or nanoparticle catalyst. The faster excitation of the catalyst to excite electrons under irradiation and the methanol to quenching the photogenerated holes helped in the effective reduction of U^{6+}.

8.4 Homogenous Photocatalysis

The homogenous photo-Fentons process is largely studied for the treatment of organic compounds. In most of the photo-Fenton pollution abatement studies, iron is used as the metal ion giving rise to Fe^{2+} ions by the photoreduction of Fe^{3+}. In aqueous medium, low-molecular-weight Fe(III)- hydroxyl complex shows a photocatalytic reduction in the presence of UV radiations. Consequently, the reduced form reoxidizes to Fe(III) in the presence of oxidizing agents and completes a basic redox cycle of Fe(III)/Fe(II). The photons help in the regeneration of Fe(II) by the reduction of Fe(III) ions and overcome the limitation of conventional Fentons process by continuously producing radicals. The major limitations of photo-Fenton processes are the pH dependence, stability of the catalyst and the separation of the catalyst after experiments. Its efficiency can be improved by extending the solubility of iron species to higher pH values as they form strong complexes with carboxylates rather than forming with water and remain active in the aqueous solution over a wide range of pH to produce oxidative species after photochemical reaction (Soares et al. 2015; Pereira et al. 2014). It increases the production of more Fe^{2+} species in the solution and extending the solar absorption spectrum up to 580 nm (Soares et al. 2015; Huang et al. 2012). Poly-carboxylate groups like citrate, malonate, and oxalate, having more than one carboxylic group, are commonly used organic ligands (Gulshan et al. 2010). In the presence of sunlight, these poly-carboxylate coordination compounds form oxidative species such as $^{\bullet}OH$, $O_2^{\bullet}$, and $HO^{\bullet}_2$, in atmospheric water and surface water by Ligand Metal Charge Transfer (LMCT) process (Seraghni et al. 2012). These complexes should absorb light in the UV–Visible region and allow the reduction of ligand by metals for making the system reactive which helps in degrading organic pollutants (Clarizia et al. 2017). These photo-Fenton process using Metal–organic complexes are more beneficial due to the following reasons (a) higher degradation rate, (b) sunlight can be used as the source of photons, (c) effective at circumneutral pH, and (d) use of easily available and cheaper reagents (Kassinos et al. 2009). The efficiency of these reactions depends on pH, initial Fe to poly-carboxylate ratio, and ligand used. Excess amount of carboxylate can also compete for hydroxyl radicals and thus decrease the reaction rate (Souza et al. 2014). Among the different complexes,

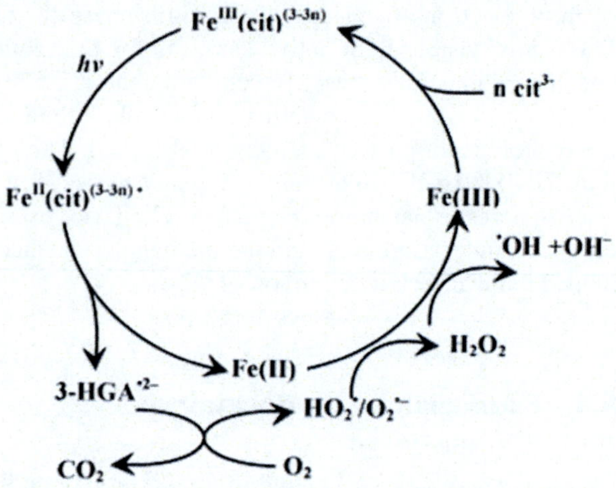

Fig. 8.1 Basic mechanism of Fe(III)- citrate complex. Reprinted with permission from Ou et al. (2008)

Fe(III)-citrate complexes hold high stability and photoreactivity, and they are theoretically stable at neutral pH (Chen et al. 2011; Sharma et al. 2019). Fe(III)-citrate complex was used for the photodegradation of atrazine under Xe lamp irradiation (Ou et al. 2008). Under irradiation, through LCMT process, the reactive radicals are formed as Fe(III)- citrate complex absorbs light and followed by the reduction to Fe(II) as shown in Fig. 8.1. The radical generation with Fe (III)- citrate complexes are dependent on the pH of the solution (Chen et al. 2011). In a study conducted by Guo et al. (2011), the maximum photodegradation of methyl orange using this complex was observed at pH 6. They reported that at this particular pH, Fe(III) and citrate coexisted together and the maximum transformation of Fe(III) to Fe(II) took place under sunlight irradiation. Souza et al. (2014) evaluated the photo-Fenton degradation of diclofenac using a ferric-oxalate complex, and complete degradation was observed at pH 6 and Fe-oxalate molar ratio of 1:9 in 90 min of treatment. A study conducted by Feng et al. (2012) examined the $^{\bullet}OH$ radical generation in comparison with the pH of the solution and Fe(III)-to-citrate ratio. They observed that if sufficient amount of Fe(III)-Citrate are available in the system near-neutral pH, Fe(II) can exist in the form of Fe(II)-carboxylate species surpassing hydroxide ions and the $^{\bullet}OH$ generation can be continued to near neutral pH (Feng et al. 2012).

8.5 Heterogeneous Photocatalysis

Heterogeneous photocatalysis has been widely investigated by the researchers for the degradation of organic contaminants (Ani et al. 2018). It is based on using materials having a wide bandgap, usually semiconductors, which generate holes and electrons when illuminated with photons. When the semiconductor catalyst undergoes UV light photolysis, a surface reaction occurs which moves the electrons from the valence band to the conduction band creating the positive holes in the valence band. Among the different semiconductor materials, TiO_2 and ZnO have received extensive consideration due to their photocatalytic performance under UV light irradiation. TiO_2 has emerged as a promising photocatalyst material for the degradation of organic pollutants because of its chemical stability, cost-effectiveness, unique photocatalytic efficiency and low toxicity (Sanches et al. 2010; Pelaez et al. 2012; Yap et al. 2010; Gao et al. 2011; Eren 2012). For catalytically activating TiO_2, it requires radiation of less than 380 nm to induce photoexcitation since it has a bandgap of ~ 3.2 eV (Wong and Chu 2003). Anatase (3.2 eV) and rutile (3 eV) are the main polymorphs of TiO_2 (Pelaez et al. 2012). A combination of anatase and rutile phases of TiO_2 were reported to have more photocatalytic activity due to higher photostability and enhanced charge separation than the pure anatase form (Li et al. 2006; Ohtani 2010). The heterogeneous photocatalysis has the advantages such as complete mineralization of pollutants with less sludge production at a rapid rate and efficiency in removing recalcitrant pollutants whereas the formation of undesirable intermediates, pH dependence and limited to laboratory scale operation are the major disadvantages.

The basic mechanism of the heterogeneous photocatalytic system is as shown in Fig. 8.2 (Bora and Mewada 2017). The process is initiated with the bombardment of photons to a semiconductor surface with energy equal or higher than the bandgap energy of the catalyst. When a photon of energy more than or equal to the bandgap energy of semiconductor falls on the surface of the semiconductor, the electron (e^-) in the valence band goes to the excited state that is in the conduction band. This creates a hole (h^+) in the valence band. The next step is the photo-reduction reaction of excited electrons (e^-) with an oxygen molecule (O_2) to form superoxide radicals ($^{\bullet}O2$), or hydro-peroxide radicals ($HO_2^{\bullet}$). These oxygen species are highly reactive

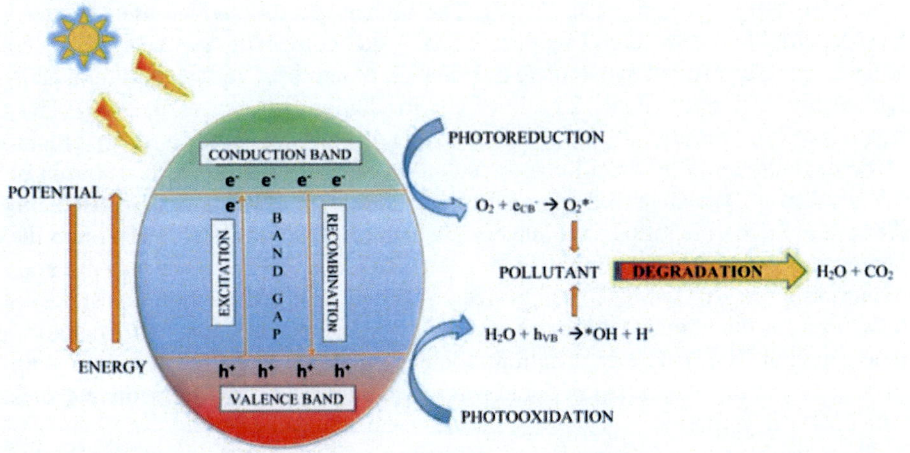

Fig. 8.2 Basic mechanism of heterogeneous photocatalysis. Reprinted with permission from Bora and Mewada (2017)

and trigger the complete degradation reaction of recalcitrant pollutants into H_2O and CO_2. The generated electrons reduce the pollutant. Photo-oxidation of H_2O with the hole (h^+) generates hydroxyl radicals ($^{\bullet}OH$) and hydrogen ions (H^+). Also, the generated holes oxidize the pollutant. Therefore, the generated hydroxyl radicals and peroxide radicals are responsible for the photodecomposition of organic pollutants (Akpan and Hameed 2009). $^{\bullet}OH$ radicals, having a standard redox potential of + 2.8 V, oxidizes the pollutant to the mineral end-products (Bora and Mewada 2017).

In TiO_2 photocatalytic reaction, hydroxyl radical is generated near the catalyst surface due to the direct oxidation by the photogenerated holes as shown in Eqs. 8.5 and 8.6 (Moctezuma et al. 2012).

$$TiO_2 + h\upsilon \rightarrow TiO_2(e^- + h^+) \tag{8.5}$$

$$TiO_2(h^+) + H_2O \rightarrow TiO_2 + \bullet OH + H^+ \tag{8.6}$$

When the catalyst particles in the reaction medium increase beyond the optimum amount, scattering, and screening phenomena commences resulting in non-uniform light distribution (Velegraki et al. 2015). The presence of TiO_2 or other catalysts in the aqueous solution beyond the optimum dose results in decreasing the degradation rate due to the decrease in light penetration, more light scattering, sedimentation of catalyst, and agglomeration (Elmolla and Chaudhuri 2010). The photocatalytic degradation of malachite green dye using TiO_2 particles with UV-LED irradiation was primarily through the hydroxyl and superoxide radicals produced by electron–hole pair formation in the catalyst surface (Natarajan et al. 2011). Supporting TiO_2 particles on activated carbon can improve the photocatalytic efficiency by reducing the electron–hole recombination rate, improving the interfacial charge transfer and promoting the pollutant transfer process by acting as a co-adsorbent (Li et al. 2006; Gao et al. 2011; Ouzzine et al. 2014). The higher specific surface area, reduced bandgap, and better structural features of TiO_2-AC composite, synthesized by the sol–gel process, resulted in complete degradation of tetracycline at an optimum catalyst loading of 1 g L^{-1} under 75 min of UV irradiation (Martins et al. 2017). TiO_2 supported on chitosan scaffolds prepared by 3D printing was used for the photocatalytic degradation of amoxicillin and resulted in 80% degradation after 180 min of UV irradiation. The enhanced degradation of residues from wastewaters was due to the high area/volume ratio of the substrate which offered more adsorption on to the photocatalyst surface (Bergamonti et al. 2019). TiO_2 immobilized on a clay mineral, montmorillonite (TiO_2-MMT) synthesized by a hydrothermal method was successfully used for the photocatalytic degradation of ciprofloxacin under UV irradiation (Hassani et al. 2015). The degradation efficiency increased with catalyst dosage up to 0.1 g L^{-1} at the natural pH of 5 and was effectively reused for 5 repetitive cycles with 55.1% degradation.

Photocatalysis based on the ZnO catalyst has engrossed wide attention because of its high electron mobility at room temperature, high interfacial charge-transfer efficiency, and similarity with the TiO_2 bandgap (Pawar and Lee 2014; Ahmad et al.

2013; Dong et al. 2014). The greatest advantage of ZnO as photocatalyst as compared to TiO_2 is that it is cheaper, effective in acidic as well as basic medium, and it can absorb more light in the UV region than TiO_2 (Boussatha et al. 2018). The photocatalytic degradation of a mixture of azo dyes using ZnO under irradiation was more favourable at acidic pH conditions with the $O_2^{\bullet -}$ radicals produced by the reduction of adsorbed O_2 predominant reactive species during the degradation (Chen et al. 2017). The enhanced photocatalytic degradation of methylene blue observed using ZnO/ montmorillonite photocatalyst was because of the improved surface area and reduced bandgap of the catalyst (Fatimah et al. 2011). Even though the catalyst has high oxidizing power for organic pollutant removal, the frequently occurring photo corrosion is a major concern (Hariharan 2006).

Apart from the conventionally used heterogeneous photocatalyst, several semiconductor catalysts such as WO_3, CdS, CdO, GaP, SnO_2, etc. with varying bandgap were utilized for the photocatalytic degradation of organic compounds and are shown in Table 8.1. Cadmium Sulphide (CdS) is one of the most potential visible light active catalysts with a narrow bandgap of ~ 2.4 eV which are extensively used for the photodegradation of organic compounds (Wang et al. 2012b). CdS microspheres synthesized by the hydrothermal process were used by Repo et al. (2013) for the degradation of a mixture of dye under UV and blue LED irradiation. Nearly complete degradation of dyes was observed after 3 h of irradiation which was due to the efficient adsorption of dyes on the catalyst surface. Further, the catalyst did not lose its photocatalytic efficiency for five reuse cycles. However, CdS particles are prone to photo corrosion leading to the leaching of Cd into the solution, and rapid recombination rate limits its photocatalytic activity. An ideal way to resolve this issue is to support the CdS catalyst on suitable support material or combining it with another semiconductor to form hybrid composites (Cui et al. 2014).

Bismuth containing materials have been broadly studied as visible light photocatalyst attributable to its unique layered structure and electronic properties facilitating effective separation of electron–hole pairs (Alansi et al. 2018; Sun et al. 2013). In a study conducted by Yan et al. (2013), $BiVO_4$ photocatalysts synthesized by a chemical method was used for the photocatalytic degradation of thiobencarb using visible light irradiation (Lai et al. 2014). They observed 97% degradation of the compound in 5 h of irradiation, and the catalyst remained stable and showed no substantial loss in the catalytic performance after three successive runs. In another study, the enhanced visible-light activity of $BiVO_4$ nanocrystal during the degradation of ciprofloxacin was due to the Pt loading which acted as charge collectors and separators on its surface (Yan et al. 2013). Furthermore, Ag-based visible light photocatalyst such as Ag_3PO_4, Ag_2CO_3, Ag_3VO_4, etc. has attracted significant attention for the degradation of organic pollutants due to its narrow bandgap and great potential in harvesting solar energy (Liu et al. 2016b; Wang et al. 2014). The higher specific surface area and reduced bandgap of 2.4 eV of Ag_3PO_4 synthesized by precipitation method resulted in the complete removal of ethylparaben under simulated solar radiation (Frontistis et al. 2017).

Table 8.1 Photocatalytic degradation studies using semiconductor catalyst

Catalyst	Pollutant	Synthesis method	Bandgap (eV)	Light source	Results	References
ZnS	Rhodamine B	One-pot hydrothermal	3.7	300 W Hg lamp	96.6% degradation in 120 min	Dong et al. (2013)
CeO_2	Xylene Milling Yellow 6 G	Sol–gel	2.52	125 W Hg lamp	Complete degradation of dye in 30 min at a catalyst dose of 1 gL^{-1}	Tambat et al. (2016)
V_2O_5	Phenol and its derivatives	Chemical precipitation	2.295	Sunlight	Surface defects on the catalyst acted electron–hole centres and enhanced the degradation	Aslam et al. (2015)
$BiVO_4$	Ibuprofen	Hydrothermal	2.4	Simulated solar light	Degradation followed the first-order kinetics with maximum degradation at pH 4.5 and catalyst dose 5 gL^{-1}	Li et al. (2016)
CuO	Methylene blue Methylene violet	Hydrothermal	2.67	UV light	89% and 96% degradation of methylene blue and methylene violet respectively after 180 min of UV irradiation	Sonia et al. (2015)
CdO	Congo Red Malachite Green Crystal Violet	Hydrothermal	1.9	500 W Hg lamp	Complete degradation of all the dyes was observed and the catalyst was stable for 3 reuse cycles	Tadjarodi et al. (2014)

8.6 Visible Light Photocatalysis

The necessity for high-energy ultraviolet (UV) radiation for the activation of TiO_2 has restricted both the practicality and ecofriendly benefits of most organic photochemical processes (Yoon et al. 2010). Any method which can decrease the electron–hole pair recombination and reduces the bandgap of the catalyst will substantially increase the photocatalytic performance. One of the primary motivations driving the advancement of progressively effective, practical, and synthetically useful visible light-mediated photocatalytic processes is to efficiently harness the energy of solar radiation for more environmentally responsible chemical processes. Different strategies such as non-metal and/or metal doping, dye sensitization, and coupling semiconductors were developed to modify photocatalyst particles for extending their absorption range into the visible-light region.

8.6.1 *Doping*

Photocatalysts should be modified to absorb visible light which is the major portion of the solar spectrum. Doping is the process of incorporation of atoms or ions into the crystal lattice of a photocatalyst to modify its photo responsiveness under visible light irradiation by creating structural imperfections (Ohtani 2010; Ahmed et al. 2011; Dong et al. 2015b; Yap et al. 2011). The effect of doping on a photocatalytic material is dependent on various factors such as the type of dopant, its concentration, doping process, and other properties of the catalyst (Dong et al. 2015b). The optimum dosage of dopant varies from catalyst to catalyst. The excess amount can inhibit the photocatalytic activity by reducing the surface area and delaying the adsorption of pollutants on to its surface (Akpan and Hameed 2009). During the doping of TiO_2 particles using anions, the oxygen atoms in the TiO_2 get replaced by other elements, introducing a mid bandgap level that supports the visible light absorption (Abdullah et al. 2016; Lin et al. 2016). Among the different anionic species, doping with carbon, nitrogen, and sulphur has the potential to form mid bandgap level (impurity level) closer to the valence band of the photocatalyst for better photocatalytic activity as shown in Fig. 8.3 (Ananpattarachai et al. 2009). Asahi et al. (2001) reported that if anionic species are used as doping materials then the bandgap can cover adequately with the band states and enable easy transfer of excited carriers to the surface of the catalyst. It is also reported that doping of TiO_2 particles with anions will be more effective as they can take the place of oxygen in the lattice (Ohno et al. 2003).

Cong et al. (2007) conducted the N doping of a homogenous anatase TiO_2 using different N sources. Among the dopants, triethylamine as N source resulted in enhanced photocatalytic activity and better degradation of 2,4-dichlorophenol. The XPS and Raman spectroscopy results confirmed that N was present in the structure as Ti–O-N and O-Ti-N, and shifted the absorption of the catalyst to the visible region. Ananpattarachai et al. (2009) reported that, among the three N dopants, doping

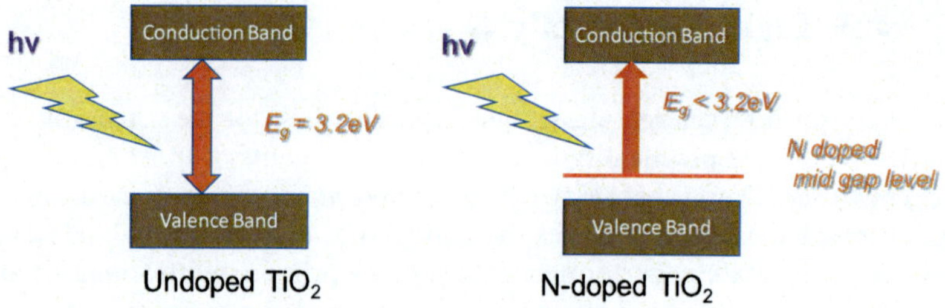

Fig. 8.3 Formation of midgap level after N doping in TiO_2

with diethanolamine resulted in enhanced visible light activity for 2-chlorophenol degradation. XPS results confirmed that nitrogen was incorporated in the TiO_2 structure interstitially and resulted in the smallest crystallite size and bandgap energy which helped in enhanced degradation. In another study, the degradation of 4-chlorophenoxyacetic acid by N-TiO_2 was studied under Visible LED as the light source (Abdelhaleem and Chu 2017). They observed a complete degradation of the pollutant with 73% mineralization and the catalyst was found effective for five successive reuse studies. Carbon is considered as one of the anionic non-metal dopants with higher stability. C doped TiO_2 particles synthesized by the sol–gel method enhanced the photocatalytic degradation of ethylene by retarding the anatase to rutile transformation by forming Ti–O-C structure using a visible light source. They observed that C atoms were incorporated in the interstitial spaces of TiO_2 lattice and it improved the crystallinity of the catalyst and hence the activity (Lin et al. 2013). The C doped ZnO nanoparticles synthesized by Bechambi et al. (2015) displayed exceptional photocatalytic activity for four reuse cycles during the degradation of BPA under UV irradiation in the presence of H_2O_2. The higher stability of the catalyst under the optimized conditions resulted in 60% mineralization of BPA after four reuse cycles.

Chen et al. (2007a) compared the photocatalytic degradation efficiency of C-TiO_2, N-TiO_2, and C-N-TiO_2 for the degradation of methylene blue. They found that C and N atoms suppressed the crystal growth of TiO_2 and exhibited higher activity by producing more reactive radicals in the aqueous solution. An N-doped $TiO_2/SiO_2/Fe_3O_4$ magnetic catalyst synthesized by the sol–gel method maintained its photocatalytic efficiency during the degradation of paraquat for eight reuse cycles with a negligible loss of 7.9% degradation efficiency owing to the reduced bandgap and enhanced specific surface area of the catalyst with doping (Pourzad et al. 2019). Ce/N co-doped $TiO_2/NiFe_2O_4$/Diatomite ternary composite exhibited excellent adsorption, visible light activity, and easy separability because of the synergistic effect of ferromagnetic properties of $NiFe_2O_4$ and competent adsorptive properties of diatomite (Chen and Liu 2017). The co-doping with Ce and N resulted in forming an intra-band structure which reduced the bandgap of the hybrid catalyst and enhanced the charge transfer with visible light all through the degradation of tetracycline. The enhanced photocatalytic degradation of naphthalene observed with La, N co-doped TiO_2 particles under visible light was due to the synergistic effect of the reduced bandgap, increased adsorption of pollutant to the volume pores and decreased recombination of photogenerated charge carriers (Liu et al. 2016a).

Metals and transition metal ions can also improve the surface properties of the catalyst for visible light absorption (Salimi et al. 2019). The bandgap of catalysts can be tailored for visible light absorption by reducing its bandgap using metals such as Co, Ni, W, V, Bi, etc. (Caglar Yılmaz et al. 2019; Salimi et al. 2019; Qin et al. 2012). Accordingly, by doping with a metal element, an impurity level is acquainted with the centre of the forbidden band of the semiconductor to enable visible light activity (Guo et al. 2020). In a study conducted by Gao et al. (2010), the transition metal Zr was doped onto TiO_2 by the sol–gel process and the efficiency of the catalyst was evaluated for the degradation of BPA under UV irradiation. Enhanced photocatalytic activity resulting in 90% TOC reduction was observed after 2 h which may be credited to the alteration in the conduction band edge towards higher potential and producing more reactive radicals. Ni-doped TiO_2 particles synthesized by microwave-assisted sol–gel process for the degradation of bisphenol A (BPA) resulted in complete degradation and 77% mineralization of the compound in 210 min (Blance-Vega et al. 2017). The incorporation of Ni into the crystal lattice inhibited the phase transformations and shifted the bandgap to the visible light region and resulted in improving the degradation of BPA. In another study, Ce metal was incorporated into the crystal lattice of ZnO by the co-precipitation method and effectively enhanced the degradation of BPA under sunlight irradiation. The metal dopant created an impurity level below the conduction band of ZnO and enhanced the electron mobility resulting incomplete degradation of BPA (Kamaraj et al. 2014).

Metal doping of $BiVO_4$ can also promote the electron–hole recombination and improve the capacity of the catalyst to absorb visible light. Pd doped $BiVO_4$ synthesized by hydrothermal and followed by the impregnation strategy improved the degradation of methyl orange using the visible light treatment. The Pd particles were dispersed on the surface of the catalyst and acted as recombination centres and enhanced charge separation (Ge 2008). Zhang et al. (2005) reported that Ag-TiO_2 supported on activated carbon exhibited higher photocatalytic activity when compared to TiO_2 owing to the enhanced charge separation and hence producing more hydroxyl radicals with the reaction of surface adsorbed OH^- with photogenerated holes. The tungsten doped TiO_2 supported on activated carbon with a narrow bandgap and higher surface area resulted in 97.3% decolourization of rhodamine B under visible light. The amount of dopant and catalyst dosage showed a strong relation to the degradation efficiency, with 0.6% W ion and 8 g L^{-1} as the optimum conditions for maximum decolourization (Qin et al. 2012). More than 90% degradation of Amoxicillin was observed under visible light degradation with Co-doped TiO_2 particles synthesized by the reflux route (Caglar Yılmaz et al. 2019). The synthesized catalyst possessed a higher surface area and enhanced the production of more electron–hole pairs due to reduced bandgap. Among the different dopants, non-metal ions are more stable when compared to metal and transition metal ions. Even though these metals reduce the bandgap and act as charge trapping centres to electron–hole recombination, their stability is a major concern (Lin et al. 2013; Ahmed et al. 2011; Yap et al. 2011).

8.6.2 Dye-Sensitization

Dye sensitization is another surface modification technique by which the dyes are chemisorbed or physisorbed on the surface of the semiconductor to increase the efficiency of the excitation processes and extend the wavelength to visible range (Dong et al. 2015a). The dyes having high redox property and visible light sensitivity can be utilized as a part of photocatalytic reactions. Under illumination by visible light, the excited dyes can inject electrons to the conduction band of semiconductors to start the catalytic reactions (Ni et al. 2007). The requirements of a good photosensitizer include an extended range of absorption spectrum, strong adsorption onto the semiconductor surface, high quantum yield, and excited states with a long lifetime, and similar band structure to decline energy loss during the electron transfer process (Narayan 2012; Saien and Mesgari 2016). The basic mechanism of the dye-sensitized photocatalytic system under visible light irradiation is shown in Fig. 8.4. In the case of visible light irradiation of the dye-sensitized catalytic system, the first step is the excitation of sensitizer which is adsorbed on TiO_2 to the singlet state with generating an electron. The produced electrons are then injected into the conduction band (CB) of TiO_2 while the valance band remains intact. The transferred electrons react with the adsorbed oxygen molecules on the surface of the TiO_2 catalyst resulting in the formation of superoxide which is a precursor for the degradation of pollutants (Youssef et al. 2018).

Several studies are there in which synthetic dyes such as eosin-Y (Chowdhury et al. 2012), thionine (Chatterjee and Mahata 2001), Nile blue-A, safranine-O, rhodamine-B and methylene blue (Chatterjee et al. 2006), porphyrin (Wei et al. 2017; Chen et al. 2008; Niu et al. 2013) ruthenium complexes (Nazeeruddin et al. 2005), etc. are being used as a sensitizer in semiconductors for photodegradation of various dyes. Chowdhury et al. (2012) developed eosin-Y dye-sensitized visible light active TiO_2 photocatalyst for the photodegradation of phenol. It was observed that eosin-Y dye-sensitized TiO_2 with Pt gave 93% phenol degradation within 90 min. Chatterjee

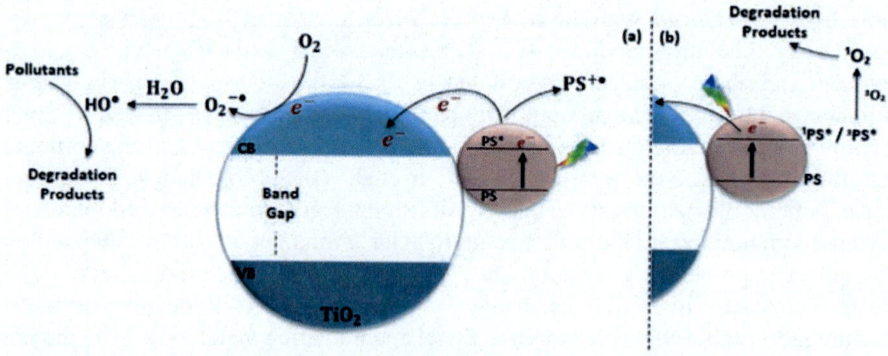

Fig. 8.4 Basic mechanism of the dye-sensitized photocatalytic system under visible light irradiation. Reprinted with permission from Youssef et al. (2018)

and Mahata (2001) developed thionine and eosin Y dye-sensitized TiO_2 for the degradation of organic pollutants such as phenol, chlorophenol, 1,2-dichloroethane, and trichloroethylene in water. It was found that 55–72% degradation of pollutants was achieved after 5 h of irradiation with a 50 W tungsten lamp. Another important work on synthetic dye-sensitization was done by Chatterjee et al. (2006). They have used Thionine, Nile blue-A, safranine-O, rhodamine-B, methylene blue, eosin dyes for the modification of TiO_2 semiconductor photocatalyst for the degradation of various halocarbons under visible light irradiation. They reported 55–72% pollutants removal within 5 h of irradiation with a 150 W Xenon lamp.

Recent studies are focusing on metal porphyrin, a compound containing a macrocyclic aromatic conjugation system. It was found that the central metal atom improves the regularity of the porphyrin molecule and it also has a stronger absorption band in the visible light (Wei et al. 2017). Chen et al. (2008) have used Zn porphyrin dye for utilizing visible light more efficiently in the photocatalytic reaction of N-doped TiO_2. It was observed that Zn porphyrin is adsorbed on the surface of TiO_2 through an O = C–O–Ti linkage. They inferred that the coupling of nitrogen doping with dye sensitization is a good method to strikingly enhance the visible-light absorption and photocatalytic action of TiO_2. Wei et al. (2017) developed the Fe(III) porphyrin sensitized TiO_2 nanotubes prepared through improved hydrothermal and heating reflux process for the photodegradation of methylene blue under visible light. It was observed that Fe(III) porphyrin played an important role in absorbing photons and expanding the absorption wavelength to the visible light region, increasing the separation of the electron–hole pairs, and thus accelerating the decomposition of organic pollutants.

The major problem with organic dyes includes their complex synthetic routes, environmental toxicity, and low yield. Whereas natural dyes such as anthocyanin, chlorophylls, etc. are found in flowers, leaves, and fruits of plants and can be extracted by simple methods and are cost-effective, non-toxic, and provide complete degradation. Kathiravan et al. (2009) have reported the use of cyanobacterial chlorophyll extracted from cyanobacteria namely *Spirulina* sp. as a sensitizer for colloidal TiO_2 and observed that the adsorption of chlorophyll on the surface of colloidal TiO_2, is through electrostatic interaction. Phongamwong et al. (2017) have developed chlorophyll and Mg co-modified P25 nanohybrid catalysts for the degradation of rhodamine B under visible light. It was observed that loading free Mg together with chlorophyll prompted the development of another intricate structure, bringing about a synergistic impact between chlorophyll-Mg and P25. Also, another study by Phongamwong et al. (2015) developed N-doped TiO_2 catalysts loaded with chlorophyll in Spirulina (Sp/N–TiO_2) in an attempt to enhance photocatalytic CO_2 reduction. It was observed that activities of catalysts Sp/N–TiO_2 is much more than pure TiO_2 and N doped TiO_2. Anthocyanins extracted from natural Caribbean species *S. cumini* were employed to sensitize TiO_2 for the photo-degradation of methylene blue dye by Díaz-Uribe et al. (2018). They have reported that the sensitization procedure enhanced the photocatalytic activity of the TiO_2 in the visible range. It was also observed that the photocatalytic activity of TiO_2/*S. cumini* was around 3 times more predominant than unmodified TiO_2. Therefore, natural dyes as sensitizers for photodegradation of pollutants are promising alternatives to synthetic organic dyes. They offer environmental friendliness, low-cost production, simple preparation technique, and wide availability as compared with other synthetic dyes.

8.6.3 Surface Adsorbates

Recently, researchers are more focusing on the development of surface modification of TiO_2 with conducting polymers for achieving visible light photocatalysis (Eskizeybek et al. 2012; Online et al. 2017; Sboui et al. 2017). Among the different polymers, polyaniline (PANI) homopolymer is having more attention among the researchers due to its ability to act as surface capping agents or stabilizers. Its properties include high conductivity due to the presence of -NH groups, easy to synthesize, good stability, and high light absorption (Eskizeybek et al. 2012; Saravanan et al. 2016). Also, the electron transfer from PANI to TiO_2 occurs because of the presence of conduction band at the more negative side and lower bandgap around 2.8 eV as compared with that of TiO_2. Therefore, the interfacial charge transfers and separation between the PANI and the semiconductor take place, thereby making it a good sensitizer for photocatalysis treatments to work under visible light (Sboui et al. 2017; Mais et al. 2019). Eskizeybek et al. (2012) have demonstrated the feasibility of solar-assisted degradation of methylene blue and malachite green dyes by PANI/ZnO nanocomposite. The maximum degradation of 99% was observed with 0.4 gL^{-1} catalyst dosage under the irradiation of 5 h of natural sunlight. A similar study by Saravanan et al. (2016) used PANI/ZnO nanocomposite for the degradation of methylene blue and methyl orange dyes. During irradiation, the PANI gets excited about $\pi \rightarrow \pi^*$ transition and produces electrons. These electrons get transferred to ZnO catalyst and stimulate the production of radicals for the degradation. Recently, Sambaza et al. (2019) modified Ag/TiO_2 by the incorporation of PANI supported by oxidative polymerization. They compared the degradation efficiencies with pure TiO_2, Ag/TiO_2, and PANI/Ag-TiO_2 and observed an enhancement in the degradation efficiency for bisphenol-A (BPA) for PANI/Ag/TiO_2 at around 99.7% under visible light with h^+ and $-O_2^-$ as the dominating reactive species. It also showed good reuse potential for 4 cycles with > 90% removal till 4th cycle. Another study by Sboui et al. (2017) demonstrated solar photocatalytic degradation of methyl orange dye and other organic pollutants by the small pieces of cork as support for PANI-TiO_2. Due to the low density and hydrophobic property of the cork, it can adsorb the contaminants from the water thereby enhancing the efficiency of the process.

Apart from PANI, other conducting polymers such as poly-o-phenylenediamine (PoPD) and poly-2-aminobenzene sulfonic acid (P2ABSA) have attracted considerable attention for developing visible light active TiO_2 catalyst. The study conducted by Online et al. (2017) has established the applicability of P2ABSA/TiO_2 nanocomposites synthesized by oxidative polymerization method for the degradation of methylene blue. It was observed that the incorporation of P2ABSA on the surface of TiO_2 does not change the lattice structure and grain size, whereas it enhanced the visible light activity of the catalyst due to the lowering of bandgap to 2.70 eV. Also, it gave six times higher rate of degradation and excellent photocatalytic stability than TiO_2. Yang et al. (2017) developed PoPD modified TiO_2 nanocomposites by in situ oxidative polymerization method and demonstrated the enhanced visible-light activity with the degradation of methylene blue dye. It was observed that the narrow bandgap of PoPD of ~ 1.89 eV showed high light absorption, thus it can act as a photosensitizer for TiO_2. Also, the maximum degradation of methylene blue was observed at pH 11.41, catalyst dosage = 1.5 g L^{-1}, and initial methylene blue concentration of 40 ppm having good reuse potential up to 5 cycles of operation.

8.6.4 Metal Deposition

Modification of semiconductor by noble metal deposition, such as palladium (Pd), platinum (Pt), silver (Ag), and gold (Au), etc., results in the efficient photocatalytic reaction since the Fermi levels of these noble metals are lower than that of a semiconductor. These metals absorb visible light due to surface plasmon resonance effect thereby promoting the photoreactions of catalysts (Gomes et al. 2017). Additionally, photo-excited electrons can be transferred from the conduction band to metal particles deposited on the surface of TiO_2, while photo-generated valence band holes remain on the TiO_2. These activities enormously diminish the possibility of electron–hole recombination, bringing higher photocatalytic reactivity (Ni et al. 2007; Safajou et al. 2017). The study on photocatalytic performance for the degradation of the paraben mixture was conducted by Gomes et al. (2017). They compared the performance of different noble metals, viz. Pt, Pd, Au, and Ag, supported TiO_2 and observed that 0.5% Ag-TiO_2 and 0.5% Pd-TiO_2 gave maximum removal efficiency for the paraben mixture wastewater. Among these noble metals, recently palladium has been extensively used by the researchers because of its high stability and catalytic activity (Safajou et al. 2017). Nguyen et al. (2018) demonstrated the complete degradation as well as mineralization of methylene blue and methyl orange dyes with Pd/TiO_2. The maximum degradation for 20 mg L^{-1} initial concentration of MB and MO dyes was observed with 0.5 wt.% Pd-TiO_2 at initial pH of 10 for MB and 2.5 for MO. Moreover, Pd has also been used as the active element for interacting with the surface of various oxides as supports. The study conducted by Safajou et al. (2017) synthesized Pd nanoparticle embedded graphene sheets on TiO_2 nanowires for the degradation of Rhodamine B. Their results showed enhanced photoactivity for graphene supported Pd/TiO_2 because the presence of Pd reduces the recombination rate whereas graphene increases the surface area, light absorption, and conductivity thereby improves the separation and transportation of the charge carriers or electrons. Another study by Yu et al. (2010) investigated the Au and Pd co-modified TiO_2 nanofilm for the degradation of pesticide malathion. The Au and Pd co-modified TiO_2 catalyst performed 1.72 times better than pure TiO_2.

8.6.5 Heterogeneous Composites

Several reports have tried to understand the stability of photocatalyst materials during organic compound removal (Ge et al. 2015). Coupling of two semiconductors with an appropriate bandgap position can improve the photocatalytic performance due to the enhanced electron–hole separation and visible light absorption (Yan et al. 2015).

Here, an extensive bandgap semiconductor is combined with a small bandgap semiconductor with a more negative conduction band level. The conduction band electrons from the small bandgap semiconductor are injected to the large bandgap semiconductor with good stability (Ni et al. 2007). TiO_2-ZnO nanocomposite synthesized by Menon et al. (2019) resulted in complete decolourization of methyl orange due to the bandgap narrowing caused by the coexistence of different phases in the composite. The synthesis of binary/ternary nanocomposites can facilitate easy charge separation when compared to single component materials (Zhang et al. 2012). ZnO-CuO heterojunction photocatalyst synthesized using a chemical method resulted in more than 90% degradation of Amoxicillin due to the effective charge transfer from CuO to ZnO and subsequent production of $^{\bullet}OH$ and $O_2^{\bullet}$ (Belaissa et al. 2016). Coupling of semiconductors can help in improving the stability of one of the photocatalyst by preventing leaching and can also improve the interfacial charge transfer (Zhu et al. 2013). Fe_2O_3/In_2O_3 nanocomposites synthesized by hydrothermal method effectively degraded rhodamine B due to the incorporation of In_2O_3 causing a defect in the crystal and reducing bandgap and prolonged the electron–hole recombination (Guo et al. 2020). In a study conducted by Lai et al. (2019), with an optimal mass ratio of 7% of CuS fabricated on a CuS/$BiVO_4$ heterojunction catalyst, complete removal of ciprofloxacin was observed under visible light. CuS enhanced the visible light absorption of heterogeneous catalyst and provided more active sites on the composite with higher photostability for reuse experiments.

Semiconductor composite photocatalyst with narrow bandgap immobilized on suitable support materials can exhibit improved photocatalytic activity. TiO_2-CdS nanocomposite decorated on carbon nanofibers prepared by an electrospinning method enhanced the removal of a mixture of dye pollutants under visible light irradiation. Complete removal was observed with 5 min of irradiation due to the enhanced adsorption characteristic of carbon nanofiber and hindered electron–hole recombination of composite (Pant et al. 2014). ZnO/MgO composite supported on activated carbon synthesized by the sol–gel method resulted in complete degradation of methylene blue at natural pH due to the improved trapping of electrons on MgO sites and holes on ZnO surface consenting charge separation (Karimi et al. 2015). The co-existence of BiOI and Zn_2SnO_4 were confirmed by XRD results as two different phases which facilitated enhanced charge separation in BiOI/Zn_2SnO_4 heterostructure enhanced the photocatalytic degradation of rhodamine B, bisphenol A, and methylene blue. (Yan et al. 2015). Under UV irradiation, both catalysts absorb light resulting in their excitation and charge separation occurs at their junction such that electron in CB of BiOI gets transferred to CB of Zn_2SnO_4 and vice versa in the VB. The carbon spheres acted as dispersing support to control the particle growth and as a photosensitizer which narrowed the bandgap in a carbon-supported CuO-BiVO_4 nanocomposite synthesized by Zhao et al. (2012) using a hydrothermal method followed by impregnation. The photocatalyst was stable for four successive reuse cycles with more than 85% degradation of methylene blue in all cycles with both $^{\bullet}OH$ and $O_2^{-\bullet}$ as active species for degradation. The hole generated in the valence band of BiOBr showed higher oxidation potential due to effective charge separation under visible light irradiation and enhanced the degradation efficiency of BPA using magnetically separable BiOBr/$SiO_2Fe_3O_4$ catalyst (Zhang et al. 2014).

8.6.6 *Hybrid Heterostructures*

Graphene-based materials with their unique structure have received considerable attention in the area of photocatalysis due to their electronic and charge separation properties (Brindha and Sivakumar 2017; Zhang et al. 2011). They are used as ideal conductive support for photocatalyst because of their capacity to stretch out the extend of light absorption to the visible region (Gao et al. 2012). TiO_2 supported on graphene synthesized by hydrothermal method resulted in degradation of methyl orange at a kinetic rate of $5.9*10^{-3}$ min^{-1} when compared to $0.9*10^{-3}$ min^{-1} for pure TiO_2. The enhanced degradation was due to higher adsorption of dye molecules on the catalyst and the electron-accepting nature of the π conjugating graphene structure which suppressed the charge recombination (Khalid et al. 2013). N-doped TiO_2 particles on reduced Graphene oxide had shown excellent visible light activity for the degradation of tetracycline hydrochloride. The synergistic effect of improved absorption of pollutants on the catalyst surface and the effective charge transfer due to the π conjugation structure of graphene resulted in degradation (Tang et al. 2018). The enhancement in the photocatalytic degradation of BPA under visible light using TiO_2/reduced GO (TiO_2-rGO) composites was due to effective charge separation in the composite which occurred through the interfacial contact and the larger adsorption of pollutant on the aromatic planes of graphene through π-π conjugation (Xu et al. 2018). Similarly, the enhanced photocatalytic activity was shown by the TiO_2-rGO composite prepared by Wang and Zhang (2011) for the degradation of rhodamine B. It also depended on the enhanced adsorption of the compound on the aromatic region of graphene, and higher charge transfer occurred in the 2-dimensional graphene network.

The photocatalytic activity of ZnO/GO composite catalyst depended on the exposure of graphene on the surface of ZnO which helped in rapid migration of photo-induced electron to the delocalized conjugated π structure of graphene (Fig. 8.5) and enhanced the production of more $O_2^{\bullet-}$ (Xu et al. 2011). In another study, the photocatalytic degradation of Bisphenol A by Ag_3PO_4/GO composite was studied by Wang et al. (2014) and observed that the electric field inside composite resulted in retransferring the excited electron from the conduction band back to the surface of GO and enhanced the photocatalytic activity. The high charge carrier mobility properties of

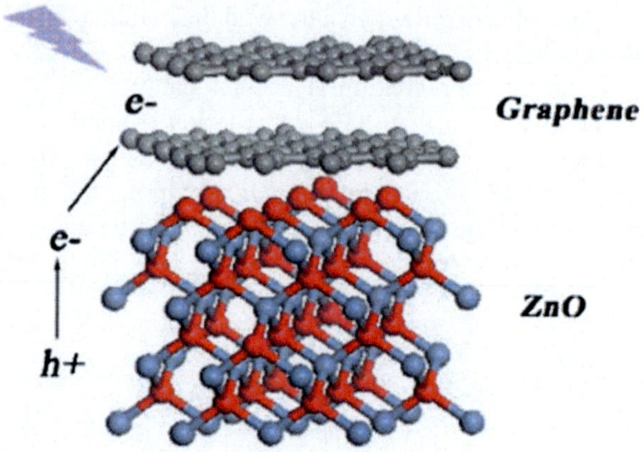

Fig. 8.5 Schematic of charge transfer in ZnO/Graphene composite. Reprinted with permission from Xu et al. (2011)

graphene which helped in accepting the photoinduced electrons from bismuth molybdate ($BiMoO_6$) in a Graphene- $BiMoO_6$ hybrid catalyst improved the degradation of reactive brilliant red dye X-3B using visible light treatment (Wang et al. 2012a). The stability of CdS/rGO composite toward the photocatalytic degradation of methylene blue was studied for three reuse cycles and only a 4% reduction in efficiency was observed after the third cycle. Supporting CdS on rGO avoided the leaching of Cd ions into the solution and suppressed the charge recombination by photoluminescence quenching by rGO (Wang et al. 2012b). The reduction in degradation of rhodamine B under visible light irradiation using $FeWO_4$-rGO composites was due to the competition between rGO and $FeWO_4$ for light absorption beyond a weight percentage of 0.3% rGO. $BiVO_4$-rGO composite showed excellent photocatalytic activity when related to pure $BiVO_4$ particles for the degradation of ciprofloxacin because of the effective separation of electron–hole pairs and enhanced visible light absorption of the hybrid catalyst (Yan et al. 2013).

Graphitic Carbon Nitride (g-C_3N_4) are π conjugated materials with higher stability and lower bandgap (2.7 eV) and are extensively used in photocatalytic applications (Gao et al. 2020; Liu et al. 2016b). The heterojunction established between mesoporous graphitic carbon nitride (MPg-C_3N_4) and ZnO at their interface in the $MPGC_3N_4$/ZnO nanocomposites resulted in the advancement of charge separation and resulted in retaining the stability of the catalyst for five reuse cycles (Le et al. 2017). They observed that the photocatalytic activity was retained over 88% after the fifth cycle as the change in edge potential at the heterojunction interfaces blocked the charge transfer at the recombination centre. Ag/TiO_2/$MPgC_3N_4$ ternary nanocomposite displayed higher photocatalytic activity due to the synergistic effect of Ag nanoparticles deposited on TiO_2/$MPgC_3N_4$ heterojunction (Gao et al. 2020). An optimum amount of Ag particles helped in inhibiting the recombination rate and the porous gC_3N_4 increased the light absorption and specific surface area for photocatalytic reaction. Further, considering the Fermi levels of each component, the electrons generated in the CB of $MPgC_3N_4$ get transferred to the surface of Ag particles through the CB of TiO_2 and suppressed the electron hole-pairs and resulted in 4.1 times more degradation of AMX than TiO_2/MPgC_3N_4. Magnetic fluorinated graphitic carbon nitride (FeGF) photocatalyst synthesized by hydrothermal method with porous structure and higher surface area had shown higher light absorption and resulted in 60% mineralization of Amoxicillin in 2 h. The particles had shown higher photocatalytic activity due to the heterojunction formed between Fe_3O_4 and fluorinated graphitic carbon nitride which reduced the charge recombination (Mirzaei et al. 2019). The coupling of Ag_3PO_4 with g-C_3N_4 improves the stability of the catalyst by protecting the dissolution of Ag_3PO_4 and improved its visible-light absorption range (Liu et al. 2016b). Ag_3PO_4@g-C_3N_4 core@shell nanostructures possess large contact area between the core and shell facilitating enhanced charge separation resulting in degrading 97% methylene blue in 30 min of irradiation.

Lamellar graphitized mesoporous carbon (GMC) nanosheets with the large surface area were used as a support material for the degradation of ciprofloxacin from water. The carbon component inhibited the agglomeration of TiO_2 particles and enhanced the adsorption of pollutants resulting in complete mineralization of 15 mg L^{-1} of CPX in 90 min of UV irradiation (Zheng et al. 2018). Apart from the above-mentioned hybrid materials, carbon nanotubes, mesoporous carbon, C60, and zeolites were used for the photocatalytic degradation of organic pollutants. These materials have a positive effect on improving photocatalytic activity by inducing synergistic effects or improving adsorption properties. Some of the recent studies conducted using these hybrid materials are shown in Table 8.2.

8.7 Catalyst Immobilization

The industrial application of photocatalytic treatment with nanoparticles is still a challenge among the researchers due to several limitations associated with the technique, including agglomeration of the particles, difficulty in separation and recovery, etc. (Shen et al. 2012). The key solution to the problem is the immobilization of nanoparticles on suitable support media. Therefore, many researchers are finding the best ways to anchor the photocatalyst onto the supports with minimum cost and without lowering photocatalytic efficiencies. The immobilizing support must be having high light absorption capabilities, good adhesion with the nanoparticles, ease in separation, chemically inert, abundant in nature, low cost, and resistant to degradation (Ola and Maroto-valer 2015; Kaur et al. 2018). The various support materials such as glass beads/plates (Shen et al. 2012), alumina (Sakthivel et al. 2002), zeolites (Liu et al. 2018b), clay beads (Kaur et al. 2018), magnetic core–shell (He et al. 2008), stainless steel fibres (Hosseini et al. 2007), activated carbon (Quiñones et al. 2014), polymers (Sboui et al. 2017), etc. has been reported for the photodegradation of organic contaminants from wastewater. The different methods of immobilization for the TiO_2 catalyst includes sol–gel, electrophoretic deposition, spray pyrolysis deposition, chemical vapour deposition, and physical vapour deposition (Ola and Maroto-valer 2015).

Table 8.2 Photocatalytic degradation using hybrid nanomaterials

Photocatalyst	Pollutant	Light source	Major findings	References
Multiwalled carbon nanotubes (CNT)TiO_2	Methylene blue	125 W mercury lamp	The charge separation and stabilization attained by carbon nanotubes improved the photocatalytic degradation process	Zhao et al. (2013)
CdS-graphene-CNT composite	Methylene blue	Visible light	CNT acted as good conducting material which reduced aggregation of graphene and suppressed electron–hole recombination	Wang et al. (2013)
Zeolite Socony Mobil-5(ZSM-5)-TiO_2	Acetaminophen	14 W UV lamp	At an optimum loading of 40% of ZSM-5 on TiO_2, 96.6% degradation of acetaminophen was observed at a catalyst dosage of 1 gL^{-1} after 180 min UV irradiation	Chang et al. (2015)
C60/$Bi_2TiO_4F_2$ Heterojunction	Rhodamine B Eosin Y	visible light ($\lambda > 420$ nm)	The higher visible light absorption and photostability of the catalyst enhanced the oxidation of dye pollutants	Li et al. (2013)
TiO_2/MoS_2@zeolite	Methyl orange	XG500 xenon long-arc lamp	The synergistic effect of better adsorption and efficient charge transfer resulted in 95% degradation of methyl orange after 60 min of visible light irradiation	Zhang et al. (2015)

(continued)

Table 8.2 (continued)

Photocatalyst	Pollutant	Light source	Major findings	References
Carbon black/Ag_3PO_4	Methyl orange	300 W xenon lamp	Enhanced charge transfer ability of carbon black resulted in 1.6 times more degradation of pollutant when compared to Ag_3PO_4	Liu et al. (2018a)
Mesoporous Carbon/Bi_2WO_6	Rhodamine B	500 W Xe lamp	Mesoporous carbon improved the crystallization of Bi_2WO_6 and increased the surface area of the composite resulted in complete degradation of Rhodamine B	Chen et al. (2012)
Mesoporous Carbon/ZnO Hybrid	Methylene Blue	300 W Hg lamp	The porosity rises in the material and reduced bandgap increased the photodegradation	Wang et al. (2017)

For instance, Hosseini et al. (2007) investigated and compared the performance of TiO_2 films supported on three different substrates such as perlite granules, glass plates, and steel fibre for the photocatalytic degradation of phenol. It was observed that the steel fibre support enhances the UV light absorption and surface area of TiO_2 whereas the glass plate forms stable bonds between the OH groups bound to the TiO_2 catalyst. Besides, the perlite granules showed a maximum porosity of 95% which makes them float on the surface of the water. The study by Alberti et al. (2019) synthesized environmental friendly zeolite having good magnetic properties from iron and steel waste by simple hydrothermal activation method and it becomes the support for TiO_2 nanoparticles. The sol–gel synthesized TiO_2 particles were immobilized on zeolite by ultrasounds and the particles showed excellent degradation of methylene blue with good recovery and reuse potential. Another study by Kaur et al. (2018) developed TiO_2 and Fe doped TiO_2 supported on clay beads by simple heat attachment and the catalyst showed enhanced degradation of Carbendazim around 88% with a Flat plate photocatalytic reactor under sunlight. The stable immobilized catalyst was formed at optimum conditions of double coating of TiO_2 on 12.42 mm diameter clay beads which are calcined at 400 °C.

8.8 Practical Limitations and Catalyst Fouling

The major limitations of photocatalysis during pollutant abatement include the lack of visible light activity, pH dependence, separation and recovery, catalyst fouling and stability, etc. As most of the conventional semiconductor photocatalyst materials require higher intensity light sources for their excitation, the scaling up of the process becomes expensive. However, the surface modifications of catalysts can improve the visible light activity, reduces electron–hole recombination and stability of the catalyst (Batalović et al. 2017; Karim and Shriwastav 2020). Even though photocatalysis has been extensively studied for the removal of pollutants, the difficulty in recovery and loss in photocatalytic activity in the consecutive cycles presents a major problem in the practical application of the treatment technique. The morphology, crystalline structure, and particle size of nanoparticles are also a problem for remediation since they can change the surface area and the recovery after the treatment. Apart from these limitations, catalyst fouling or decay may occur due to various causes such as oxidation of metal catalysts to inactive oxides and loss of catalyst sites by adsorption of impurities on the surface (Argyle and Bartholomew 2015). For instance, Li et al. (2009) reported that zinc oxide is transformed into zinc carbonate and soluble carboxyl zinc by reaction with carbon dioxide or intermediates products such as carboxylic acid. They have observed regeneration of catalyst by soft-mechanochemical ion exchange method without producing any harmful byproducts. The deactivated catalyst can be regenerated by various methods including chemical and thermal regeneration methods. The study by Wang et al. (2019) investigated the regeneration of deactivated TiO_2 photocatalyst caused by sulfur poisoning. It was observed that the active sites of the catalyst were adsorbed with H_2SO_4/SO_4^{2-} and the thermal treatment was found to be effective by increasing the oxygen vacancy thereby improving the catalyst activity.

8.9 Challenges and Future Direction

Intense research in the field of photocatalytic degradation of organic and inorganic pollutants will be needed to address several key technical constraints before commercializing the technology. The commercialization of the photocatalytic technology suffers several challenges mainly due to rapid electron–hole recombination, restricted light absorption in the visible region, stability, difficulty in the separation, and reuse of catalyst. The photocatalyst should have a smaller bandgap to allow it to use the broad range of the solar spectrum. To overcome these issues, several surface modification techniques have been developed and greater research interest has been focused in the field of visible light active photocatalysis. The modification of catalyst surface should be able to deliver the photostability of the catalyst for multiple experimental cycles without causing dissolution of functional groups or metal ions. It should also increase the incidence of photons on the active sites of the catalyst to ensure the efficient charge separation and transportation of electrons in the catalyst. Multifunctional photocatalyst with higher stability and excellent visible light activity can maintain the long-term activity of the catalyst without reducing the surface-active sites.

Developing an efficient immobilization technique in slurry based photoreactor or designing reactors which can immobilize photocatalytic systems can overcome the limitation of recovery and aggregation of particles. The mass transfer limitation associated with heterogeneous catalysis can be improved to absorb all kinds of pollutants without considering their partition coefficient. Furthermore, fate and transport of the nanoparticles reaching the aquatic bodies have to be investigated concerning the safety and toxicity to other levels of organisms. Besides, there are limited studies on the toxicity of the final or intermediate product of the photocatalysis process or the photocatalyst itself. Because in various degradation processes the final product may be more dangerous than the parent compound. Therefore, there must be more research on the development of an environmentally friendly catalyst for the degradation and complete mineralization of emerging contaminants, so that it can be applied in real life.

8.10 Conclusion

Photocatalytic treatment of wastewater has been widely studied due to its superior photocatalytic activity under irradiation. The present chapter provides the basic review of the application of the various homogeneous and heterogeneous photocatalytic system and its mechanism for the degradation of different contaminants in the wastewater. Major factors affecting the photocatalytic degradation include pH of the aqueous solution, catalyst dose, source of light and intensity, presence of oxidants, and radical scavengers. The degradation of an organic compound by semiconductor photocatalyst involves the absorption of a photon of energy greater than the bandgap energy of the catalyst to form electron–hole pairs. The electron–hole pairs eventually recombine or initiate the oxidation–reduction reaction. However, these methods have some drawbacks such as lack of visible light activity, faster electron–hole recombination, photostability, etc. Several modifications of catalyst have been developed such as metal and non-metal doping, heterogeneous composites, hybrid heterostructures, polymers, metal deposition, and dye-sensitization to extend the visible light response and stability of the catalyst. To enhance the adsorption of contaminants and recovery potential of catalyst, various immobilization techniques of TiO_2 nanoparticles on support structures and magnetic separation has been adopted. However, there exist some technical barriers to the large-scale commercialization of the process. Further detailed research and development in this area are required to design one novel multifunctional catalyst having well visible light absorption, high recovery potential, high adsorption, and stability.

References

Abdelhaleem A, Chu W (2017) Photodegradation of 4-chlorophenoxyacetic acid under visible LED activated N-doped TiO_2 and the mechanism of stepwise rate increment of the reused catalyst. J Hazard Mater 338:491–501. https://doi.org/10.1016/j.jhazmat.2017.05.056

Abdullah AM, Al-Thani NJ, Tawbi K, Al-Kandari H (2016) Carbon/nitrogen-doped TiO_2: new synthesis route, characterization and application for phenol degradation. Arab J Chem 9:229–237. https://doi.org/10.1016/j.arabjc.2015.04.027

Ahmad M, Ahmed E, Hong ZL et al (2013) A facile one-step approach to synthesizing ZnO/graphene composites for enhanced degradation of methylene blue under visible light. Appl Surf Sci 274:273–281. https://doi.org/10.1016/j.apsusc.2013.03.035

Ahmed S, Rasul MG, Martens WN et al (2011) Advances in heterogeneous photocatalytic degradation of phenols and dyes in wastewater: a review. Water Air Soil Pollut 215:3–29. https://doi.org/10.1007/s11270-010-0456-3

Akpan UG, Hameed BH (2009) Parameters affecting the photocatalytic degradation of dyes using TiO_2-based photocatalysts: a review. J Hazard Mater 170:520–529. https://doi.org/10.1016/j.jhazmat.2009.05.039

Alansi AM, Al-Qunaibit M, Alade IO et al (2018) Visible-light responsive BiOBr nanoparticles loaded on reduced graphene oxide for photocatalytic degradation of dye. J Mol Liq 253:297–304. https://doi.org/10.1016/j.molliq.2018.01.034

Alberti S, Caratto V, Peddis D et al (2019) Synthesis and characterization of a new photocatalyst based on TiO_2 nanoparticles supported on a magnetic zeolite obtained from iron and steel industrial waste. J Alloys Compd 797:820–825. https://doi.org/10.1016/j.jallcom.2019.05.098

Aleboyeh A, Kasiri MB, Aleboyeh H (2012) Influence of dyeing auxiliaries on AB74 dye degradation by UV/H_2O_2 process. J Environ Manage 113:426–431. https://doi.org/10.1016/j.jenvman.2012.10.008

Ananpattarachai J, Kajitvichyanukul P, Seraphin S (2009) Visible light absorption ability and photocatalytic oxidation activity of various interstitial N-doped TiO_2 prepared from different nitrogen dopants. J Hazard Mater 168:253–261. https://doi.org/10.1016/j.jhazmat.2009.02.036

Anderson JA (2011) Photocatalytic nitrate reduction over Au/TiO_2. Catal Today 175:316–321. https://doi.org/10.1016/j.cattod.2011.04.009

Ani IJ, Akpan UG, Olutoye MA, Hameed BH (2018) Photocatalytic degradation of pollutants in petroleum refinery wastewater by TiO_2—and ZnO-based photocatalysts: Recent development. J Clean Prod 205:930–954. https://doi.org/10.1016/j.jclepro.2018.08.189

Ao X, Liu W (2017) Degradation of sulfamethoxazole by medium pressure UV and oxidants: peroxymonosulfate, persulfate, and hydrogen peroxide. Chem Eng J 313:629–637. https://doi.org/10.1016/j.cej.2016.12.089

Ao X, Liu W, Sun W et al (2018) Medium pressure UV-activated peroxymonosulfate for ciprofloxacin degradation: kinetics, mechanism, and genotoxicity. Chem Eng J 345:87–97. https://doi.org/10.1016/j.cej.2018.03.133

Argyle MD, Bartholomew CH (2015) Heterogeneous catalyst deactivation and regeneration: a review. Catalysts 5:145–269. https://doi.org/10.3390/catal5010145

Asahi R, Morikawa T, Aoki K, Taga Y (2001) Visible-Light Photocatalysis in Nitrogen-Doped Titanium Oxides. Science (80-) 293:269–271. https://doi.org/10.1126/science.1061051

Aslam M, Ismail IMI, Salah N et al (2015) Evaluation of sunlight induced structural changes and their effect on the photocatalytic activity of V_2O_5 for the degradation of phenols. J Hazard Mater 286:127–135. https://doi.org/10.1016/j.jhazmat.2014.12.022

Avisar D, Lester Y, Mamane H (2010) pH induced polychromatic UV treatment for the removal of a mixture of SMX, OTC and CIP from water. J Hazard Mater 175:1068–1074. https://doi.org/10.1016/j.jhazmat.2009.10.122

Bahnemann D (2004) Photocatalytic water treatment: solar energy applications. Sol Energy 77:445–459. https://doi.org/10.1016/j.solener.2004.03.031

Batalović K, Bundaleski N, Radaković J et al (2017) Modification of N-doped TiO_2 photocatalysts using noble metals (Pt, Pd)—a combined XPS and DFT study. Phys Chem Chem Phys 19:7062–7071. https://doi.org/10.1039/c7cp00188f

Bechambi O, Sayadi S, Najjar W (2015) Photocatalytic degradation of bisphenol A in the presence of C-doped ZnO: effect of operational parameters and photodegradation mechanism. J Ind Eng Chem 32:201–210. https://doi.org/10.1016/j.jiec.2015.08.017

Belaissa Y, Nibou D, Assadi AA et al (2016) A new hetero-junction p-CuO/n-ZnO for the removal of amoxicillin by photocatalysis under solar irradiation. J Taiwan Inst Chem Eng 68:254–265. https://doi.org/10.1016/j.jtice.2016.09.002

Benitez FJ, Acero JL, Real FJ et al (2013) Photolysis of model emerging contaminants in ultra-pure water: kinetics, by-products formation and degradation pathways. Water Res 47:870–880. https://doi.org/10.1016/j.watres.2012.11.016

Bergamonti L, Bergonzi C, Graiff C et al (2019) 3D printed chitosan scaffolds: a new TiO_2 support for the photocatalytic degradation of amoxicillin in water. Water Res 163:114841.https://doi.org/10.1016/j.watres.2019.07.008

Blance- Vega MP, Guzman- Mar JL, Villanuve-Rodriguez M et al (2017) Photocatalytic elimination of bisphenol A under visible light using Ni-doped TiO_2 synthesized by microwave assisted sol-gel method. Mater Sci Semicond Process 71:275–282. https://doi.org/10.1016/j.mssp.2017.08.013

Bora LV, Mewada RK (2017) Visible / solar light active photocatalysts for organic effluent treatment : Fundamentals, mechanisms and parametric review. Renew Sustain Energy Rev 76:1393–1421. https://doi.org/10.1016/j.rser.2017.01.130

Boussatha N, Gilliot M, Ghoualem H, Martin J (2018) Formation of nanogranular ZnO ultrathin films and estimation of their performance for photocatalytic degradation of amoxicillin antibiotic. Mater Res Bull 99:485–490. https://doi.org/10.1016/j.materresbull.2017.11.053

Brindha A, Sivakumar T (2017) Visible active N, S co-doped TiO_2/graphene photocatalysts for the degradation of hazardous dyes. J Photochem Photobiol A Chem 340:146–156. https://doi.org/10.1016/j.jphotochem.2017.03.010

Caglar Yılmaz H, Akgeyik E, Bougarrani S et al (2019) Photocatalytic degradation of amoxicillin using Co-doped TiO_2 synthesized by reflux method and monitoring of degradation products by LC–MS/MS. J Dispers Sci Technol 0:1–12. https://doi.org/10.1080/01932691.2019.1583576

Canterino M, Di Somma I, Marotta R, Andreozzi R (2008) Kinetic investigation of Cu(II) ions photoreduction in presence of titanium dioxide and formic acid. Water Res 42:4498–4506. https://doi.org/10.1016/j.watres.2008.07.035

Chakrabarti S, Chaudhuri B, Bhattacharjee S et al (2009) Photo-reduction of hexavalent chromium in aqueous solution in the presence of zinc oxide as semiconductor catalyst. Chem Eng J 153:86–93. https://doi.org/10.1016/j.cej.2009.06.021

Chang CT, Wang JJ, Ouyang T et al (2015) Photocatalytic degradation of acetaminophen in aqueous solutions by TiO_2/ZSM-5 zeolite with low energy irradiation. Mater Sci Eng B Solid-State Mater Adv Technol 196:53–60. https://doi.org/10.1016/j.mseb.2014.12.025

Chatterjee D, Dasgupta S, Rao N, N, (2006) Visible light assisted photodegradation of halocarbons on the dye modified TiO_2 surface using visible light. Sol Energy Mater Sol Cells 90:1013–1020. https://doi.org/10.1016/j.solmat.2005.05.016

Chatterjee D, Mahata A (2001) Demineralization of organic pollutants on the dye modified TiO_2 semiconductor particulate system using visible light. Appl Catal B Environ 33:119–125. https://doi.org/10.1016/S0926-3373(01)00170-9

Chekir N, Benhabiles O, Tassalit D et al (2016) Photocatalytic degradation of methylene blue in aqueous suspensions using TiO_2 and ZnO. Desalin Water Treat 57:6141–6147. https://doi.org/10.1080/19443994.2015.1060533

Chen D, Jiang Z, Geng J et al (2007) Carbon and nitrogen co-doped TiO_2 with enhanced visible-light photocatalytic activity. Ind Eng Chem Res 46:2741–2746. https://doi.org/10.1021/ie061491k

Chen D, Yang D, Geng J et al (2008) Improving visible-light photocatalytic activity of N-doped TiO_2 nanoparticles via sensitization by Zn porphyrin. Appl Surf Sci 255:2879–2884. https://doi.org/10.1016/j.apsusc.2008.08.032

Chen SH, Yin Z, Luo SL et al (2012) Photoreactive mesoporous carbon / Bi_2WO_6 composites : synthesis and reactivity. Appl Surf Sci 259:7–12. https://doi.org/10.1016/j.apsusc.2012.05.075

Chen X, Wu Z, Liu D, Gao Z (2017) Preparation of ZnO photocatalyst for the efficient and rapid photocatalytic degradation of azo dyes. Nanoscale Res Lett 12:4–13. https://doi.org/10.1186/s11671-017-1904-4

Chen Y, Liu K (2017) Fabrication of magnetically recyclable Ce/N co-doped $TiO_2/NiFe_2O_4$/diatomite ternary hybrid: improved photocatalytic efficiency under visible light irradiation. J Alloys Compd 697:161–173. https://doi.org/10.1016/j.jallcom.2016.12.153

Chen Y, Liu Z, Wang Z et al (2011) Photodegradation of propranolol by Fe(III)-citrate complexes: kinetics, mechanism and effect of environmental media. J Hazard Mater 194:202–208. https://doi.org/10.1016/j.jhazmat.2011.07.081

Chiou C, Wu C, Juang R (2008) Influence of operating parameters on photocatalytic degradation of phenol in UV / TiO_2 process. 139:322–329. https://doi.org/10.1016/j.cej.2007.08.002

Chong MN, Jin B, Chow CWK, Saint C (2010) Recent developments in photocatalytic water treatment technology: a review. Water Res 44:2997–3027. https://doi.org/10.1016/j.watres.2010.02.039

Chowdhury P, Moreira J, Gomaa H, Ray AK (2012) Visible-solar-light-driven photocatalytic degradation of phenol with dye-sensitized TiO_2: parametric and kinetic study. Ind Eng Chem Res 51:4523–4532. https://doi.org/10.1021/ie2025213

Clarizia L, Russo D, Di Somma I et al (2017) Homogeneous photo-Fenton processes at near neutral pH: a review. Appl Catal B Environ 209:358–371. https://doi.org/10.1016/j.apcatb.2017.03.011

Cong Y, Zhang J, Chen F, Anpo M (2007) Synthesis and characterization of nitrogen-doped TiO_2 nanophotocatalyst with high visible light activity. J Phys Chem C 111:6976–6982. https://doi.org/10.1021/jp0685030

Cui B, An W, Liu L et al (2014) Synthesis of CdS/BiOBr composite and its enhanced photocatalyticdegradation for Rhodamine. Appl Surf Sci 319:298–305. https://doi.org/10.1016/j.apsusc.2014.05.179

Deng J, Shao Y, Gao N et al (2013) Degradation of the antiepileptic drug carbamazepine upon different UV-based advanced oxidation processes in water. Chem Eng J 222:150–158. https://doi.org/10.1016/j.cej.2013.02.045

Díaz-Uribe C, Vallejo W, Campos K et al (2018) Improvement of the photocatalytic activity of TiO_2 using Colombian Caribbean species (Syzygium cumini) as natural sensitizers: experimental and theoretical studies. Dye Pigment 150:370–376. https://doi.org/10.1016/j.dyepig.2017.12.027

Djellabi R, Ghorab MF (2015) Photoreduction of toxic chromium using TiO_2-immobilized under natural sunlight: effects of some hole scavengers and process parameters. Desalin Water Treat 55:1900–1907. https://doi.org/10.1080/19443994.2014.927335

Dong F, Guo Y, Zhang J et al (2013) Size-controllable hydrothermal synthesis of ZnS nanospheres and the application in photocatalytic degradation of organic dyes. Mater Lett 97:59–63. https://doi.org/10.1016/j.matlet.2013.01.029

Dong H, Zeng G, Tang L et al (2015) An overview on limitations of TiO_2-based particles for photocatalytic degradation of organic pollutants and the corresponding countermeasures. Water Res 79:128–146. https://doi.org/10.1016/j.watres.2015.04.038

Dong S, Feng J, Fan M et al (2015) Recent developments in heterogeneous photocatalytic water treatment using visible light-responsive photocatalysts: a review. RSC Adv 5:14610–14630. https://doi.org/10.1039/C4RA13734E

Dong S, Sun J, Li Y et al (2014) $ZnSnO^3$ hollow nanospheres/reduced graphene oxide nanocomposites as high-performance photocatalysts for degradation of metronidazole. Appl Catal B Environ 144:386–393. https://doi.org/10.1016/j.apcatb.2013.07.043

Doudrick K, Monzón O, Mangonon A et al (2012) Nitrate reduction in water using commercial titanium dioxide photocatalysts (P25, P90, and Hombikat UV100). J Environ Eng (united States) 138:852–861. https://doi.org/10.1061/(ASCE)EE.1943-7870.0000529

Dozzi MV, Saccomanni A, Selli E (2012) Cr(VI) photocatalytic reduction: effects of simultaneous organics oxidation and of gold nanoparticles photodeposition on TiO_2. J Hazard Mater 211–212:188–195. https://doi.org/10.1016/j.jhazmat.2011.09.038

Duan X, He X, Wang D et al (2017) Decomposition of iodinated pharmaceuticals by UV-254 nm-assisted advanced oxidation processes. J Hazard Mater 323:489–499. https://doi.org/10.1016/j.jhazmat.2016.04.022

Elmolla ES, Chaudhuri M (2010) Photocatalytic degradation of amoxicillin ampicillin and cloxacillin antibiotics in aqueous solution using UV/TiO_2 and UV/H_2O_2/TiO_2 photocatalysis. Desalination 252:46–52. https://doi.org/10.1016/j.desal.2009.11.003

Eren Z (2012) Degradation of an azo dye with homogeneous and heterogeneous catalysts by sonophotolysis. Clean—Soil Air Water 40:1284–1289. https://doi.org/10.1002/clen.201100384

Eskizeybek V, Sari F, Gülce H et al (2012) Preparation of the new polyaniline/ZnO nanocomposite and its photocatalytic activity for degradation of methylene blue and malachite green dyes under UV and natural sun lights irradiations. Appl Catal B Environ 119–120:197–206. https://doi.org/10.1016/j.apcatb.2012.02.034

Fatimah I, Wang S, Wulandari D (2011) ZnO/montmorillonite for photocatalytic and photochemical degradation of methylene blue. Appl Clay Sci 53:553–560. https://doi.org/10.1016/j.clay.2011.05.001

Feng J, Yang Z, He S et al (2018) Photocatalytic reduction of Uranium(VI) under visible light with Sn-doped In2S3 microspheres. Chemosphere 212:114–123. https://doi.org/10.1016/j.chemosphere.2018.08.070

Feng X, Wang Z, Chen Y et al (2012) Effect of Fe(III)/citrate concentrations and ratio on the photoproduction of hydroxyl radicals: application on the degradation of diphenhydramine. Ind Eng Chem Res 51:7007–7012. https://doi.org/10.1021/ie300360p

Frontistis Z, Antonopoulou M, Petala A et al (2017) Photodegradation of ethylparaben using simulated solar radiation and Ag_3PO_4 photocatalyst. J Hazard Mater 323:478–488. https://doi.org/10.1016/j.jhazmat.2016.04.017

Gao B, Lim TM, Subagio DP, Lim TT (2010) Zr-doped TiO_2 for enhanced photocatalytic degradation of bisphenol A. Appl Catal A Gen 375:107–115. https://doi.org/10.1016/j.apcata.2009.12.025

Gao B, Wang J, Dou M et al (2020) Enhanced photocatalytic removal of amoxicillin with Ag/TiO_2 / mesoporous g-C_3N_4 under visible light: property and mechanistic studies. Environ Sci Pollut Res 27:7025–7039

Gao B, Yap PS, Lim TM, Lim TT (2011) Adsorption-photocatalytic degradation of Acid Red 88 by supported TiO_2: Effect of activated carbon support and aqueous anions. Chem Eng J 171:1098–1107. https://doi.org/10.1016/j.cej.2011.05.006

Gao Z, Liu N, Wu D et al (2012) Graphene-CdS composite, synthesis and enhanced photocatalytic activity. Appl Surf Sci 258:2473–2478. https://doi.org/10.1016/j.apsusc.2011.10.075

Gaya UI, Abdullah AH (2008) Heterogeneous photocatalytic degradation of organic contaminants over titanium dioxide: A review of fundamentals, progress and problems. J Photochem Photobiol C Photochem Rev 9:1–12. https://doi.org/10.1016/j.jphotochemrev.2007.12.003

Ge L (2008) Novel Pd/$BiVO_4$ composite photocatalysts for efficient degradation of methyl orange under visible light irradiation. Mater Chem Phys 107:465–470. https://doi.org/10.1016/j.matchemphys.2007.08.016

Ge M, Chen Y, Liu M, Li M (2015) Synthesis of magnetically separable Ag_3PO_4/$ZnFe_2O_4$ composite photocatalysts for dye degradation under visible LED light irradiation. J Environ Chem Eng 3:2809–2815. https://doi.org/10.1016/j.jece.2015.10.011

Giraldo-Aguirre AL, Erazo-Erazo ED, Flórez-Acosta OA et al (2015) TiO_2 photocatalysis applied to the degradation and antimicrobial activity removal of oxacillin: Evaluation of matrix components, experimental parameters, degradation pathways and identification of organics by-products. J Photochem Photobiol A Chem 311:95–103. https://doi.org/10.1016/j.jphotochem.2015.06.021

Gomes JF, Leal I, Bednarczyk K et al (2017) Detoxification of parabens using UV-A enhanced by noble metals—TiO_2 supported catalysts. J Environ Chem Eng 5:3065–3074. https://doi.org/10.1016/j.jece.2017.06.010

Gulshan F, Yanagida S, Kameshima Y et al (2010) Various factors affecting photodecomposition of methylene blue by iron-oxides in an oxalate solution. Water Res 44:2876–2884. https://doi.org/10.1016/j.watres.2010.01.040

Guo J, Du Y, Lan Y, Mao J (2011) Photodegradation mechanism and kinetics of methyl orange catalyzed by Fe(III) and citric acid. J Hazard Mater 186:2083–2088. https://doi.org/10.1016/j.jhazmat.2010.12.112

Guo N, Liu H, Fu Y, Hu J (2020) Preparation of Fe_2O_3 nanoparticles doped with In_2O_3 and photocatalytic degradation property for rhodamine B. Optik (Stuttg) 201:163537https://doi.org/10.1016/j.ijleo.2019.163537

Hariharan C (2006) Photocatalytic degradation of organic contaminants in water by ZnO nanoparticles: revisited. Appl Catal A Gen 304:55–61. https://doi.org/10.1016/j.apcata.2006.02.020

Hassani A, Khataee A, Karaca S (2015) Photocatalytic degradation of ciprofloxacin by synthesized TiO_2 nanoparticles on montmorillonite: effect of operation parameters and artificial neural network modeling. J Mol Catal A Chem 409:149–161. https://doi.org/10.1016/j.molcata.2015.08.020

He Q, Zhang Z, Xiong J et al (2008) A novel biomaterial—Fe_3O_4:TiO_2 core-shell nano particle with magnetic performance and high visible light photocatalytic activity. Opt Mater (amst) 31:380–384. https://doi.org/10.1016/j.optmat.2008.05.011

Hosseini SN, Borghei SM, Vossoughi M, Taghavinia N (2007) Immobilization of TiO_2 on perlite granules for photocatalytic degradation of phenol. 74:53–62. https://doi.org/10.1016/j.apcatb.2006.12.015

Huang W, Brigante M, Wu F et al (2012) Development of a new homogenous photo-Fenton process using Fe(III)-EDDS complexes. J Photochem Photobiol A Chem 239:17–23. https://doi.org/10.1016/j.jphotochem.2012.04.018

Idris A, Majidnia Z, Nor Kamarudin KS (2016) Photocatalyst treatment for lead(II) using titanium oxide nanoparticles embedded in PVA-alginate beads. Desalin Water Treat 57:5035–5044. https://doi.org/10.1080/19443994.2015.1006256

Jagannath R, Central T, Researc MC (2009) Photocatalytic degradation of methylene blue dye using ultraviolet light emitting diodes photocatalytic degradation of methylene blue dye using ultraviolet light. Ind Eng Chem Res 48:10262–10267. https://doi.org/10.1021/ie9012437

Jo W, Tayade RJ (2014) New generation energy-efficient light source for photocatalysis : LEDs for environmental applications. Ind Eng Chem Res 53:2073–2084. https://doi.org/10.1021/ie404176g

Joshi M, Kamble SP, Labhsetwar NK et al (2009) Chlorophyll-based photocatalysts and their evaluations for methyl orange photoreduction. J Photochem Photobiol A Chem 204:83–89. https://doi.org/10.1016/j.jphotochem.2009.01.016

Kamaraj M, Ranjith KS, Sivaraj R et al (2014) Photocatalytic degradation of endocrine disruptor Bisphenol-A in the presence of prepared CexZn1-xO nanocomposites under irradiation of sunlight. J Environ Sci (china) 26:2362–2368. https://doi.org/10.1016/j.jes.2014.09.022

Karim A V, Shriwastav A (2020) Degradation of ciprofloxacin using photo, sono, and sonophotocatalytic oxidation with visible light and low-frequency ultrasound: degradation kinetics and pathways. Chem Eng J 392:124853.https://doi.org/10.1016/j.cej.2020.124853

Karimi MA, Mehrjard AH, Kabir AA, Zaydabadi M (2015) ZnO nanocomposite supported on activated carbon for photocatalytic degradation of methylene blue. Res Chem Intermed 41:6157–6168. https://doi.org/10.1007/s11164-014-1729-z

Kassinos D, Varnava N, Michael C, Piera P (2009) Homogeneous oxidation of aqueous solutions of atrazine and fenitrothion through dark and photo-fenton reactions. Chemosphere 74:866–872. https://doi.org/10.1016/j.chemosphere.2008.10.008

Kathiravan A, Chandramohan M, Renganathan R, Sekar S (2009) Cyanobacterial chlorophyll as a sensitizer for colloidal TiO_2. Spectrochim Acta—Part A Mol Biomol Spectrosc 71:1783–1787. https://doi.org/10.1016/j.saa.2008.06.031

Kaur T, Sraw A, Wanchoo RK, Toor AP (2018) Solar assisted degradation of carbendazim in water using clay beads immobilized with TiO_2 & Fe doped TiO_2. Sol Energy 162:45–56. https://doi.org/10.1016/j.solener.2017.11.033

Khalid NR, Ahmed E, Hong Z et al (2013) Enhanced photocatalytic activity of graphene-TiO_2 composite under visible light irradiation. Curr Appl Phys 13:659–663. https://doi.org/10.1016/j.cap.2012.11.003

Khataee AR, Kasiri MB (2010) Photocatalytic degradation of organic dyes in the presence of nanostructured titanium dioxide: influence of the chemical structure of dyes. J Mol Catal A Chem 328:8–26. https://doi.org/10.1016/j.molcata.2010.05.023

Kim I, Yamashita N, Tanaka H (2009) Performance of UV and UV/H_2O_2 processes for the removal of pharmaceuticals detected in secondary effluent of a sewage treatment plant in Japan. J Hazard Mater 166:1134–1140. https://doi.org/10.1016/j.jhazmat.2008.12.020

Lai C, Zhang M, Li B et al (2019) Fabrication of CuS/$BiVO_4$ (0 4 0) binary heterojunction photocatalysts with enhanced photocatalytic activity for Ciprofloxacin degradation and mechanism insight. Chem Eng J 358:891–902. https://doi.org/10.1016/j.cej.2018.10.072

Lai HF, Chen CC, Chang YK et al (2014) Efficient photocatalytic degradation of thiobencarb over $BiVO_4$ driven by visible light: parameter and reaction pathway investigations. Sep Purif Technol 122:78–86. https://doi.org/10.1016/j.seppur.2013.10.049

Le S, Jiang T, Li YY et al (2017) Highly efficient visible-light-driven mesoporous graphitic carbon nitride/ZnO nanocomposite photocatalysts. Appl Catal B Environ 200:601–610. https://doi.org/10.1016/j.apcatb.2016.07.027

Li F, Kang Y, Chen M et al (2016) Photocatalytic degradation and removal mechanism of ibuprofen via monoclinic $BiVO_4$ under simulated solar light. Chemosphere 150:139–144. https://doi.org/10.1016/j.chemosphere.2016.02.045

Li G, Jiang B, Li X et al (2013) C60/$Bi_2TiO_4F_2$ heterojunction photocatalysts with enhanced visible-light activity for environmental remediation. ACS Appl Mater Interfaces 5:7190–7197. https://doi.org/10.1021/am401525m

Li P, Wang J, Peng T et al (2019) Heterostructure of anatase-rutile aggregates boosting the photoreduction of U(VI). Appl Surf Sci 483:670–676. https://doi.org/10.1016/j.apsusc.2019.03.330

Li XJ, Wang DY, Yang K, Zhong ZX (2013) Photoreduction of Hg(II) in the presence of SO_4^{2-} under artificial solar radiation. Adv Mater Res 821–822:917–921. https://doi.org/10.4028/www.scientific.net/AMR.821-822.917

Li Y, Jiao Z, Yang N, Gao H (2009) Regeneration of nano-ZnO photocatalyst by the means of soft-mechanochemical ion exchange method. J Environ Sci 21:S69–S72. https://doi.org/10.1016/S1001-0742(09)60040-1

Li Y, Li X, Li J, Yin J (2006) Photocatalytic degradation of methyl orange by TiO_2-coated activated carbon and kinetic study. Water Res 40:1119–1126. https://doi.org/10.1016/j.watres.2005.12.042

Liang PL, Yuan LY, Deng H et al (2020) Photocatalytic reduction of uranium(VI) by magnetic $ZnFe_2O_4$ under visible light. Appl Catal B Environ 267:118688. https://doi.org/10.1016/j.apcatb.2020.118688

Lin Y, Hsueh H, Chang C, Chu H (2016) The visible-light-driven photodegradation of dimethyl sulfide on S-doped TiO_2: Characterization kinetics and reaction pathways. Appl Catal B Environ 199:1–10. https://doi.org/10.1016/j.apcatb.2016.06.024

Lin YT, Weng CH, Lin YH et al (2013) Effect of C content and calcination temperature on the photocatalytic activity of C-doped TiO_2 catalyst. Sep Purif Technol 116:114–123. https://doi.org/10.1016/j.seppur.2013.05.018

Liu D, Wu Z, Tian F et al (2016) Synthesis of N and la co-doped TiO_2/AC photocatalyst by microwave irradiation for the photocatalytic degradation of naphthalene. J Alloys Compd 676:489–498. https://doi.org/10.1016/j.jallcom.2016.03.124

Liu D, Zhao X, Song L, Zhang S (2018) Synthesis, performance and action mechanism of carbon black/Ag_3PO_4 photocatalysts. Ceram Int 44:13712–13719. https://doi.org/10.1016/j.ceramint.2018.04.212

Liu L, Qi Y, Lu J et al (2016) A stable Ag_3PO_4@g-C_3N_4 hybrid core@shell composite with enhanced visible light photocatalytic degradation. Appl Catal B Environ 183:133–141. https://doi.org/10.1016/j.apcatb.2015.10.035

Liu X, Liu Y, Lu S et al (2018) Performance and mechanism into TiO_2/Zeolite composites for sulfadiazine adsorption and photodegradation. Chem Eng J 350:131–147. https://doi.org/10.1016/j.cej.2018.05.141

Liu Y, He X, Fu Y, Dionysiou DD (2016) Degradation kinetics and mechanism of oxytetracycline by hydroxyl radical-based advanced oxidation processes. Chem Eng J 284:1317–1327. https://doi.org/10.1016/j.cej.2015.09.034

Mais L, Mascia M, Palmas S, Vacca A (2019) Photoelectrochemical oxidation of phenol with nanostructured TiO_2-PANI electrodes under solar light irradiation. Sep Purif Technol 208:153–159. https://doi.org/10.1016/j.seppur.2018.03.074

Martins AC, Cazetta AL, Pezoti O et al (2017) Sol-gel synthesis of new TiO_2/activated carbon photocatalyst and its application for degradation of tetracycline. Ceram Int 43:4411–4418. https://doi.org/10.1016/j.ceramint.2016.12.088

Menon NG, Tatiparti SSV, Mukherji S (2019) Synthesis, characterization and photocatalytic activity evaluation of TiO_2–ZnO nanocomposites: elucidating effect of varying Ti: Zn molar ratio. Colloids Surf A Physicochem Eng Asp 565:47–58. https://doi.org/10.1016/j.colsurfa.2018.12.053

Miller JS, Olejnik D (2001) Photolysis of polycyclic aromatic hydrocarbons in water. Water Res 35:233–243. https://doi.org/10.1016/S0043-1354(00)00230-X

Mirzaei A, Chen Z, Haghighat F, Yerushalmi L (2019) Magnetic fluorinated mesoporous g-C_3N_4 for photocatalytic degradation of amoxicillin: transformation mechanism and toxicity assessment. Appl Catal B Environ 242:337–348. https://doi.org/10.1016/j.apcatb.2018.10.009

Moctezuma E, Leyva E, Aguilar CA et al (2012) Photocatalytic degradation of paracetamol: intermediates and total reaction mechanism. J Hazard Mater 243:130–138. https://doi.org/10.1016/j.jhazmat.2012.10.010

Monteagudo JM, Durán A, San Martín I (2014) Mineralization of wastewater from the pharmaceutical industry containing chloride ions by UV photolysis of H_2O_2/Fe(II) and ultrasonic irradiation. J Environ Manage 141:61–69. https://doi.org/10.1016/j.jenvman.2014.03.020

Murruni L, Conde F, Leyva G, Litter MI (2008) Photocatalytic reduction of Pb(II) over TiO_2: new insights on the effect of different electron donors. Appl Catal B Environ 84:563–569. https://doi.org/10.1016/j.apcatb.2008.05.012

Narayan MR (2012) Review: dye sensitized solar cells based on natural photosensitizers. Renew Sustain Energy Rev 16:208–215. https://doi.org/10.1016/j.rser.2011.07.148

Nasr O, Mohamed O, Al-Shirbini AS, Abdel-Wahab AM (2019) Photocatalytic degradation of acetaminophen over Ag, Au and Pt loaded TiO_2 using solar light. J Photochem Photobiol A Chem 374:185–193. https://doi.org/10.1016/j.jphotochem.2019.01.032

Natarajan K, Natarajan TS, Bajaj HC, Tayade RJ (2011) Photocatalytic reactor based on UV-LED/TiO_2 coated quartz tube for degradation of dyes. Chem Eng J 178:40–49. https://doi.org/10.1016/j.cej.2011.10.007

Nazeeruddin MK, Klein C, Liska P, Grätzel M (2005) Synthesis of novel ruthenium sensitizers and their application in dye-sensitized solar cells. Coord Chem Rev 249:1460–1467. https://doi.org/10.1016/j.ccr.2005.03.025

Nguyen CH, Fu CC, Juang RS (2018) Degradation of methylene blue and methyl orange by palladium-doped TiO_2 photocatalysis for water reuse: efficiency and degradation pathways. J Clean Prod 202:413–427. https://doi.org/10.1016/j.jclepro.2018.08.110

Ni M, Ã MKHL, Leung DYC, Sumathy K, (2007) A review and recent developments in photocatalytic water-splitting using TiO_2 for hydrogen production. Renew Sustain Energy Rev 11:401–425. https://doi.org/10.1016/j.rser.2005.01.009

Niu J, Yao B, Chen Y et al (2013) Enhanced photocatalytic activity of nitrogen doped TiO_2 photocatalysts sensitized by metallo Co, Ni-porphyrins. Appl Surf Sci 271:39–44. https://doi.org/10.1016/j.apsusc.2012.12.175

Nu Hoai Nguyen V, Amal R, Beydoun D (2005) Photocatalytic reduction of selenium ions using different TiO_2 photocatalysts. Chem Eng Sci 60:5759–5769. https://doi.org/10.1016/j.ces.2005.04.085

Ohno T, Mitsui T, Matsumura M (2003) Photocatalytic activity of S-doped TiO_2 photocatalyst under visible light. Chem Lett 32:364–365. https://doi.org/10.1246/cl.2003.364

Ohtani B (2010) Photocatalysis A to Z-What we know and what we do not know in a scientific sense. J Photochem Photobiol C Photochem Rev 11:157–178. https://doi.org/10.1016/j.jphotochemrev.2011.02.001

Ola O, Maroto-valer MM (2015) Review of material design and reactor engineering on TiO_2 photocatalysis for CO_2 reduction. J Photochem Photobiol C Photochem Rev 24:16–42. https://doi.org/10.1016/j.jphotochemrev.2015.06.001

Online VA, Yang C, Dong W et al (2017) Highly e ffi cient photocatalytic degradation of synergetic e ff ect of TiO_2 and P2ABSA. RSC Adv 7:23699–23708. https://doi.org/10.1039/c7ra02423a

Ou X, Quan X, Chen S et al (2008) Photocatalytic reaction by Fe(III)-citrate complex and its effect on the photodegradation of atrazine in aqueous solution. J Photochem Photobiol A Chem 197:382–388. https://doi.org/10.1016/j.jphotochem.2008.02.001

Ouzzine M, Romero-Anaya AJ, Lillo-Ródenas MA, Linares-Solano A (2014) Spherical activated carbon as an enhanced support for TiO_2/AC photocatalysts. Carbon N Y 67:104–118. https://doi.org/10.1016/j.carbon.2013.09.069

Pant B, Barakat NAM, Raj H et al (2014) Synthesis and photocatalytic activities of CdS/TiO_2 nanoparticles supported on carbon nanofibers for high efficient adsorption and simultaneous decomposition of organic dyes. J Colloid Interface Sci 434:159–166. https://doi.org/10.1016/j.jcis.2014.07.039

Pawar RC, Lee CS (2014) Single-step sensitization of reduced graphene oxide sheets and CdS nanoparticles on ZnO nanorods as visible-light photocatalysts. Appl Catal B Environ 144:57–65. https://doi.org/10.1016/j.apcatb.2013.06.022

Pelaez M, Nolan N, Seery M (2012) A review on the visible light active titanium dioxide photocatalysts for environmental applications. Appl Catal B Environ 125:331–349. https://doi.org/10.1016/j.apcatb.2012.05.036

Pereira JHOS, Queirós DB, Reis AC et al (2014) Process enhancement at near neutral pH of a homogeneous photo-Fenton reaction using ferricarboxylate complexes: application to oxytetracycline degradation. Chem Eng J 253:217–228. https://doi.org/10.1016/j.cej.2014.05.037

Pereira VJ, Linden KG, Weinberg HS (2007) Evaluation of UV irradiation for photolytic and oxidative degradation of pharmaceutical compounds in water. Water Res 41:4413–4423. https://doi.org/10.1016/j.watres.2007.05.056

Phongamwong T, Chareonpanich M, Limtrakul J (2015) Role of chlorophyll in Spirulina on photocatalytic activity of CO_2 reduction under visible light over modified N-doped TiO_2 photocatalysts. Appl Catal B Environ 168–169:114–124. https://doi.org/10.1016/j.apcatb.2014.12.022

Phongamwong T, Donphai W, Prasitchoke P et al (2017) Novel visible-light-sensitized Chl-Mg / P25 catalysts for photocatalytic degradation of rhodamine B. Appl Catal B Environ 207:326–334. https://doi.org/10.1016/j.apcatb.2017.02.042

Pourzad A, Sobhi HR, Behbahani M et al (2019) Efficient visible light-induced photocatalytic removal of paraquat using N-doped TiO_2@SiO_2@Fe_3O_4 nanocomposite. J Mol Liq 299. https://doi.org/10.1016/j.molliq.2019.112167

Qin S, Chen W, Li Y et al (2012) Photodecolorization of Rhodamine B on tungsten-doped TiO_2/activated carbon under visible-light irradiation. J Hazard Mater 227–228:25–33. https://doi.org/10.1016/j.jhazmat.2012.04.071

Quiñones DH, Rey A, Álvarez PM et al (2014) Enhanced activity and reusability of TiO_2 loaded magnetic activated carbon for solar photocatalytic ozonation. Appl Catal B Environ 144:96–106. https://doi.org/10.1016/j.apcatb.2013.07.005

Rauf MA, Ashraf SS (2009) Fundamental principles and application of heterogeneous photocatalytic degradation of dyes in solution. Chem Eng J 151:10–18. https://doi.org/10.1016/j.cej.2009.02.026

Repo E, Rengaraj S, Pulkka S et al (2013) Photocatalytic degradation of dyes by CdS microspheres under near UV and blue LED radiation. Sep Purif Technol 120:206–214. https://doi.org/10.1016/j.seppur.2013.10.008

Sabaté J, Bayona JM, Solanas AM (2001) Photolysis of PAHs in aqueous phase by UV irradiation. Chemosphere 44:119–124. https://doi.org/10.1016/S0045-6535(00)00208-3

Safajou H, Khojasteh H, Salavati-Niasari M, Mortazavi-Derazkola S (2017) Enhanced photocatalytic degradation of dyes over graphene/Pd/TiO_2 nanocomposites: TiO_2 nanowires versus TiO_2 nanoparticles. J Colloid Interface Sci 498:423–432. https://doi.org/10.1016/j.jcis.2017.03.078

Saien J, Mesgari Z (2016) Highly efficient visible-light photocatalyst of nitrogen-doped TiO_2 nanoparticles sensitized by hematoporphyrin. J Mol Catal A Chem 414:108–115. https://doi.org/10.1016/j.molcata.2015.12.027

Sakthivel S, Shankar MV, Palanichamy M et al (2002) Photocatalytic decomposition of leather dye. Comparative study of TiO_2 supported on alumina and glass beads. J Photochem Photobiol A Chem 148:153–159. https://doi.org/10.1016/S1010-6030(02)00085-0

Salimi M, Behbahani M, Sobhi HR et al (2019) A new nano-photocatalyst based on Pt and Bi co-doped TiO_2 for efficient visible-light photo degradation of amoxicillin. New J Chem 43:1562–1568. https://doi.org/10.1039/c8nj05020a

Sambaza S, Maity A, Pillay K (2019) Enhanced degradation of BPA in water by PANI supported Ag/TiO_2 nanocomposite under UV and visible light. J Environ Chem Eng 7:102880https://doi.org/10.1016/j.jece.2019.102880

Sanches S, Barreto Crespo MT, Pereira VJ (2010) Drinking water treatment of priority pesticides using low pressure UV photolysis and advanced oxidation processes. Water Res 44:1809–1818. https://doi.org/10.1016/j.watres.2009.12.001

Sanches S, Leitão C, Penetra A et al (2011) Direct photolysis of polycyclic aromatic hydrocarbons in drinking water sources. J Hazard Mater 192:1458–1465. https://doi.org/10.1016/j.jhazmat.2011.06.065

Saravanan R, Sacari E, Gracia F et al (2016) Conducting PANI stimulated ZnO system for visible light photocatalytic degradation of coloured dyes. J Mol Liq 221:1029–1033. https://doi.org/10.1016/j.molliq.2016.06.074

Sboui M, Nsib MF, Rayes A et al (2017) TiO_2–PANI/Cork composite: a new floating photocatalyst for the treatment of organic pollutants under sunlight irradiation. J Environ Sci 60:3–13. https://doi.org/10.1016/j.jes.2016.11.024

Seraghni N, Belattar S, Mameri Y et al (2012) Fe(III)-citrate-complex-induced photooxidation of 3-methylphenol in aqueous solution. Int J Photoenergy 2012:1–10. https://doi.org/10.1155/2012/630425

Shaban YA (2013) Effective photocatalytic reduction of Cr(VI) by carbon modified (CM)-n-TiO_2 nanoparticles under solar irradiation. World J Nano Sci Eng 03:154–160. https://doi.org/10.4236/wjnse.2013.34018

Shaban YA, El Maradny AA, Al Farawati RK (2016) Photocatalytic reduction of nitrate in seawater using C/TiO_2 nanoparticles. J Photochem Photobiol A Chem 328:114–121. https://doi.org/10.1016/j.jphotochem.2016.05.018

Sharma CP, Karim AV, Shriwastav A (2019) Decolorization of methylene blue using Fe(III)-citrate complex in a solar photo-Fenton process: impact of solar variability on process optimization. Water Sci Technol 80:2047–2057. https://doi.org/10.2166/wst.2019.411

Shayeghi M, Dehghani MH, Alimohammadi M, Goodini K (2012) Using ultraviolet irradiation for removal of malathion pesticide in water. J Arthropod Borne Dis 6:45–53

Shen C, Wang YJ, Xu JH, Luo GS (2012) Facile synthesis and photocatalytic properties of TiO_2 nanoparticles supported on porous glass beads. Chem Eng J 209:478–485. https://doi.org/10.1016/j.cej.2012.08.044

Sichel C, Garcia C, Andre K (2011) Feasibility studies: UV/chlorine advanced oxidation treatment for the removal of emerging contaminants. Water Res 45:6371–6380. https://doi.org/10.1016/j.watres.2011.09.025

Soares PA, Batalha M, Souza SMAGUAGU et al (2015) Enhancement of a solar photo-Fenton reaction with ferric-organic ligands for the treatment of acrylic-textile dyeing wastewater. J Environ Manage 152:120–131. https://doi.org/10.1016/j.jenvman.2015.01.032

Son HS, Ko G, Zoh KD (2009) Kinetics and mechanism of photolysis and TiO_2 photocatalysis of triclosan. J Hazard Mater 166:954–960. https://doi.org/10.1016/j.jhazmat.2008.11.107

Sonia S, Poongodi S, Kumar PS et al (2015) Hydrothermal synthesis of highly stable CuO nanostructures for efficient photocatalytic degradation of organic dyes. Mater Sci Semicond Process 30:585–591. https://doi.org/10.1016/j.mssp.2014.10.012

Souza BM, Dezotti MWC, Boaventura RAR, Vilar VJP (2014) Intensification of a solar photo-Fenton reaction at near neutral pH with ferrioxalate complexes: a case study on diclofenac removal from aqueous solutions. Chem Eng J 256:448–457. https://doi.org/10.1016/j.cej.2014.06.111

Sowmya A, Meenakshi S (2015) Photocatalytic reduction of nitrate over Ag-TiO_2 in the presence of oxalic acid. J Water Process Eng 8:e23–e30. https://doi.org/10.1016/j.jwpe.2014.11.004

Sun B, Reddy EP, Smirniotis PG (2005) Visible light Cr(VI) reduction and organic chemical oxidation by TiO_2 photocatalysis. Environ Sci Technol 39:6251–6259. https://doi.org/10.1021/es0480872

Sun Y, Wang W, Zhang L, Sun S (2013) The photocatalysis of Bi_2MoO_6 under the irradiation of blue LED. Mater Res Bull 48:4357–4361. https://doi.org/10.1016/j.materresbull.2013.07.015

Tadjarodi A, Imani M, Kerdari H et al (2014) Preparation of CdO rhombus-like nanostructure and its photocatalytic degradation of Azo dyes from aqueous solution. Nanomater Nanotechnol 4https://doi.org/10.5772/58464

Tambat S, Umale S, Sontakke S (2016) Photocatalytic degradation of milling yellow dye using sol-gel synthesized CeO_2. Mater Res Bull 76:466–472. https://doi.org/10.1016/j.materresbull.2016.01.010

Tan C, Gao N, Deng Y et al (2013) Degradation of antipyrine by UV, UV/H_2O_2 and UV/PS. J Hazard Mater 260:1008–1016. https://doi.org/10.1016/j.jhazmat.2013.06.060

Tang X, Wang Z, Wang Y (2018) Visible active N-doped TiO_2/reduced graphene oxide for the degradation of tetracycline hydrochloride. Chem Phys Lett 691:408–414. https://doi.org/10.1016/j.cplett.2017.11.037

Tezcanli-Güyer G, Ince NH, Tezcanli-g G (2004) Individual and combined effects of ultrasound, ozone and UV irradiation: a case study with textile dyes. Ultrasonics 42:603–609. https://doi.org/10.1016/j.ultras.2004.01.096

Velegraki T, Hapeshi E, Fatta-Kassinos D, Poulios I (2015) Solar-induced heterogeneous photocatalytic degradation of methyl-paraben. Appl Catal B Environ 178:2–11. https://doi.org/10.1016/j.apcatb.2014.11.022

Vignesh K, Priyanka R, Rajarajan M, Suganthi A (2013) Photoreduction of Cr(VI) in water using Bi_2O_3- ZrO_2 nanocomposite under visible light irradiation. Mater Sci Eng B Solid-State Mater Adv Technol 178:149–157. https://doi.org/10.1016/j.mseb.2012.10.035

Wang C, Cao M, Wang P, Ao Y (2013) Preparation, characterization of CdS-deposited graphene-carbon nanotubes hybrid photocatalysts with enhanced photocatalytic activity. Mater Lett 108:336–339. https://doi.org/10.1016/j.matlet.2013.06.102

Wang C, Zhu J, Wu X et al (2014) Photocatalytic degradation of bisphenol A and dye by graphene-oxide/Ag_3PO_4 composite under visible light irradiation. Ceram Int 40:8061–8070. https://doi.org/10.1016/j.ceramint.2013.12.159

Wang F, Zhang K (2011) Reduced graphene oxide-TiO_2 nanocomposite with high photocatalytic activity for the degradation of rhodamine B. J Mol Catal A Chem 345:101–107. https://doi.org/10.1016/j.molcata.2011.05.026

Wang H, Liu Q, You C (2019) Regeneration of sulfur-deactivated TiO_2 photocatalysts. Appl Catal A Gen 572:15–23. https://doi.org/10.1016/j.apcata.2018.12.031

Wang L, Wang N, Zhu L et al (2008) Photocatalytic reduction of Cr(VI) over different TiO_2 photocatalysts and the effects of dissolved organic species. J Hazard Mater 152:93–99. https://doi.org/10.1016/j.jhazmat.2007.06.063

Wang P, Ao Y, Wang C et al (2012) A one-pot method for the preparation of graphene-Bi_2MoO_6 hybrid photocatalysts that are responsive to visible-light and have excellent photocatalytic activity in the degradation of organic pollutants. Carbon N Y 50:5256–5264. https://doi.org/10.1016/j.carbon.2012.06.063

Wang X, Tian H, Yang Y et al (2012) Reduced graphene oxide/CdS for efficiently photocatalytic degradation of methylene blue. J Alloys Compd 524:5–12. https://doi.org/10.1016/j.jallcom.2012.02.058

Wang Z, Liu S, Zhang J et al (2017) Photocatalytic active mesoporous carbon/ZnO hybrid materials from block copolymer tethered ZnO nanocrystals. Langmuir 33:12276–12284. https://doi.org/10.1021/acs.langmuir.7b02492

Wei M, Wan J, Hu Z et al (2017) Preparation, characterization and visible-light-driven photocatalytic activity of a novel Fe(III) porphyrin-sensitized TiO_2 nanotube photocatalyst. Appl Surf Sci 391:267–274. https://doi.org/10.1016/j.apsusc.2016.05.161

Wols BA, Hofman-Caris CHM (2012) Review of photochemical reaction constants of organic micropollutants required for UV advanced oxidation processes in water. Water Res 46:2815–2827. https://doi.org/10.1016/j.watres.2012.03.036

Wong CC, Chu W (2003) The direct photolysis and photocatalytic degradation of alachlor at different TiO_2 and UV sources. Chemosphere 50:981–987. https://doi.org/10.1016/S0045-6535(02)00640-9

Wu Q, Zhao J, Qin G et al (2013) Photocatalytic reduction of Cr(VI) with TiO_2 film under visible light. Appl Catal B Environ 142–143:142–148. https://doi.org/10.1016/j.apcatb.2013.04.056

Xu L, Yang L, Johansson EMJ et al (2018) Photocatalytic activity and mechanism of bisphenol a removal over TiO_{2-x}/rGO nanocomposite driven by visible light. Chem Eng J 350:1043–1055. https://doi.org/10.1016/j.cej.2018.06.046

Xu T, Zhang L, Cheng H, Zhu Y (2011) Significantly enhanced photocatalytic performance of ZnO via graphene hybridization and the mechanism study. Appl Catal B Environ 101:382–387. https://doi.org/10.1016/j.apcatb.2010.10.007

Yan T, Liu H, Sun M et al (2015) Efficient photocatalytic degradation of bisphenol A and dye pollutants over BiOI/Zn_2SnO_4 heterojunction photocatalyst. RSC Adv 5:10688–10696. https://doi.org/10.1039/c4ra13990a

Yan Y, Sun S, Song Y et al (2013) Microwave-assisted in situ synthesis of reduced graphene oxide-$BiVO_4$ composite photocatalysts and their enhanced photocatalytic performance for the degradation of ciprofloxacin. J Hazard Mater 250–251:106–114. https://doi.org/10.1016/j.jhazmat.2013.01.051

Yang C, Dong W, Cui G et al (2017) Highly-efficient photocatalytic degradation of methylene blue by PoPD-modified TiO_2 nanocomposites due to photosensitization-synergetic effect of TiO_2 with PoPD. Sci Rep 7:1–12. https://doi.org/10.1038/s41598-017-04398-x

Yap PS, Lim TT, Lim M, Srinivasan M (2010) Synthesis and characterization of nitrogen-doped TiO_2/AC composite for the adsorption-photocatalytic degradation of aqueous bisphenol-A using solar light. Catal Today 151:8–13. https://doi.org/10.1016/j.cattod.2010.01.012

Yap PS, Lim TT, Srinivasan M (2011) Nitrogen-doped TiO_2/AC bi-functional composite prepared by two-stage calcination for enhanced synergistic removal of hydrophobic pollutant using solar irradiation. Catal Today 161:46–52. https://doi.org/10.1016/j.cattod.2010.09.024

Yoon TP, Ischay MA, Du J (2010) Visible light photocatalysis as a greener approach to photochemical synthesis. Nat Chem 2:527–532. https://doi.org/10.1038/nchem.687

Youssef Z, Colombeau L, Yesmurzayeva N, Baros F (2018) Dye-sensitized nanoparticles for heterogeneous photocatalysis : Cases studies with TiO_2, ZnO, fullerene and graphene for water puri fication. Dye Pigment 159:49–71. https://doi.org/10.1016/j.dyepig.2018.06.002

Yu H, Wang X, Sun H, Huo M (2010) Photocatalytic degradation of malathion in aqueous solution using an Au-Pd-TiO_2 nanotube film. J Hazard Mater 184:753–758. https://doi.org/10.1016/j.jhazmat.2010.08.103

Zhang H, Xu P, Du G et al (2011) A facile one-step synthesis of TiO_2/graphene composites for photodegradation of methyl orange. Nano Res 4:274–283. https://doi.org/10.1007/s12274-010-0079-4

Zhang L, Wang W, Sun S et al (2014) Elimination of BPA endocrine disruptor by magnetic BiOBr@SiO_2@Fe_3O_4 photocatalyst. Appl Catal B Environ 148–149:164–169. https://doi.org/10.1016/j.apcatb.2013.10.053

Zhang L, Zhang H, Huang H et al (2012) Ag_3PO_4/SnO_2 semiconductor nanocomposites with enhanced photocatalytic activity and stability. New J Chem 36:1541–1544. https://doi.org/10.1039/c2nj40206h

Zhang W, Xiao X, Zheng L, Wan C (2015) Fabrication of TiO_2/MoS_2@zeolite photocatalyst and its photocatalytic activity for degradation of methyl orange under visible light. Appl Surf Sci 358:468–478. https://doi.org/10.1016/j.apsusc.2015.08.054

Zhang X, Zhou M, Lei L (2005) Preparation of an Ag-TiO_2 photocatalyst coated on activated carbon by MOCVD. Mater Chem Phys 91:73–79. https://doi.org/10.1016/j.matchemphys.2004.10.058

Zhao D, Yang X, Chen C, Wang X (2013) Enhanced photocatalytic degradation of methylene blue on multiwalled carbon nanotubes-TiO_2. J Colloid Interface Sci 398:234–239. https://doi.org/10.1016/j.jcis.2013.02.017

Zhao W, Wang Y, Yang Y et al (2012) Carbon spheres supported visible-light-driven CuO-$BiVO_4$ heterojunction : preparation, characterization, and photocatalytic properties. Appl Catal B Environ 115–116:90–99. https://doi.org/10.1016/j.apcatb.2011.12.018

Zheng X, Xu S, Wang Y et al (2018) Enhanced degradation of ciprofloxacin by graphitized mesoporous carbon (GMC)-TiO_2 nanocomposite: strong synergy of adsorption-photocatalysis and antibiotics degradation mechanism. J Colloid Interface Sci 527:202–213. https://doi.org/10.1016/j.jcis.2018.05.054

Zhu H, Jiang R, Xiao L et al (2013) CdS nanocrystals/TiO_2/crosslinked chitosan composite: facile preparation, characterization and adsorption-photocatalytic properties. Appl Surf Sci 273:661–669. https://doi.org/10.1016/j.apsusc.2013.02.106

Chapter 9
Technological Advancement in Photocatalytic Degradation of Dyes Using Metal-Doped Biopolymeric Composites—Present and Future Perspectives

Palliyalil Sirajudheen, Sivakumar Vigneshwaran, Perumal Karthikeyan, Chettithodi Poovathumkuzhi Nabeena, and Sankaran Meenakshi

Abstract The rapid and extensive industrialization and the unrestrained growth of modern textile industries together with the lack of appropriate wastewater treatment facilities led to the discharge of effluents into water bodies, thereby causing a serious threat to the environment. The presence of such pollutants in the water bodies deteriorates the water quality and makes it unfit for use. From an environmental perspective, it is essential to develop new technologies for the wastewater treatment and recycling of dye-contaminated water. The metal-doped polymeric and biopolymeric materials, especially chitosan and cellulose composites based photocatalysts, have a prominent role in the removal of toxic organic dyes from water. Since adsorption is the key step of photocatalysis, it can be assumed that doping of metals with chitosan and cellulose biopolymers can show a synchronous effect in improving photocatalytic activity. The surface modification of biopolymers with metals produces more active sites at the surface of the adsorbent, which enhances dye and semiconductor interaction as well as suppresses electron–hole recombination rates during the photocatalytic process. Herein, the chapter brought in the thought of the application of various metal activated composites in wastewater photodegradation and suggested the versatility in composites for the development of rapid, selective, and effective degradation processes for the removal of a variety of organic dyes. It further emphasized the existing obstruction and impending prediction for the deprivation of dyes via photocatalytic techniques.

Keywords Biopolymer · Cellulose · Chitosan · Photocatalysis · Mineralization

---

P. Sirajudheen · S. Vigneshwaran · P. Karthikeyan · S. Meenakshi (✉)
Department of Chemistry, The Gandhigram Rural Institute—Deemed to be University, Gandhigram, Dindigul, Tamil Nadu 624302, India

P. Sirajudheen · C. P. Nabeena
Department of Chemistry, Pocker Sahib Memorial Orphanage College, Tirurangadi, Malappuram, Kerala 676306, India

S. P. Singh et al. (eds.), *Nanomaterials and Nanocomposites for Environmental Remediation*, Energy, Environment, and Sustainability,
https://doi.org/10.1007/978-981-16-3256-3_9

9.1 Introduction

The life of any kind does not exist without water as it is an essential and inimitable natural resource on the earth. Water plays a crucial role in preserving the earth ecology and quality of life. Recently, scarcity of water is observed all over the world, since the percentage of freshwater is inadequate to meet up the need of existing population (Singh et al. 2020). The growing water demand, increase in population, uncontrolled water usage, inadequate rainfall, sudden climate change, and most prominently, water pollution lead to water scarcity. In the modern era, water contamination and pollution are regarded as one of the most substantial and annoying problems that demand an instantaneous and practical solution. Extensive researches have been carried out to develop efficient treatment methods for removing pollutants from industrial effluents. The discharge of the untreated industrial effluents, fertilizers used in agriculture, pharmaceutical wastes, disposal of pesticides, and dumping of domestic waste into the water bodies are the primary source of water pollution (Sirajudheen and Meenakshi 2019a). The industrial effluents contain various organic compounds such as biphenyls (PCBs), dibenzofurans, dioxins, polychlorinated, chlorophenols, and dyes. Among these compounds, dyes are regarded as one of the most rigorous contributors to water pollution. Most of the industries use dyes, and the untreated industrial effluents contain numerous quantity of organic dyes (Natarajan et al. 2018).

Dyes are the colored complex structures that show affinity to the substrate on to which it is being applied. Dyes appear colored as they absorb light in the visible region of the electromagnetic spectrum. They are mostly aromatic, ionizing compounds having diverse chromophore molecules which render them colored. Chromophores present in dyes are responsible for the absorption of radiation. The dyes can be classified into various classes such as acid, basic, direct, disperse, metallic, mordant, pigment, reactive, solvent, sulfur, and vat dyes, which reveals their macroscopic properties and also their existing functionalities. More than one lakh commercial dyes are presently obtainable in the market, and all over the world more than 7×10^5 tons of dyestuff are produced per annum. It is estimated that 10–15% of dyes are disposed of in water during their synthesis and application processes (Natarajan et al. 2018). The type, nature, and properties of the dyes reported in some literature are enlisted in Table 9.1.

The dyes interact either physically or chemically with various substrates and form strong bonds with them. Figure 9.1 shows the classification of various dyes based on their method of synthesis, nature of chromophores present, starting material used, and the type of electronic excitation. Varieties of synthetic dyestuffs are available for various industrial applications and are listed in Table 9.2. However, the dyes have been found to be noxious pollutants which are also highly carcinogenicand mutagenic (Sirajudheen and Meenakshi 2019b). In addition, their persistence deteriorates the fertile agriculture land and also poses a threat to aquatic life. Hence, the elimination of dye molecules from wastewater is of primary importance before it is released into the water bodies. Dyes are resistant to biodegradation, and their removal from aquatic

Table 9.1 Types and properties of dyes used in industries for various applications (Natarajan et al. 2018). *Source* (Natarajan et al. 2018). Reprinted with permission from Elsevier Copyright © 2018 Elsevier Ltd.

Name of dye	Type	Color	Molecular formula	Molecular weight (g/mol)	Absorption maxima (nm)
Acid Red 114	Azo	Dark red powder	$C_{37}H_{28}N_4Na_2O_{10}S_3$	830.81	514
Acid Red 27	Azo	Dark red to purple	$C_{20}H_{11}N_2Na_3O_{10}S_3$	604.47	520
Acid Orange 7,	Azo	Orange	$C_{16}H_{11}N_2NaO_4S$	350.32	483
Acid Orange 8	Azo	Red orange	$C_{17}H_{13}N_2NaO_4S$	364.35	490
Acid Yellow 17	Azo	Yellow	$C_{16}H_{10}C_{12}N_4Na_2O_7S_2$	551.29	400
Allura Red AC	Azo	Red	$C_{18}H_{14}N_2Na_2O_8S_2$	496.42	504
Acid Red 14,	Azo	Dark red	$C_{20}H_{12}N_2Na_2O_7S_2$	502.43	515
Acid Yellow 23	Azo	Yellow	$C_{16}H_9N_4Na_3O_9S_2$	534.37	455
Acid Black 1	Azo	Dark brown	$C_{22}H_{14}N_6Na_2O_9S_2$	616.49	620
Acid Red 73	Azo	Yellow light red	$C_{22}H_{14}N_4Na_2O_7S_2$	556.48	507
Acid Brown 14	Azo	Red light brown	$C_{26}H_{16}N_4Na_2O_8S_2$	622.54	465
Acid Orange 10	Azo	Bright orange	$C_{16}H_{10}N_2Na_2O_7S_2$	452.37	478
Acid Red 186	Azo	Purplish red	$C_{20}H_{14}N_4O_8S_2.2Na$	548.46	455
Acid Orange 6	Azo	Yellow orange	$C_{12}H_9N_2NaO_5S$	316.27	387
Alizarin Red S	Anthrq-uinone	Orange-yellow	$C_{14}H_7NaO_7S$	342.26	517
Acid Blue 80	Anthraqui-none	Red light blue	$C_{32}H_{28}N_2Na_2O_8S_2$	678.68	626
Acid Blue 25	Anthraqui-none	Dark blue	$C_{20}H_{13}N_2NaO_5S$	416.38	600
Acid Blue 7	Triaryl-methane	Blue	$C_{37}H_{35}N_2NaO_6S_2$	690.81	625

(continued)

Table 9.1 (continued)

Name of dye	Type	Color	Molecular formula	Molecular weight (g/mol)	Absorption maxima (nm)
Acid Blue 1	Triaryl-methane	Dark blue	$C_{27}H_{31}N_2NaO_6S_2$	566.67	630
Acid Blue 9	Triaryl-methane	Violet	$C_{37}H_{42}N_4O_9S_3$	787.9	625
Acid Yellow 73	Xanthene	Brilliant yellow	$C_{20}H_{10}Na_2O_5$	376.27	490
Basic Red 46	Azo	Dark red	$C_{18}H_{21}BrN_6$	401.3	530
Basic Violet 3	Triaryl-methane	Bright blue purple	$C_{25}H_{30}C_lN3$	407.98	590
Brilliant Green	Triaryl-methane	Blue-green	$C_{27}H_{33}N_2.HO_4S$	482.63	624
Basic Blue 41	Azo	Dark violet	$C_{20}H_{26}N_4O_6S_2/C_{21}H_{27}C_lN_4O_3S$	482.57/ 450.98	600
Congo Red	Azo	Red	$C_{32}H_{22}N_6Na_2O_6S_2$	696.66	510
Crocein Orange G	Azo	Bright orange	$C_{16}H_{11}N_2NaO_4S$	350.32	488
Direct Blue 160	Azo	Dark grey		1373	570
Direct Red 80	Azo	Purple	$C_{45}H_{26}N_{10}Na_6O_{21}S_6$	1373.08	529
Direct Red 81	Azo	Red	$C_{29}H_{19}N_5Na_2O_8S_2$	675.6	510
Direct Red 23	Azo	Purple	$C_{35}H_{25}N_7Na_2O_{10}S_2$	813.73	508
Direct Yellow 27	Azo	Yellow	$C_{25}H_{20}N_4Na_2O_9S_3$	662.63	398
Direct Yellow 50	Azo	Red light yellow	$C_{35}H_{24}N_6Na_4O_{13}S_4$	956.82	395
Direct Violet 17	Azo	Dark blue purple	$C_{38}H_{24}N_5Na_3O_{11}S_3$	891.79	546
Direct Yellow 12	Azo	Deep yellow	$C_{30}H_{26}N_4Na_2O_8S_2$	680.66	395
Gentian Violet	Triaryl-methane	Violet	$C_{25}N_3H_{30}C_l$	407.98	536
Indigo Carmine	Indigotine	Indigo to dark blue	$C_{16}H_8N_2Na_2O_8S_2$	466.35	610

(continued)

Table 9.1 (continued)

Name of dye	Type	Color	Molecular formula	Molecular weight (g/mol)	Absorption maxima (nm)
Janus Green B	Azo	Violet	$C_{30}H_{31}C_lN_6$	511.07	544–550
Methylene Blue	Aniline	Dark green	$C_{16}H_{18}N_3SCl$	319.85	664
Methyl Orange	Azo	Orange-yellow	$C_{14}H_{14}N_3NaO_3S$	327.33	460
Methyl Red	Azo	Dark red	$C_{15}H_{15}N_3O_2$	269.3	540
Orange G	Azo	Orange	$C_{16}H_{10}N_2Na_2O_7S_2$	452.38	495
Ponceau S	Azo	Light red	$C_{22}H_{16}N_4O13S_4$	672.63	514
Reactive Black 5	Azo	Black	$C_{26}H_{21}N_5Na_4O_{19}S_6$	991.82	597
Rhodamine B	Azo	Reddish-violet	$C_{28}H_{31}C_lN_2O_3$	479.02	554
Reactive Yellow 2	Azo	Bright yellow	$C_{25}H_{15}C_{l3}N_9Na_3O_{10}S_3$	872.97	400
Reactive Yellow 14	Azo	Yellow	$C_{20}H_{19}ClN_4Na_2O_{11}S_3$	669	410
Reactive Yellow 17	Azo	Yellow	$C_{20}H_{20}K_2N_4O_{12}S_3$	682.79	426
Reactive Yellow 145	Azo	Orange	$C_{28}H_{20}ClN_9Na_4O_{16}S_5$	1026.2	419
Reactive Red 22	Azo	Red	$C_{19}H_{16}N_2Na_2O_{11}S_3$	590.51	511
Reactive Red 15	Azo	Big red	$C_{25}H_{14}ClN_7Na_4O_{13}S_4$	876.09	500
Reactive Blue 4	Anthraqui-none	Dark blue	$C_{23}H_{12}Cl_2N_6Na_2O_8S_2$	681.39	596
Remazol Brilliant Blue R	Anthraqui-none	Blue-black	$C_{22}H_{16}N_2Na_2O_{11}S_3$	626.54	592
Reactive Orange 16	Azo	Red	$C_{20}H_{17}N_3Na_2O_{11}S_3$	617.54	494
Reactive Red 2	Azo	Purplish red	$C_{19}H_{10}Cl_2N_6Na_2O_7S_2$	615.34	538
Reactive Yellow 84	Azo	Yellow	$C_{56}H_{38}Cl_2N_{14}Na_6O_20S_6$	918.4	420
Reactive Orange 4	Azo	Orange	$C_{24}H_{13}Cl_2N_6Na_3O_{10}S_3$	781.47	490
Sudan 1 V	Azo	Reddish brown	$C_{24}H_{20}N_4O$	380.44	520
Sulforhodamine-B	Azo	Red	$C_{27}H_{30}N_2O_7S_2$	558.67	565

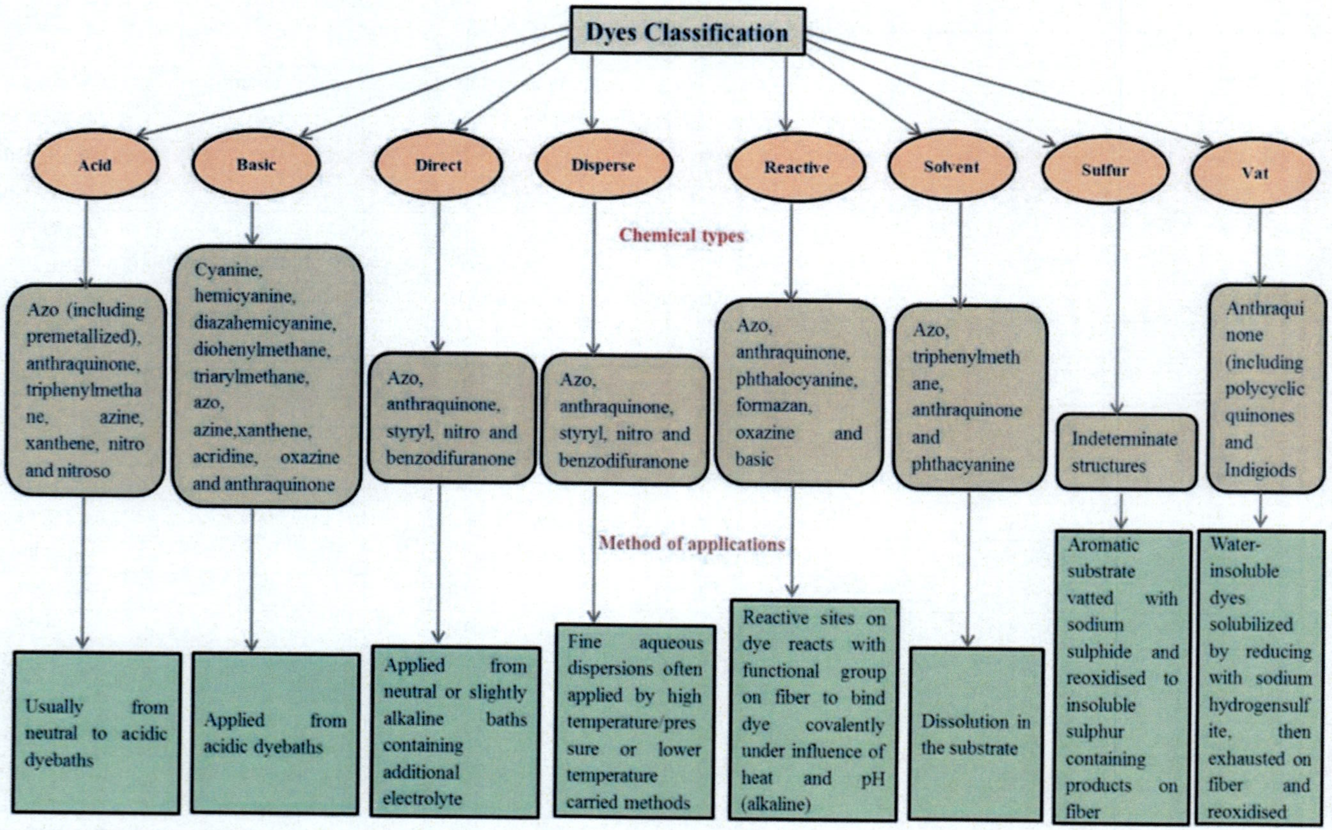

Fig. 9.1 Classification, nature, and method of applications of dyes (Natarajan et al. 2018). *Source* Natarajan et al. (2018). Reprinted with permission from Elsevier Copyright © 2018 Elsevier Ltd

Table 9.2 Various applications of dyes (Natarajan et al. 2018). *Source* (Natarajan et al. 2018). Reprinted with permission from Elsevier. Copyright © 2018 Elsevier Ltd.

No.	Types of dyes	Applications
1.	Leather dyes:	In leather, fat soluble components from these dyes by applying high temperature have an affinity to migrate upwards into the layers to give the smooth finish to leather surface
2.	Oxidation dyes:	These dyes are used as colourant material for hair
3.	Solvent dyes:	In wood staining and production of coloured lacquers, solvent inks, waxes and colouring oils etc. They are also used in making distempers, soaps, spirit inks, lacquers, plastic, rubber, ink for High Density Polyethylene (HDPE) woven bags, wood gloss finish, metallised polyester film, paddle dyeing of wool skins and furs, etc
4.	Fluorescent dyes:	It is used as penetrant liquids for crack detection, plastics, synthetic resins, printing inks, non-destructive testing, and sports ball dyeing. Merantine Brilliant Yellow 8G fluorescent dye is used in the textile applications
5.	Fuel dyes:	Used in fuels
6.	Smoke dyes:	Coloured smoke used for smoke signals in a military context. It can be produced by smoke grenades, or by various other pyrotechnical devices
7.	Sublimation dyes:	These dyes have printing applications. These are readily available in the market as inkjet ink, toner for laser printers, or as ribbons for the thermal-transfer printing
8.	Inkjet dyes:	Writing industry including the inkjet printers
9.	Leuco dyes:	It is used in the security printing, novelty bath toys, swimming pool toys, battery testers
10.	Direct dyes:	Direct dyes are widely used on cotton, paper, leather, wool, silk and nylon. They are also used as pH indicators and as biological stains
11.	Vat dyes:	These dyes are applied in the cotton, cellulosic and blends
12.	Sulphur dyes:	Sulphur dyes are water insoluble dyes and have no affinity for the cellulosics as such, but solubilized when treated with reducing agent to form a leuco compound. These leuco compounds are water soluble and have affinity for the cellulosic materials such as cotton, viscose, jute and flex etc
13.	Organic pigments	Used in cotton, cellulosic, blended fabrics and paper

(continued)

Table 9.2 (continued)

No.	Types of dyes	Applications
14.	Reactive dyes:	Reactive dyeing is the important method for the coloration of cellulosic fibres. It can also be applied on wool and nylon
15.	Dispersed dyes:	It is commonly used to dye polyester fibres and nylon
16.	Acid dyes:	Acid dyes are used for dyeing protein fibres such as wool, angora, cashmere and silk. Apart these, milk protein fibres like Silk Latte, Soya Protein etc., can also be used
17.	Azoic dyes:	These dyes are used in printing inks and pigments
18.	Basic dyes:	For dyeing acrylic fibres, basic dyes are used and also for dyeing of jute, cut flowers, dried nflower, coir, etc
19.	Developed dyes:	Developed dyes are applied in the cellulosic fibres and fabric
20.	Mordant dyes	These water soluble dyes are affinity for silk, wool and polyamides

systems is challenging (Sirajudheen et al. 2020). The non-degradability of dyes and their stability toward the light and/or oxidizing agents complicate the selection of a suitable dye removal method. In order to meet the required water quality, several physical and/or chemical processes consisting adsorption, coagulation (Mittal et al. 2010), electrochemical methods (Xia et al. 2013), and photocatalysis (Saravanan et al. 2014), etc., are required.

Among these methods, photocatalysis is the most effective, simple, and known for the treatment of effluents containing dyes. Photodegradation processes are perhaps the best method for the removal of the toxic, non-biodegradable pollutants and for removing pathogens. Unlike photodegradation process, other physical methods do not degrade the dye molecule but only reduce the dye concentration in water bodies by converting it from one form to another, thus creating secondary pollution. (Sirajudheen and Meenakshi 2019a). Recent researches on the effluent treatment process in the dyeing industry specified the use of photocatalysis for both degradation and detoxification. The final products formed during photocatalytic degradation of organic dye molecule are water and CO_2 (Chen et al. 2010), which will not produce any further toxicity to water.

This paper focuses on discussing some recent articles allied to the synthesis, properties, and application of the chitosan/cellulose-based semiconductor catalyst composites reported for the dye degradation process via visible-light photocatalysis. The plausible mechanism for the dye degradation process is critically examined, and underlying mechanisms for the photocatalysis have been explored. The methods of synthesizing chitosan cellulose composites and their surface morphologies are discussed. Their photocatalytic efficiencies toward a variety of cationic anionic and

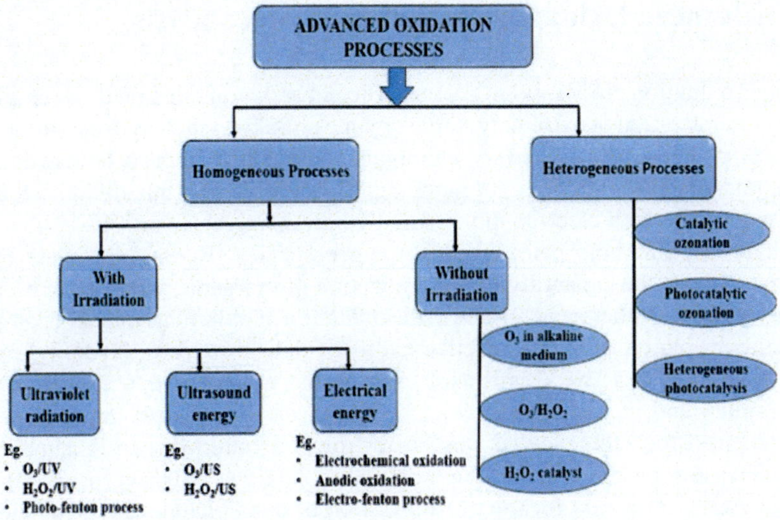

Fig. 9.2 Classification of advanced oxidation processes (Singh et al. 2020). *Source* Singh et al. (2020). Reprinted with permission from Elsevier Copyright © 2020 Elsevier Ltd.

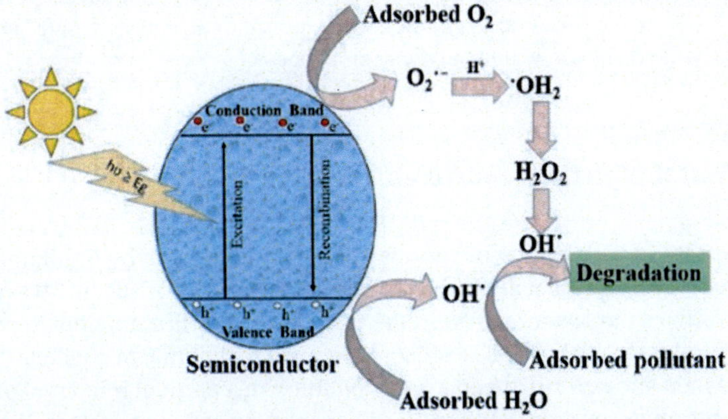

Fig. 9.3 Mechanistic view of semiconductor-based photocatalysis (Singh et al. 2020). *Source* Singh et al. (2020). Reproduced with permission from Copyright © 2020 Elsevier Ltd

azo dyes mineralization are also explained. The existing limitations, their solutions, and need for further research in the areas of treatment techniques of dye contaminant of interest are also discussed. Some related scientific hypothesis and perspectives are put forth for further studies in this field.

9.2 Advanced Oxidation Process in Photocatalysis

Advanced oxidation processes (AOPs) are efficient and promising methods employed for the mineralization of extremely stable organic pollutants such as dyes, surfactants, pesticides, phenolic wastes, and some inorganic pollutants which are hard to degrade. The photodegradation process got more significant attraction due to their ability to create photogenerated electron–hole pairs by harnessing solar or UV–Visible radiation. The electron–hole pairs produce oxygen species (ROSs) which are highly reactive and have the capability of mineralization of inorganic and organic effluents from water. On the other hand, in AOPs, the oxidation of pollutants mainly takes place due to the formation of highly reactive hydroxyl radical oxidant. Hydroxyl radical ($OH^{\cdot}$), a strongly reactive electrophile, having the oxidation potential of 2.8 eV, reacts rapidly and non-selectively with nearly all electron-rich compounds. AOPs have enhanced the efficiency of wastewater mineralization due to their ability to utilize solar energy directly for chemical detoxification (Singh et al. 2020). The different AOPs employed for the mineralization of dyes include chemical, electrochemical, sonochemical, and photochemical processes. Depending upon the phase of the catalysts used, AOPs can be classified into homogeneous and heterogeneous processes. The different catalysis methods used for dye degradation are shown in Fig. 9.2. In homogeneous catalysis, the catalyst recovery after photodegradation is difficult. However, heterogeneous catalysis is economical, easy, simple, and could be practically used for the water purification process.

9.2.1 *Photocatalytic Experiments*

The photocatalytic degradation of various dyes are investigated by irradiating a definite mg/L of dye solutions with different irradiation sources of specific wavelengths and under different exposure times. In the process, before illumination, a particular amount of catalyst in milligrams is added into a specific volume of aqueous dye solutions. It has then allowed stirring in a dark condition in a particular interval of time to create adsorption/desorption equilibrium. The sampling is done at specific time intervals. The progress of the degradation processes is monitored spectrophotometrically. The percentage of degradation is estimated using Eq. (9.1) as

$$\text{Percentage removal} = \left(\frac{C_0 - C_t}{C_0}\right) 100\% \tag{9.1}$$

Here, C_0 and C_t are the concentrations at the beginning of light irradiation and at time t, respectively. In some experiments, visible light was used as a light source, to study the effect of essential parameters like initial concentration of dye, catalyst dosage, pH of solution, and existence of co-ions.

The chemical oxygen demand (COD) analysis serves to quantify the amount of organic effluents present in the wastewater. The COD value is the measure of the total quantity of oxygen required for the oxidation of organic matter contained in effluents, to CO_2 and water. The effectiveness of the photocatalyst composite was calculated using COD measurement and thereby assessing the efficacy of the photocatalyst in wastewater treatment (Sirajudheen and Meenakshi 2019a). The degree of mineralization in terms of COD reduction was determined using Eq. (9.2)

$$\text{COD reduction} = \left(\frac{\text{COD}_{\text{initial}} - \text{COD}_{\text{final}}}{\text{COD}_{\text{initial}}}\right) \times 100\% \tag{9.2}$$

The efficiency of decolorization is inversely related to the dye concentration. For measuring COD, a specific amount dye solution in (mL) and photocatalyst in (mg) were irradiated using light source of suitable wavelength using a photoreactor, and the corresponding COD values were measured before and after the irradiation. The decrease in the COD values of the dye solution after irradiation indicates the substantial reduction of organic matter via photocatalytic degradation. The greater extent of photodegradation together with lower toxicity of the products shows the higher potential of the photocatalyst in the elimination of dyes from wastewater (Meenakshi and Farzana 2013).

9.2.2 Mechanism of Photocatalysis

Dispersion of the biopolymeric composite in the dye solution followed by irradiation using visible light leads to the ultimate electron transfer from the surface of biopolymers to the dyes. Hence, the electrons (e^-) in the valence band (VB) of the metal can be excited to the conduction band (CB), thereby generating an equal number of holes in the VB. The excited electron and created hole in the semiconductor are responsible for the degradation of the dye. The schematic view of the photocatalytic mechanism is represented in Fig. 9.3. Moreover, the photo-induced holes confined by OH^- result in the formation of hydroxyl radical species ($OH^{\bullet}$). These act as a powerful oxidant for the complete or partial mineralization of organic dye pollutants (Saravanan et al. 2013).

$$\text{Semiconductors} + \text{photons}(hv) \rightarrow e^-(\text{CB}) + h^+(\text{VB}) \tag{9.3}$$

$$h^+ + H_2O \rightarrow H^+ + OH^- \tag{9.4}$$

$$h^+ + OH^- \rightarrow OH^{\cdot} \tag{9.5}$$

$$e^+ + O_2 \rightarrow O_2^{\cdot} \tag{9.6}$$

$$e^- + O_2 \rightarrow O_2^{\cdot} \quad (9.6)$$

$$2e^- + O_2 + H_2O \rightarrow H_2O_2 \quad (9.7)$$

$$2e^- + O_2 + H_2O \rightarrow H_2O_2 \quad (9.8)$$

$$e^- + H_2O_2 \rightarrow OH^{\cdot} + OH^- \quad (9.9)$$

$$\text{Dyes} + OH^{\cdot} + O_2 \rightarrow CO_2 + H_2O + \text{other products} \quad (9.10)$$

The mechanism explained here is general, and depending on the experimental conditions, additional steps may be included. It should be considered that the surface alteration of chitosan and cellulose with metal oxides or sulfides would create a suitable environment for the formation of more active sites at the surface of the composite. This results in efficient dye semiconductor interaction, and consequent enhancement of photodegradation. Further, biopolymers and polymers themselves adsorb some dye molecules which will be continually supplied to the semiconductors for mineralization. This process has a significant role in the increase in the efficiency of dye degradation. The attached semiconductors can enhance the photocatalytic degradation of dyes. This is attributed to the reduction in band gap owing to the coupling of metals with metal oxides and sulfides. The decrease in band gap, in turn, increases the wavelength range corresponding to electron–hole pair separation. There is a resultant increase in the recombination time, and a higher photocatalytic activity is thus achieved.

9.2.3 *Metallic Photocatalysts*

In the early stages of dye degradation and detoxification of water, metallic conductors and semiconductors were widely used as photocatalysts. For instance, TiO_2 was used for oxygen photo-adsorption (Bickley and Stone 1973), and ZnO was used for photo-adsorption/desorption of oxygen (Barry and Stone 1960). Phototocatalysts gain high energy by absorbing photons of the appropriate wavelength. This energy is used to transfer the electrons from the valence band (VB) to the conduction band (CB), leading to a redox reaction (Selloni 2008), which causes dye degradation. Apart from TiO_2 and ZnO, various other semiconducting systems, capable of photo splitting of water upon visible-light irradiation include sulfides or oxides like CdS, ZrO_2, and WO_3; titanates like nickel titanate, bismuth titanate, and strontium titanate; photocatalysts co-doped with antimony and chromium, etc., (Sirajudheen et al. 2016); various compounds of bismuth like bismuth tantalate (Bi_4TaO_8I, Bi_3TaO_7); bismuth niobates {Bi_2MNbO_7, (M = Ga^{3+}, In^{3+}, Al^{3+}), Bi_2MNbO_7, (M = Fe, Al, In, Sm), Bi_3TiNbO};, and oxides of bismuth like $(BiO)_2CO_3$ (Sivakumar et al. 2014), were also found to have employed in the photocatalytic degradation of dyes. Similar to metal oxides, most of these have the drawback of having a high value of band gap, and they require UV photon sources for decolorizing wastewater (Viswanathan 2017). Among these catalysts, zinc oxide has been widely used because of its wide availability, and zinc is one of the essential micronutrients for humans.

9.2.3.1 Limitations of Metallic Photocatalysts

Use of titanium and other metal dioxides as photocatalysts have some limitations in terms of low quantum efficiency (Plantard et al. 2011) and low sunlight utilization (Wang et al. 2018), that necessitated researches toward establishing an efficient photocatalyst. These photocatalysts are novel materials with large surface area, rich surface sites, and figurative morphology (Zhang et al. 2011a). However, agglomeration of these nano-photocatalysts during the reaction (Zhang et al. 2011b; Li et al. 2017a), difficulty in recovery after usage, and formation of secondary pollutant usually result in limits the development and utilization of these photocatalysts (Mohamed et al. 2015). However, as these materials have the ability to absorb light in the visible region and (Melinte et al. 2019) induce the electron shifting, they can be made more feasible by the following methods (i) by altering their morphologies with a subsequent change in chemical compositions, (ii) changing the synthesis methodologies, and (iii) imprinting or doping the natural compounds with suitable metals, metal oxides, and/or metal hydroxide compounds.

9.3 Biopolymer-Supported Composites

The most effective and suitable method for the fabrication hybrid composite is to impregnate or coat the fine metal or metal oxide particles onto solid particles of larger size. The materials like silica, granular-activated carbon, biopolymer, and polymers were used widely as host materials for the composite preparation (Nithya et al. 2014). The shape and surface morphology of the composite material has a significant role in its mineralization performance. The change in morphology of the composite matrix makes available a large specific surface area, ion-exchange capability, and numerous size defects, which will induce the sorption capacity of the composite matrix. The increase in surface area of the photocatalytic materials affects the increase in higher adsorption of pollutant viz. the dye molecules. If a semiconductor imprinted with the biopolymer, it will continuously supply the pollutants to the semiconductor for degradation. This process leads to an increase in the efficiency of photocatalyst significantly.

Biopolymer materials are most attractive due to their convenient pore size and surface morphology. Their exceptional mechanical strength which permits their long term use and the ability to act as a good host for metals attracted the researchers.

Polymers are synthetic materials, and their biodegradability and biocompatibility are much less than those of biopolymers such as β-cyclodextrin, chitin, chitosan, cellulose, and their derivatives. The cellulose, β-cyclodextrin, and chitosan are recommended as suitable functional materials because these natural polymers have desirable properties such as biodegradability, biocompatibility, adsorption properties, and non-toxicity.

9.3.1 *Chitosan-Supported Composite*

Chitosan is a fibrous biopolymer derived by N-deacetylation of Chitin, a major component of the shells of crustacean such as crab, shrimp, and crawfish. The structure of chitosan is depicted in Fig. 9.4. Chitosan is a linear polymer of N-acetyl-2-amino-2-deoxy-D-glucopyranose units with ß-(1–4) bonds. Unlike plant fibers, chitosan possesses unique properties including optical, structural characteristics and can form films, composites, etc., (Nithya et al. 2014). It is one of the most promising materials that can be used as an adsorbent for the removal of dyes (Ashori et al. 2013). It finds application as bio-adsorbent for removal of cationic and/or anionic dyes due to the simultaneous presence of reactive functional –OH and $-NH_2$ groups, (Sirajudheen and Meenakshi 2019b), which can provide dynamic doping sites for metals and metal oxides. Chitosan can also form intermolecular hydrogen bonding with the pollutants (Huang et al. 2015; Sirajudheen and Meenakshi 2019b). Most of the characteristic properties of chitosan are due to the high content of primary amino groups. It also possesses bacteriostatic and fungistatic effects, can separate proteins, and chelate with many transitional metal ions. The properties of chitosan are greatly influenced by the process environment that controls the de-acetylating process to occur. The degree of deacetylation controls the amount of free amino group, which gives a positive charge to chitosan. The amino group along with hydroxyl group allows chitosan to be a highly reactive polysaccharide (Nithya et al. 2014).

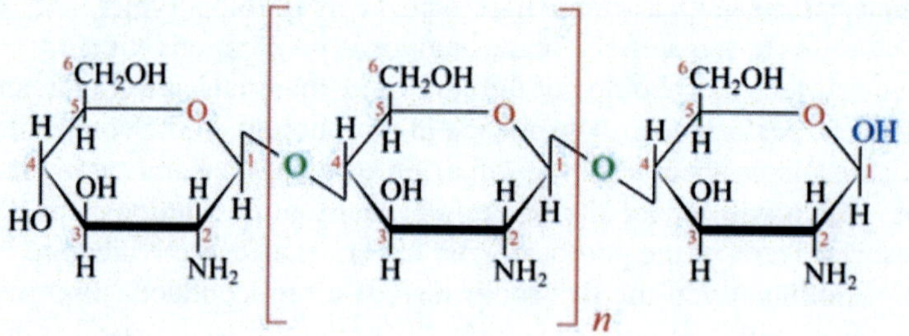

Fig. 9.4 Structure of chitosan

9.3.2 Cellulose-Supported Composites

Cellulose, another biopolymer which is the most abundant natural biopolymer having high thermal, physical, and chemical stability, can also be used as a dopant for metals in the photocatalyst fabrication process. The cellulose molecules are formed by D-pyran glucose connecting via β-1, 4 glycosidic bonds, and its structure is shown in Fig. 9.5. Hydroglucose is the structural unit of cellulose. There are three hydroxyl groups in a hydroglucose unit, wherein esterification and oxidation reactions can occur (Liang and Hu 2016). Cellulose is an ideal matrix substituted for petrochemical products (Kim et al. 2015; Sheng et al. 2017) in view of its high flexibility, hydrophilic nature, transparency, and permeability (Jayalakshmi et al. 2015; Zhang et al. 2017). Cellulose molecules are abundant in hydroxyl groups that can be altered with various functional groups such as carboxyl, amino, sulfo, or cyclodextrin. Imprinting of cellulose-based materials with photocatalysts produces new bio-hybrid matrixes for hydrogen photocatalysis and various other renewable energy applications (Saeed et al. 2019). These advantages of cellulose probably motivated various research groups to fabricate semiconductor/cellulose composites for the photocatalytic degradation of dyes (Jiang et al. 2020).

9.4 Biopolymer-Based Photocatalysts

The shape and surface morphology of the photocatalytic material has a significant role in its mineralization performance. The change in morphology of the catalyst increases the specific surface area, improve ion-exchange capability, and impart numerous size defects, which will induce the photodegradation capacity of the photocatalyst. The increase in surface area of the photocatalytic materials increases the quantity of pollutant dye molecules, which will be released continuously to semiconductor materials for degradation. The efficiency of the photocatalyst is thus increased significantly.

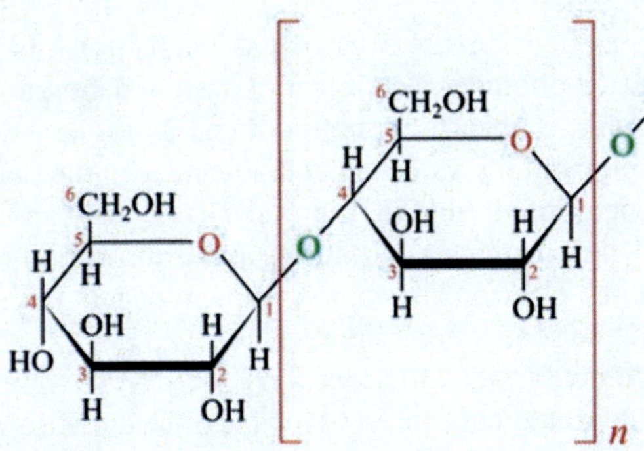

Fig. 9.5 Structure of cellulose

Doping of metals and metal oxides with biopolymers and polymers enhances photocatalytic ability by increasing the adsorption of organic compounds as well as suppressing electron–hole recombination rates during the process of photocatalytic reaction (Li et al. 2004; Nešić et al. 2013). Since adsorption is the key step of photocatalysis (Saleh and Gupta 2011), it can be assumed that doping the metals with biopolymers and/or polymers can produce a synergic effect in improving photocatalytic activity. Here, all the existing features of individual components of the compound are preserved, and at the same time, it could overcome drawbacks like rapid recombination of photogenerated electrons, low absorptivity, and hindrance effect of the photocatalyst (Sirajudheen and Meenakshi 2019a). Hence, the electrons and holes present in the catalyst have redox property which leads to the formation of active hydroxyl radicals and O_2 anion. This hydroxide radical, which is formed by the neutralization of surface –OH with photo-hole h^+ and superoxide anion, is carried out by the dissolved oxygen and surface electron (Bickley and Stone 1973). The bond energy of -OH is reasonably high compared to the bond energies of C–C, C–N, C–H, C–O, and N–H bonds in organic compounds, and –OH can effortlessly break these bonds present in impurities that contaminate water (Sivakumar et al. 2014). Investigation of a visible-light-mediated photocatalyst is warranted highly, and the band gap of the materials determines the energy of light absorbed (Kudo et al. 1999). The photocatalytic activity can be improved by doping metal oxides on the biopolymeric or polymeric surface, which leads to the charge transfer between the photocatalyst and the loaded oxides. It could delay the recombination of electron–hole pairs (Nasr et al. 1996; Kuo and Ho 2006). The size of the particles and changes in surface morphologies have a positive influence on the photocatalytic efficiency, which can be achieved by making appropriate modifications (Han et al. 2009). Figure 9.6 demonstrates the images of various photocatalytic materials such as nanosphere nanotube, nanobelt, nanoribbon, and nanoflower with different morphologies which are used in the dye mineralization process.

9.4.1 Chitosan-Based Photocatalyst

The existence of narrow band gap between the valence and conduction bands makes the semiconductors suitable for the photodegradation of organic and inorganic pollutants. When a semiconductor particle absorbs a photon of energy higher or equal to the band gap energy, an electron from the valence band advances to the conduction band with an instantaneous generation of hole (h^+) in the valence band and electron (e^-) in the conduction band. The hole generated during light absorption can have used as an oxidant, and the excited electron can act as a reducing agent in the photocatalytic process. In this regard, the semiconductors such as TiO_2, ZnO, CuO, Cu_2O, MgO, WO_3, Fe_2O_3, ZnS, CdS, double metal oxides, and g-C_3N_4, have been used as photocatalysts in the pollutant treatment process. However, the quick recombination of the photo-excited electron and consequent hole generated during the process lessen the activity of the catalyst in the visible region of the solar spectrum, increase

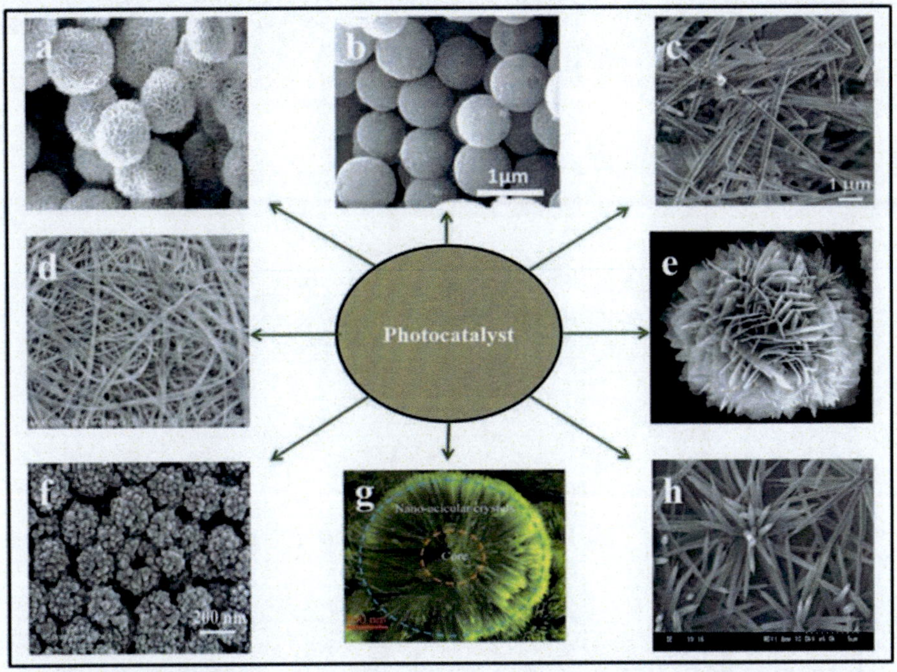

Fig. 9.6 **a** Nanoflakes, **b** spheres, **c** nanotubes, **d** nanowires, **e** flower-like, **f** hollow spheres, **g** microspheres, and **h** nano-rod photocatalytic materials (Singh et al. 2020). *Source* Singh et al. (2020) Reproduced with permission from Copyright © 2018 Elsevier Ltd

complexity in tuning of the photocatalyst, and add toxicity to biological systems. Also, use of bare binary, ternary, and quaternary metal oxide and sulfide nanoparticle for water treatment is rather limited, due to their various constraints such as partial mineralization, reduced adsorption of pollutant, highly unstable nature, and tendency to get agglomerated, difficulty in separation, low reusability, and high toxicity. To overcome these difficulties, these metal oxides and sulfide have been subjected to structural and electronic alterations using many naturally occurring and engineered materials like chitin, chitosan, starch, lignin, bacterial cellulose, activated carbon, etc., which raise the stability and photocatalytic activity of metallic photocatalysts (Lightcap et al. 2010; Xu et al. 2011). Among all the above-said materials, chitosan is found to be a highly reactive polysaccharide due to the presence of highly reactive amino group along with hydroxyl functional group on its surface structure and hence is used as supporting material for the synthesis of photocatalysts. The metal ions were found to occupy on the surfaces of chitosan, which contains several active functional groups, i.e., carboxyl and hydroxyl groups, and tend to coordinate the metal ions. The literature data of some of the major chitosan-based photocatalysts were listed in Table 9.3.

Table 9.3 The literature data of chitosan based photocatalysts for dyes degradation

Name of composite	Preparation method	Name of dye degraded	Nature of dye	Concentration of dye (mg/L)	pH	Efficiency (%)	Reference
TiO_2/Chitosan	Precipitation	Methyl orange	Anionic Azo		5	100	Dhanya and Aparna (2016)
		Alizarin red	Anthraquinone				
Chitosan/ZnO	microwave irradiation	Malachite Green	Cationic	5	2 to 7	90.3	Saad et al. (2020)
Chitosan/Ce–ZnO	microwave irradiation	Malachite Green	Cationic	5		95.4	
Cross-Linked CS/Nano-CdS	Precipitation	Methyl orange	Anionic Azo	10	4	99	Szeto et al. (2014)
Amphoteric Chitosan/TiO_2	Ultrasonic	Malachite Green	Cationic	10	7	91.4	Bahal et al. (2019)
Chitosan Membrane/ZnO/CuO	Wet inpregnation	Brilliant Blue	Aninic	30		91.21	Alzahrani (2018)
Chitosan/Titanium Dioxide		Reactive red	Anionic Azo	50		99.9	Meenakshi and Farzana (2013)
		Methylen blue	Cationic		3 to 11	95.6	
		Rhodamine B	Zwitterionic			85.04	
ZnO impregnated chitosan		Reactive red 2	Anionic azo	50	7	83	Farzana and Meenakshi (2015)
(RGO)/CS/Ag nanoparticle	Hydrogel	Methylene blue	Anionic	10		Nearly 100	Jiao et al. (2015)
		Rhodamine B	Zwitterionic	4		89	

(continued)

Table 9.3 (continued)

Name of composite	Preparation method	Name of dye degraded	Nature of dye	Concentration of dye (mg/L)	pH	Efficiency (%)	Reference
CS-g-poly(acrylamide)/ZnS	Micrwave	Congo red	Anionic azo	15	4.2	75	Pathania et al. (2016)
		Methylorange	Anionic azo	15	4.2	69	
CS/PVA/TiO_2 nanocomposite	Precipitation	Acid red 14	Anionic azo	10	3	85.4	Rasoulifard et al. (2016)
Ag/AgCl/chitosan	One-step	Rhodamine B	Zwitterionic	50		96	H. Wang et al. (2017)
MPAM-g-CMC	Solvo thermal	Congo red	Anionic Azo	20	1 to 6	99.2	Abdelwahab and Morsy (2018)
CS/CdO/NiO nano composite	Precipitation	Rhodamine B	Zwitterionic	5		90	Linda et al. (2016)
MPyTMChi	Precipitation	Methylebe blue	Cationic	10		98.9	Abdelwahab and Morsy (2018)
chitosan-TiO_2	Precipitation	Rose bengal	Cationic	30	5	87	Ahmed et al. (2018)
nano-ZnO/CS microspheres	alternate-feed spray drying	Methyl orange	anionic azo	10		80.83	Zhong et al. (2020)
Cs/ZnO	Precipitation	Methylene blue	Cationic	25	8.4	34	Aziz et al. (2020)
CS/AgCl	Precipitation	Methylene blue	Cationic	25	8.4	41	
CS/AgCl/ZnO	Precipitation	Methylene blue	Cationic	25	8.4	98	
green DNSA@CS@$MnFe_2O_4$	microwave	Methylene blue	Cationic	10	9	98.90%	Shoueir et al. (2018)

(continued)

Table 9.3 (continued)

Name of composite	Preparation method	Name of dye degraded	Nature of dye	Concentration of dye (mg/L)	pH	Efficiency (%)	Reference
cl-Ch-pMAc@ZnO/CdSQDs	microwave	Malachite Green	Cationic	100	6	97.1	Midya et al. (2019)
		Safranin	Cationic	100	6	98.4	
ZMChi/PVA	Hydrothermal	Rhodamine B	Zwitterionic	3	10	98	Abdelwahab and Ghoneim (2018)
CS-supported Cu_2O/CuO	Solid state	Methylene blue	Cationic	10	6 to 8	>90	Zayed et al. (2020)
		Methyl orange	Anionic Azo	10	6 to 8	100	
		reactive red	Anionic Azo	10	6 to 8	>90	
ZnTAPc–AgNPs	coacervation	Rhodamine 6G	Cationic	2.5	2 to 9		Khoza and Nyokong (2015)
$CoFe_{1.98}Sm_{0.02}O_4$@CS-ECH	Sol-gel	Orange 2	Anionic Azo	50	2 to 5	96.31	Humelnicu et al. (2017)
CS-ZnS-NPs	Co-precipitation	Acid black	Anionic Azo	30	3	96.7	Humelnicu et al. (2017)
	Co-precipitation	Acid brown	Anionic double azo	30	3 to 8	92.6	
g-C_3N_4–chitosan beads	Blend crosslinking	Methylene blue	Anionic	5	7	97	C. Zhao et al. (2018)
CdS/TiO_2/CSC	Co-precipitation	Methyl orange	Anionic azo	15	5 to 6	99.1	Zhu et al. (2013)

(continued)

Table 9.3 (continued)

Name of composite	Preparation method	Name of dye degraded	Nature of dye	Concentration of dye (mg/L)	pH	Efficiency (%)	Reference
CuS-CB	Co-precipitation	Malachite green	Cationic	50	8	95	Khan et al. (2019)
CS/MoO_3–TiO_2 nanocomposite	Sol-gel	Methyl orange	Anionic azo	25		85	Magesan et al. (2015)
Chitosan-Copper Oxide	Co-precipitation	Methylene blue	Anionic	0.2	7	84	Haldorai and Shim (2013)
CuO/Chitosan	Wet chemical	Rhodamine B	Zwitterionic	10		99	Haldorai and Shim (2013)
ZnO/chitosan/graphene	Hydrothermal	acid orange 7	Anionic acid	20	1 to 4	97.5	Sheshmani and Nejabat Ghamsari (2019)
Ag/Ag_3PO_4/Chitosan	Water bath heating	Rhodamine B	Zwitterionic	10		99	Wu et al. (2020)
Cu_2O/CS NCs	Precipitation-reduction	brilliant red X-3B	Anionic azo	50	5 to 6	99.8	Cao et al. (2013)
CS/n-CdS	biomineralization	Congo red	Anionic azo	20	6	85.9	Zhu et al. (2009)

9.4.1.1 Metal Oxide-Based Chitosan Composite

The most common metal oxide semiconductor used to blend with chitosan is TiO_2. It has two major crystalline forms, anatase and rutile. The band gap energy of rutile and anatase phases is 3.0 eV and 3.2 eV, respectively. TiO_2 absorbs radiation below the visible range of the electromagnetic spectrum. So, many attempts have been made to activate the TiO_2 in the visible region.

Meenakshi and Farzana (2013) reported chitosan/TiO_2 composite (CTC) and was utilized for the photodegradation of three dyes like an anionic Reactive Red 2 (RR2), dye, a cationic Methylene Blue (MB) dye, and a zwitterionic dye, Rhodamine B (RB), respectively. The addition of CTC along with H_2O_2 considerably improved the percentage of degradation to about 99.9%, 95.6%, and 85.04% for RR, MB, and RB dyes, respectively. The degradation of the pollutants is mainly influenced by the pH of the medium and synchronous role of the catalyst (Meenakshi and Farzana 2013). Generally, in acidic medium, there exist a high electrostatic force of attraction between the positively charged surface of the sorbents, composite, and the negatively charged anionic RR dye molecules. As the pH of the system increases, the number of negatively charged sites in the sorbent also increases which results in a decrease in degradation of RR but increases the degradation of MB due to its cationic in nature. In the case of RhB dye, if the pH is greater than 4, the dye molecule exists in zwitterionic form as $RB^{\pm}$, if the pH is between 1 and 3, the dye is in RhB^{+} form, and if the pH less than 1, $RhBH_2{}^{2+}$ form is adopted. It is found that the uptake of the RhB was decreased at pH = 4 because of zwitterion formation of RhB molecule. These ions formed induce the agglomeration of dye molecules to lead to the formation of dimers, which are unable to enter into the adsorbent pore. At the pH = 8 and pH = 10, the uptake was increased to around 80% due to the formation of excess OH^-, which generates competition between –N+ and –COO−, so that the agglomeration of RhB molecules will diminish. The influence of pH along with the synchronous role of CTC favors both adsorption and photodegradation activities (Sirajudheen and Meenakshi 2020).

Bahal et al. (2019) explain the green synthesis and characterization of novel acrylic acid grafted amphoteric chitosan/TiO_2 (CAT) bio-nanocomposites using ultrasonic radiations. The unvarying distribution of metal oxide on CA/TiO_2 nanocomposites using ultrasonication technique was attained by grafting acrylic acid/chitosan, which has a weak anionic –COOH group. The prepared composite was utilized for analyzing photocatalytic degradation of Malachite Green (MG) in visible light. Since the dye was cationic, the dye degradation occurs at neutral pH, and the paper does not provide an accurate explanation regarding the ambiguity that exists upon the influence of pH in dye photodegradation process. The X-ray diffraction patterns of the fabricated bio-nanocomposite showed sharp peaks at $2\theta = 25.4329°$ 28.8921°, 39.5873°, 48.3106°, and 51.8367°, which justified the incorporation of TiO_2 into chitosan/acrylic-grafted biopolymer. Further, The SEM image emphasized the formation of the spherical globular interlinked structure as compared with the cloudy morphology of acrylic acid grafted bio-nanocomposites. It might be due to the transformation of inorganic/polymeric material into composite material with an interlinked structure.

Dhanya and Aparna (2016) studied the photocatalytic degradation of MO and AR using a composite of TiO_2/chitosan in which chitosan act as both a chelating as well as an immobilization agent. The experimental result reveals that TiO_2 nanoparticles embedded chitosan can remove the dyes from wastewater almost completely. The predictable particle size from the SEM image was found to be about 40 nm for this composite. The mineralization capacity was 100%, but COD removal was only 80% for Methyl Orange (MO), which indicates the transformation of some of the dyes into their respective intermediate products. The absorbance values of MO at various pH range showed that the decoloration efficiency increased as for pH 11 < pH 9 < pH 7 < pH 5 < pH 3. The efficiency of decoloration is higher in the low pH region and lower in the region of high pH. This observation is obvious from the fact that the dyes used here for the photodegradation studies were anionic in nature. In the case of Alizarin dye, the photodecoloration effectiveness seems to be increased from pH 3–7 and then decreases from pH 7–11. The efficiency is higher toward the neutral region and lowers toward both alkaline and acidic region.

Zhong et al. (2020) investigated the photodegradation of Methyl Orange dye using alternate-feed spray drying which was implemented to synthesize microspheres nano-ZnO/chitosan composite. The nZnO retained its hexagonal phase even after the dispersion and doping the ZnO onto CS microspheres. SEM report of the paper depicts the nZnO/CS microspheres size, in the order of 2–8 μm. The TEM clearly confirms that the nZnO were attached to the surface of CS. In this report, chitosan is being regarded as a metal ion chelating agent as well as a natural adhesive (Wan Ngah et al. 2011). Thus, during composite formation, the CS chelated and adhered with Zn^{2+}, and the nanoparticles facilitate the enhancement of biocompatibility to the prepared nZnO/CS (Q. Zhong et al. 2018). The study revealed that nZnO/CS exhibits Methyl Orange (MO) decolorization rate of more than 80%. The fabricated nZnO/CS finds promising biocompatible applications in dye wastewater treatment.

Farzana and Meenakshi (2015) examined and differentiated the photocatalytic mineralization of Reactive Red 2 (RR2) dye using zinc oxide (ZnO) and zinc-oxide-incorporated chitosan beads (ZCB) under UV and visible-light irradiations. The SEM images show that the non-homogenous surface of ZCB becomes smooth after its fabrication. The band gap energies of ZnO and ZCB are found to be 3.19 and 2.8 eV, respectively. This decrease in band gap energy of ZCB from 3.19 to 2.8 eV enables the catalyst to absorb more photons and ultimately progresses the photoresponse of ZnO to visible-light region. The paper reported that the pH at which the net charge of total particle surface is equal to zero (pHzcp) of ZCB was 5.2. At pH < pHzcp, the surface of ZCB gets positively charged and thereby adsorption of anionic RR2 dye is favored. This results in a higher percentage of degradation since more and more of RR dye molecules adsorbed at this pH eventually enhance the photodegradation synergetic effect. In addition, at this pH level, $HOO^{\bullet}$ can form H_2O_2 and gives rise to OH• radicals which in turn decolorize dye molecules.

Senthil Kumar et al. (2015) reported an effortless hand picking method for the 100% revival of the photocatalyst. They used a simple wet-chemical method for the fabrication of CuO nanospheres which was subsequently imprinted onto the matrix of biopolymer (chitosan) in a mild state by solution casting method. The photocatalytic ability of CuO nanospheres toward the mineralization of organic effluents was also reported for the first time. They discussed the surface and bulk morphology, optical properties, and crystal structure were in detail. ICP-OES analysis revealed 3.025% copper was entrenched on the CS matrix. Effectiveness of the CuO/chitosan was examined against the degradation of Rhodamine B (RhB) dye. The CuO nanosphere-chitosan combination provided a higher efficiency of up to 99% toward the degradation of the dye within 60 min of irradiation. This may be ascribed to the synchronous role of the composite in having the large surface area of CuO, slow electron-hole pair recombination rate of nanosized CuO in the biopolymer matrix, and high absorption efficiency of the chitosan. The major advantage of this work is that it not only controls the azo type dyes but could also be adopted for the mineralization of diverse kinds of organic pollutants. They tested the efficiency of the CuO/CS NCF photocatalyst in several pollutants, and all the type of organic contaminants showed excellent degradation activity.

Cao et al. (2013) fabricated Cu_2O/crosslinked-chitosan nanocomposites (Cu_2O/CS NCs) in situ, were using a simple one-step liquid-phase precipitation–reduction method. They reported that Cu_2O/CS NCs were approximately similar, spherical or ellipsoidal, and the surface was porous and rough due to the wrapping of Cu_2O on chitosan matrix. The chitosan layer improves the adsorption ability of the dye and molecular oxygen and prevents the rapid recombination of electrons–holes pair. The brilliant red X-3B (X-3B) was used as a model pollutant for the investigation of visible-light photocatalytic mineralization performance of Cu_2O/CS NCs. The report says that the photodegradation process of Cu_2O/CS NCs followed an apparent pseudo-first-order kinetics model. Since the dye is negatively charged, the dye X-3B degraded more in the acidic medium than in alkaline medium. Cu_2O/CS NCs exhibited enhanced visible-light photocatalytic activity compared to other photocatalysts reported before under similar experimental conditions.

9.4.1.2 Bimetallic Double Oxide Loaded Chitosan Composites

In recent years, researchers have focused on the use of semiconductor photocatalysts to photodegrade harmful dyes because they demonstrate the ability to completely degrade such colorants. Usually, many metal oxides are used as photocatalysts due to their exceptional physical and chemical properties such as high chemical stability, non-toxicity, broad region of absorption, high photostability, and cost-effectiveness. Further, their wide direct energy band makes them appropriate for use as a photocatalyst (Xie et al. 2011; Tian et al. 2012; He et al. 2018). However, the absorption ranges of some of these metal oxides are limited to a narrow visible range; hence exhibits low photocatalytic degradation. To surpass this, some researchers tried to load metals, metal oxides, and metal sulfides onto the surface of some other metal oxides. This doping will contribute to the total photocatalytic efficiency of the metal oxides and also enhance the ability by trapping the photo-excited electrons and subsequent transference of these electrons to the metal oxides from the visible-light region (Tang et al. 2008; Ma et al. 2016). The simple doping of metal on metal oxides will cause the agglomeration that eventually reduces the specific surface area of the composite resulting in a decrease in the photodegradation of these double metal oxides. In order to overcome this, a matrix can be embedded with these compounds as a support that can prevent the reduction of the specific surface area and restrain the agglomeration of these materials (Ma et al. 2016). Specifically, chitosan is used as the locking surface for these materials, whereby doping the M–O on M–O to a chitosan polymer can upgrade its properties.

Zayed et al. (2020) reported a new solid-state preparation route of chitosan-supported copper oxide using ascorbic acid, namely chitosan-supported Cu_2O/CuO nanocomposite. The synthesis and photodegradation of Methylene Blue (MB), Methyl Orange (MO), and Reactive Red (RR) dyes are schematically represented in Fig. 9.7. The X-ray diffraction emphasized the cubic structure of Cu_2O. Also, the presence of CuO is confirmed by the presence of the diffraction peaks at $2\theta = 32, 48,$ 58, and 65 which are related to the planes (110), (202), (202), and (310). They found out the optical band gap energy of Cu_2O/CuO to be 1.85 eV. For this photocatalyst, the reaction rate is increased in the pH range of 6–9. The photocatalytic degradation

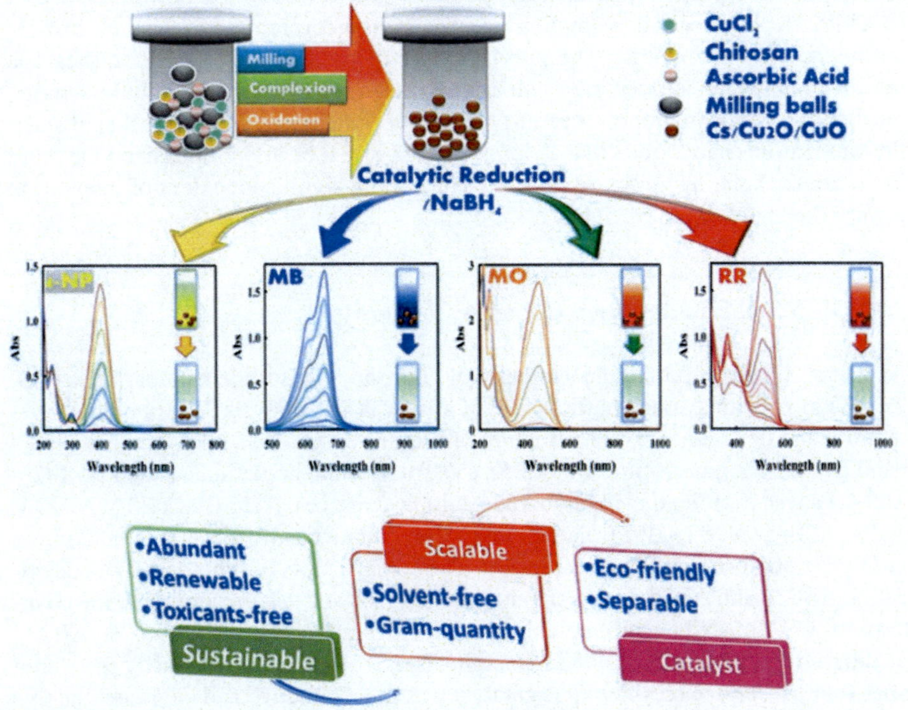

Fig. 9.7 Ball-milling preparation of $Cs/Cu_2O/CuO$ nanocomposite (Zayed et al. 2020). *Source* Zayed et al. (2020). Reproduced with permission from Copyright © 2020 Elsevier Ltd

of MB, MO, and RR was investigated using the Cu_2O/CuO and was found to be more than 90% for all the three examined dyes.

Alzahrani (2018) introduced a method to mineralize Fast Green (FCF) dye using a membrane of chitosan (CS), packed with ZnO or ZnO/CuO-heterostructured nanocomposites. The CS membrane was used as a supporting matrix, and the nanomaterials were used as photocatalysts. The percentage of photodegradation of the dye was determined by monitoring the absorbance at 623 nm, the λ_{max} of FCF, for different irradiation times. The mineralization percentages of the dyes in solar light, using the CS-ZnO and CS-ZnO/CuO membranes were found to be 57.90% and 60.23%, respectively. When irradiated with UV light, CS-ZnO and CS-ZnO/CuO exhibited photodegradation percentages, 71.45% and 91.21%, respectively. These studies pointed out that the best photocatalytic system for the degradation of FCF dye was CS-ZnO/CuO membrane in combination with UV light irradiation.

Linda et al. (2016a) fabricated a CdO/NiO/Chitosan nanocomposite and extensively examined the photocatalytic degradation of Rhodamine B (RhB) dye. Irradiation using visible light degrades nearly 90% of RhB dye in a short time. The cubic structure was observed for the both CdO and NiO particles, and in CdO the utmost orientation plane was shifted to some extent toward higher angle due to doping of NiO. The optical microscopic studies revealed the porous nature of the composite. It is a desirable property, as the nanoparticles that get adsorbed on the surface of the chitosan polymer matrix implies the successful incorporation of the nanoparticle within the chitosan matrix. The porous nature of the particles, smaller crystalline size, and enhanced surface area will contribute to the total photocatalytic activity. Further, the chitosan biopolymer surface behaves like a capping agent that absorbs the nanoparticles and looks like a cloud in the space. It clearly emphasized the more exceptional chelating ability of the chitosan, a more significant extent of interaction toward the metal oxide particles.

9.4.1.3 Metal Sulfide-Based Chitosan Composite

Aziz et al. (2020) reported the synthesis of chitosan-zinc sulfide nanoparticles (CS-ZnS-NPs) with an average particle size of 40 nm. XRD analysis illustrated the crystallinity and hexagonal crystal structure of the ZnS in CS-ZnS-NPs. The calculated band gap for the nanocomposite is 3.55 eV. The photodegradation efficiency of CS-ZnS-NPs was examined using two toxic anionic azo dyes, Acid Black 234 (AB234) and Acid Brown 98 (AB98). The sample was irradiated using a UV lamp at 254 nm during the photo-mineralization process. The contour plots for photocatalytic degradation AB234 and AB98 dyes were given in Figs. 9.8 and 9.9. According to the counterplots, CS-ZnS-NPs exhibited enhanced dye degradation of about 96.7% AB234 in 100 min and 92.6% for AB98 in 165 min. The CS-ZnS-NPs were easily recovered after four successive recycles. The catalyst was most effective and degraded the dye predominantly at the pH of 6 and 7.

Khan et al. (2019) fabricated copper sulfide nanoparticle-chitosan beads (CuS-CB), by dispersing CuS nanoparticles on chitosan beads (CB). The prepared particles

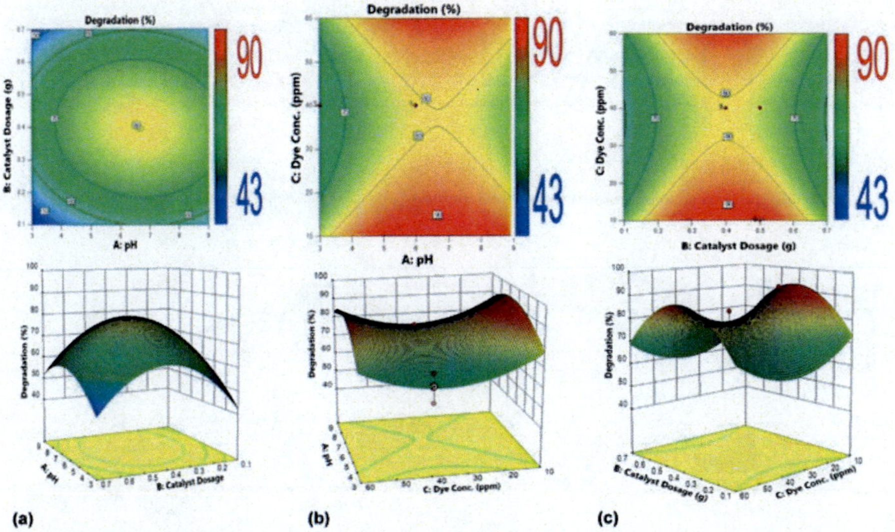

Fig. 9.8 Response surface and contour plots for photocatalytic degradation of AB98 as a function of (**a**) C: dose (**g**) and B: initial dye concentration ppm (pH = 7.0, reaction time= 150 min), **b** A:pH and B: dose (**g**) (initial dye concentration = 50 ppm, reaction time = 150 min), and **c** A: initial dye concentration ppm and B: pH, reaction time = 150 min (Aziz et al. 2020). *Source* Aziz et al. (2020) Reproduced with permission from Copyright © 2020 Elsevier Ltd

showed a smooth surface with an average of 735 μm bead size and band gap of 2.1 eV. The beads were used to analyze the photo-mineralization of one of the cationic dye, Malachite Green (MG). PH 8 was maintained for degradation. Since the increase of pH beyond 8 brings no significant change in degradation and at pH higher than 8, the Malachite Green is present in neutral form that has less attraction for negatively charged photocatalyst. The paper confirms the hexagonal structure of the CuS but does not say anything about the morphology of the CuS-CB. The CuS-CB exhibited 95% of photo-mineralization at the enhanced visible-light environment.

Zhu et al. (2009) reported the fabrication of crosslinked chitosan/nano-CdS (CS/n-CdS) composite catalyst through the biomineralization process and extensively studied the mineralization of Congo Red dye. Results of XRD in this study specified the successful development of the hexagonal phase of CdS on raw chitosan. Recycling experiments of this study confirmed the relative stability of the catalyst. The structure of the composite consists of many pleats on the surface of CS/n-CdS that provide a much larger surface area for degradation reaction. The CS/n-CdS had a spherical morphology with about diameters of 25–35 nm. The CS/n-CdS composite exhibited better thermal stability even at 800^0 C, and the compound mostly has CdS at this temperature range. The study revealed that the existence of co-anion NO_3^- would enhance the photodegradation of Congo Red (CR) dye. The decomposition rate by NO_3^- may be linked to the direct or indirect formation of hydroxyl radical that

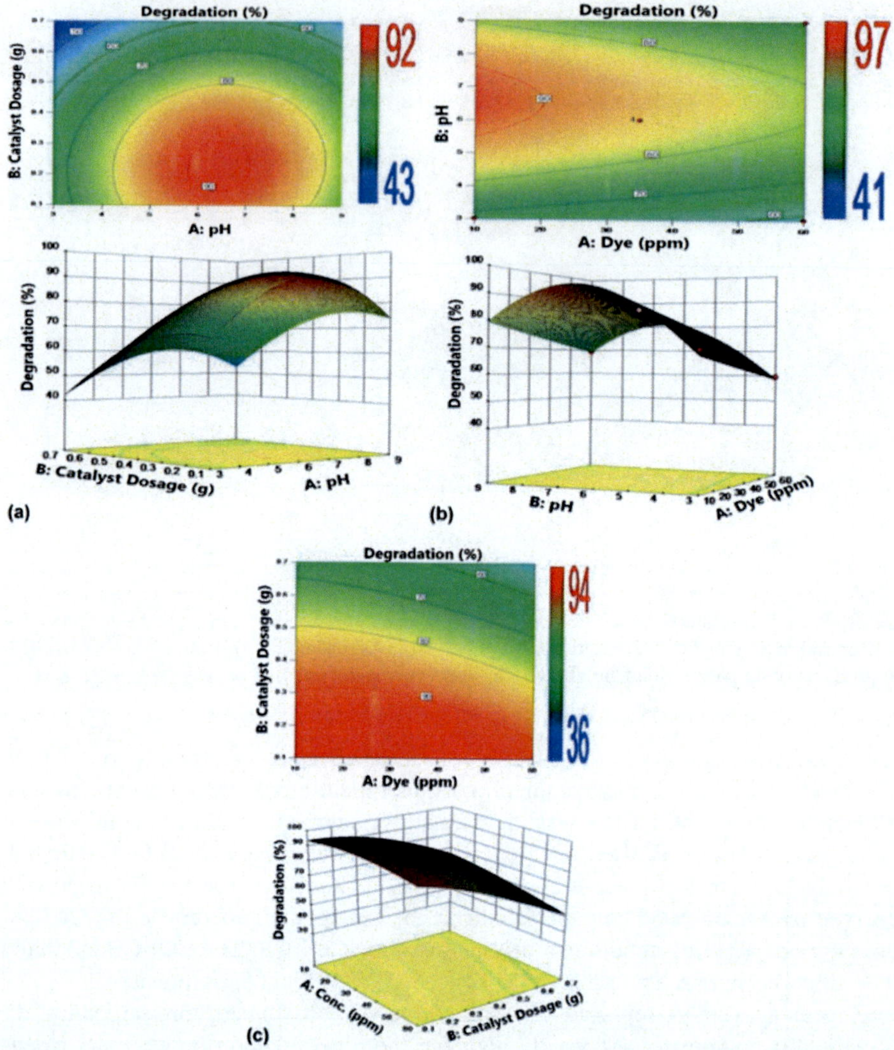

Fig. 9.9 Response surface and contour plots for photocatalytic degradation of AB234 as a function of (**a**) C: dose (**g**) and B: initial dye concentration ppm (pH = 7.0, reaction time= 150 min), **b** A: pH and B: dose (**g**) (initial dye concentration = 50 ppm, reaction time = 150 min), and **c** A: initial dye concentration ppm and B: pH, reaction time = 150 min (Aziz et al. 2020). *Source* Aziz et al. (2020) Reproduced with permission from Copyright © 2020 Elsevier Ltd

will induce the photodegradation process. Conversely, other anions had an inhibitory effect on dye degradation process.

9.4.1.4 Silver-Based Chitosan Composites

Currently, a new type of plasmonic photocatalyst Ag/AgX (X = Cl, Br, I) due to their surface plasmon resonance displays outstanding photocatalytic efficiency in the mineralization of organic effluents in the visible-light region. Conversely, there are some reasons that are disrupting the use of this catalyst. Drawbacks of Ag/AgX photocatalysts that would lead to a remarkable decrease in the photocatalytic activity are quick recombination of photogenerated electron–hole pair, easy agglomeration (Sohrabnezhad et al. 2014; Gao et al. 2014), difficulty in isolation and recovery from the reaction medium etc. (An et al. 2012). To overcome these problems, a lot of research has been carried out to fabricate suitable immobilizing/supporting materials that can effectively prevent the agglomeration of the silver particles and thereby increase photogenerated electron–hole recombination of the photocatalytic elements during the reaction. As known, chitosan (CS) is a hydrophilic biopolymer carrying cationic charge derived from partial N-deacetylation of chitin. CS contains a plentiful hydroxyl (–OH) and amino (–NH_2), which can interact with the metals via adsorption and chelation. It can act as a support and effectively prevent the agglomeration of metals in the growth of NPs and solve the problems of separation and recovery (An et al. 2012; Senthil Kumar et al. 2015).

Wang et al. (2017a) reported a simple one-step method of preparation of novel Ag/AgCl/chitosan composite photocatalyst. Chitosan played a dual role in the preparation process such as a reductant to reduce Ag^+ to Ag^0 species and acted as a supporter for Ag/AgCl nanoparticles as well. Under visible-light irradiation, the nanocomposite exhibited high photocatalytic activity for the degradation of Rhodamine B (RhB). The result of photocatalytic degradation experiment revealed that 20% the mass ratio of AgCl to chitosan was the optimum for yielding 96% degradation of the dye molecules. The XRD showed the Ag/AgCl/CS sample characteristic diffraction peaks at 2θ at 27.92°, 32.41°, 46.41°, 54.99°, 57.66°, 67.60°, 74.58°, and 76.93°, which were in good agreement with the primary diffractions of (111), (200), (220), (311), (222), (400), (331), and (420) crystalline planes of AgCl. On the other hand, no distinct peak for metallic Ag was found in the XRD, which may be due to its low content (Gupta and Saleh 2013; Jeevanantham et al. 2018).

Wu et al. (2020) synthesized plasmonic Ag/Ag_3PO_4/Chitosan composite photocatalyst using low-temperature process. As per the authors, chitosan played three crucial roles in this composite. Firstly, it was committed for the in-situ reductions of Ag^+ ions of Ag_3PO_4 into metallic Ag. Also, it acts as a transporter of Ag/Ag_3PO_4 nanoparticles and lastly prevents the aggregation of metallic particle. Besides, the prepared hydrophilic chitosan-based composite could be easily recovered from the solution. The photodegradation experiments indicated that the Ag/Ag_3PO_4/CS composites possess the superior photocatalytic activities in the Rhodamine B (RhB) degradation under visible light, but needs to improve the stability of the composite. A body-centered cubic phase of Ag_3PO_4 was identified in the XRD, but there is no distinct peak for metallic Ag which was found in the XRD, maybe due to its low content.

Taghizadeh et al. (2020) synthesized a nanocomposite chitosan/AgCl/ZnO (CS/AgCl/ZnO) hydrogel beads system and used for the photocatalyst degradation of Methylene Blue (MB) as well as *Escherichia coli* (*E. coli*) and *Staphylococcus aureus* (*S. aureus*) as gram negative and positive bacteria under visible-light irradiation. The CS/ZnO, CS/AgCl, chitosan/AgCl/ZnO degraded about 34%, 41%, and 98% of Methylene Blue (MB) dye, respectively, during the photodegradation in the ideal pH 11. This emphasized the synergetic effect in the photocatalysis. Based on the SEM image analysis, researchers found the particle size to range from 20 to 40 nm. It was also revealed that the formation of zinc oxide and silver chloride nanoparticles range, AgCl/ZnO, with 0.2 mol fraction of silver chloride. The increase in AgCl causes no perceptible change in the surface morphology of the nanocomposites. Based on the XRD, the authors claimed the wurtzite hexagonal crystalline structure for ZnO. Cubic crystalline phase of AgCl and hydrated crystalline and amorphous peaks which related to the semi-crystalline nature of pure chitosan were visible at 10.9° and 19.8°. Besides, the literature data of some of the major chitosan-based photocatalysts were listed in Table 9.3.

9.4.2 Cellulose Composite-Based Photocatalyst

As a regenerated bioresource, cellulose has been extensively explored in recent years. Being an assemble of microfibers with full of hydroxyl groups, these materials can easily combine with other materials via hydrogen bonding. Though normal cellulose does not possess functional or photo/electrochemical property, it has many other advantages, such as cheap availability, natural origin, ease of processing and recovery. Moreover, its enhanced electron–hole separation due to their ability to immobilize photocatalysts to avoid secondary contaminations (Kim et al. 2015), making cellulose as a suitable substrate candidate for preparing cellulose-based composites. Thus, cellulose composite has attracted extensive attention of researchers, and many related works have been reported (Zou et al. 2019). The semiconductor photocatalysts are playing an essential role in addressing many environmental and energy challenges. Extensive studies have been made to investigate the usage of different types of semiconductors in the synthesis of the cellulose-based photocatalyst composites of metal oxides like (ZnO, TiO_2, MnO_2, Cu_2O, WO_3), metal sulfides including (ZnS, CuS, CdS, $CuInS_2$, Ag_2S-ZnS), bismuth-based semiconductors such as (BiOCl, BiOBr, BiOI, $Bi_4O_5Br_2$), silver-based semiconductors like (AgBr, AgI, Ag_3PO_4, $AgCrO_4$, and $AgVO_4$), and non-metallic semiconductors such as (graphene, carbon nitride). The photocatalytic activities of these composites need a detailed discussion. Table 9.4 represents the summary of the photocatalytic activities of some of the cellulose-based composites.

Table 9.4 The literature data of cellulose based photocatalysts for dyes degradation

Name of composite	Preparation method	Name of dye degraded	Nature of dye	Concentration of dye (20 mg/L)	pH	Efficiency	Reference
CuS/cellulose	Aerogel	Methylene blue	Cationic	20	9	94.6	Saeed et al. (2019)
Cu_2O/GO/cellulose	in situ	Methyl orange	Anionic azo	10		72	Tu et al. (2014)
Cellulose/BiOCl	Hydrothermal	Rhodamine B	Zwitter ionic	20		60	Tian et al. (2019)
ZnO/PU/cellulose acetate	Phase separation	Reactive red 11	Anionic azo	50 to 250	7	96	Rajeswari et al. (2017)
		Reactive orange	Cationic azo	51 to 250	7	92	
ZnO/PVC/cellulose	pPhase separation	Crysta violot	triarylmethane	3		Nearly 90	Jiang et al. (2020)
		C0ngo red	Anionic azo	3		Nearly 91	
RGO/g-C_3N_4/cellolose acetate	Vacuum filtration	Rhodamine B	Zwitter ionic	10	1 to 7	More than 60	H. Zhao et al. (2016)
Ag_3PO_4/cellulose	n situ	Rhodamine B	Zwitter ionic	10		More than 50	Q. Wang et al. (2014)
Ag@AgCl/cellulose acetate	Electrospinning and in situ	Methyl orange	Anionic azo	10		More tham 73	Z. Zhou et al. (2016)
Ag@AgCl/PVP/cellulose	In situ	Methyl orange	Anionic azo	10		97	Zhang et al. (2018)
BiOBr/micro crystalline cellulose	In situ	Rhodamine B	Zwitter ionic	30	6 to 8	More than 90	W. Zhou et al. (2019)
Bacterial cellulose/ TiO_2	Sol–gel	Reactive Brilliant Red	Anionic azo	15	2.to 7	Almost 100	W. Zhou et al. (2019)

(continued)

Table 9.4 (continued)

Name of composite	Preparation method	Name of dye degraded	Nature of dye	Concentration of dye (20 mg/L)	pH	Efficiency	Reference
Cellulose-Supported $CuInS_2$	Hydrthermal	Rhodamine B	Zwitter ionic	10	5	89	Tavker et al. (2019)
Nanocellulose/ZnO	co-precipitation	Rhodamine B	Zwitter ionic	10	5	76	Tavker and Sharma (2018)
PTA/ZIF-8@Cellulose	Aerogel	Rhodamine B	Zwitter ionic	10	5	99.7	Wen et al. (2020)
P(Py-co-An)-TiO_2/NCC	Chemical oxidation	Eosin yellow	Anionic	10	4 to 5	92.3	Anirudhan and Rejeena (2015)
CA–GO/TiO_2–NH_2	electrospinning	Indigo caramine	Anionic	10	2 to 8	99.8-72	Aboamera et al. (2018)
		Methylene blue	Cationic	10	2 to 8	98.3-51	
Ag_2S–ZnS/cellulose	Vacuum filtration	Rhodamine B	Zwitter ionic	10-100	6	98	Aboamera et al. (2018)
MnO_2@Cellulose	Hydrothermal and sol-gel	Acid orange	Anionic azo	30	2.	76	Peng et al. (2019)
AuNPs/CNC	microwave irradiation	Congo red	Anionic azo	10	6 to 7	Almost 100	Alle et al. (2020)
		Rhodamine B	Zwitter ionic	10	6 to 7	Almost 100	
		Amaranth	Anionic azo	10	6 to 7	Almost 100	
Carbon/ZnO/TEMPO oxidized cellulose	Precipitation	Methyl orange	Anionic azo	5	7	96.1	Xiao et al. (2018)
C-PDA-Ag–WO_3	Self assembly	Reactive blue	Anionic	50	7	98	Fan et al. (2019)

9.4.2.1 Metal Oxide-Based Cellulose Composite

The researchers exploited the advantages of metal oxides such as large band gap, high thermal stability, non-toxicity, wide availability, and compatibility for blending with the cellulose composite. The imprinting of metal oxides with cellulose would provide preferential active sites for molecular adsorption and dissociation and also act as traps for photo-excited charge carriers and lag the recombination of the photo-excited electrons and holes (Dette et al. 2014). Possibly these advantages were explored (Wittmar et al. 2015), and cellulose–TiO_2 nanocomposites were fabricated successfully by non-solvent-induced phase separation. The optimum ratio of wet cellulose acetate–TiO_2 was found to be 15:1 and was used for photodegradation 10 mg/L of Methylene Blue (MB) and Rhodamine B (RhB). However, the increase of the dose of TiO_2 will retard the photocatalysis since the agglomeration of the composite will result in a decrease in surface area. Li et al. (2018) prepared a cellulose nanofiber/TiO_2 aerogel (CNFT) through a facile hydrothermal method and utilized it for the photocatalytic degradation of Methylene Blue (MB) solution. The synthetic route of the composite was shown in Fig. 9.10.

TiO_2 nanoparticles adhered on to the surface of cellulose nanofibers as scaffold through a hydrogen bond and were well-distributed. The particle size distribution data showed that the mean diameter of TiO_2 nanoparticles was around 6.8 nm. In the system of the as-prepared CNFT, CNF played essential roles in adsorption of pollutant and in stopping the electron–hole derived from TiO_2 from recombining. The mechanism of photodegradation is shown in Fig. 9.11. The mineralization result

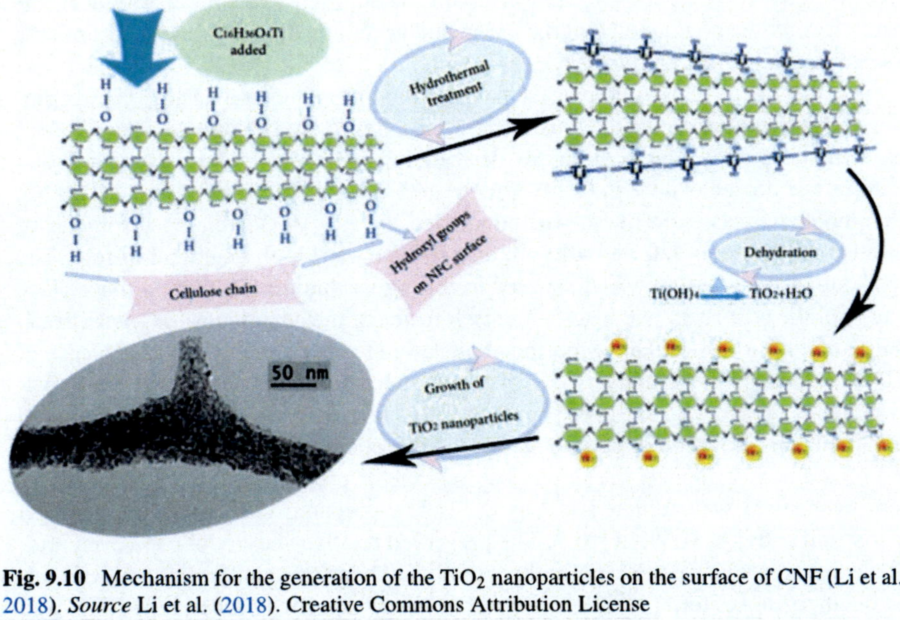

Fig. 9.10 Mechanism for the generation of the TiO_2 nanoparticles on the surface of CNF (Li et al. 2018). *Source* Li et al. (2018). Creative Commons Attribution License

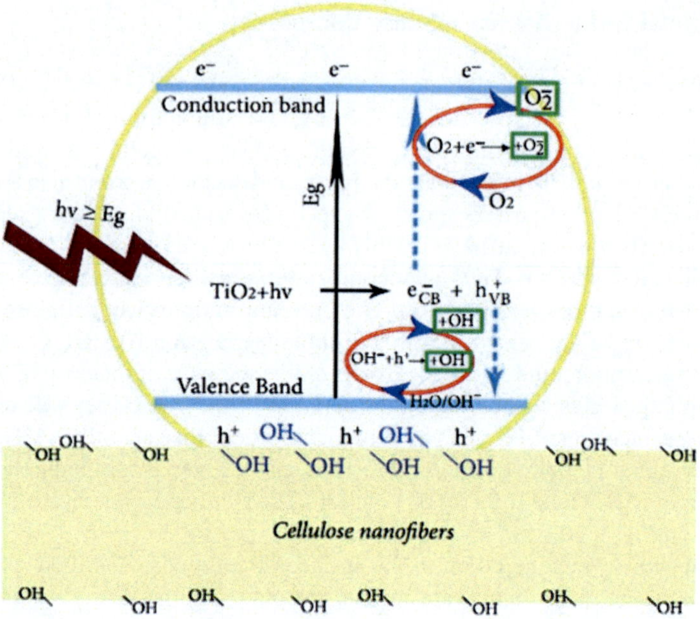

Fig. 9.11 Schematic illustrating the strategies to develop $^{\bullet}O_2{}^-$ and $^{\bullet}OH$ (Li et al. 2018). *Source* Li et al. (2018). Creative Commons Attribution License

showed that the composite aerogel showed enhanced photodegradation of pollutant (MB) in ultraviolet light (UV) as well as in the sunlight irradiation. Based on the result, it could be concluded that the green and portable photocatalyst has promising wastewater treatment application.

Aboamera et al. (2019) prepared CA/GO composite nanofibers using electrospinning techniques and crosslinked using TiO_2. The fabricated composite was stable in adsorption capacity throughout many degradation cycles, had a fast adsorption rate, and use the power of low intensity for the photodegradation of IC and MB dyes. The photocatalysis experiment was conducted using UV– Visible spectrometer of wavelength between 320 and 400 nm and showed very high degradation rate. The pH value of the solution was discussed in this paper due to its effect on the surface charge of the adsorbent and also on the new forms of the adsorbate along with availability of sorption sites. However, the explanation for the effect of pH in the case of IC was satisfactory while that of MB was somewhat ambiguous since MB was positively charged. Li et al. (2017b) prepared OBC/TiO_2-Lac composites by dissolved in 100 mM in a solution of sodium acetate and acetic acid buffer solution (pH 3.45) and mixed homogenously for 0.5 h. The scheme of preparation and photodegradation mechanism was depicted in Fig. 9.12. The prepared OBC/TiO_2-Lac showed nearly unit activity, 100% at pH 7. The preserved relative activity of 67% even after 10 cycles is proof of its exceptional reusability. Therefore, synthesis of functionalized composite bacterial cellulose nanofiber membranes with a combined bio- and

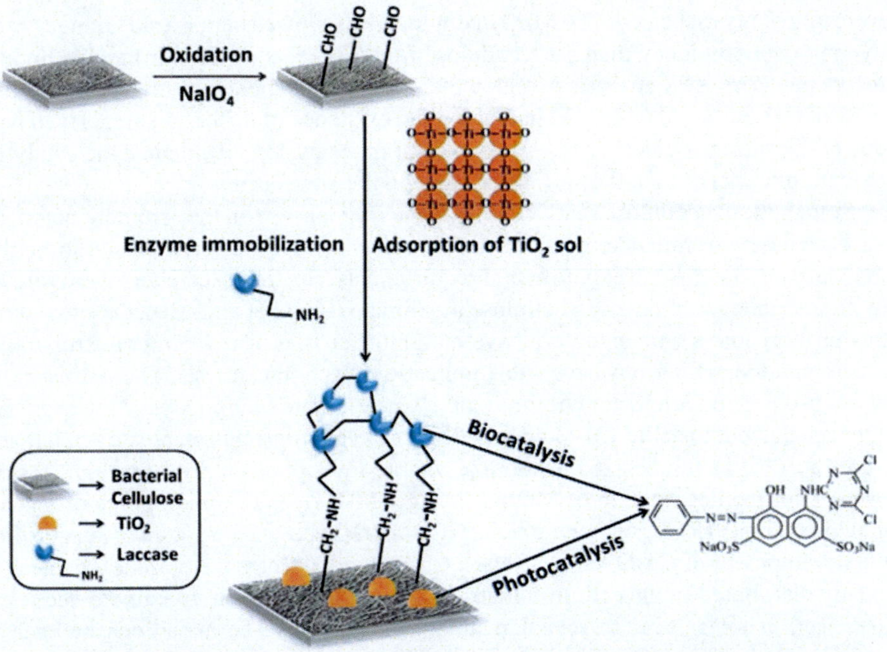

Fig. 9.12 Schematic representation of chemical modification of BC followed by the subsequent preparation of OBC/TiO_2-Lac composite functionalized membranes deployed for dye degradation (Li et al. 2017a). *Source* Li et al. (2017a). Reproduced with permission from Copyright © 2017 Elsevier Ltd

photocatalytic properties are a potentially valid approach for industrial textile dye degradation.

It has been observed that ZnO is relatively more economical than TiO_2 and other conventional semiconductor nanoparticles (Liang et al. 2012). This may be the reason for Tavker and Sharma (2018) the fabrication of Zinc oxide nanoparticles from zinc acetate dihydrate and nanocellulose via the *in-situ* method by adding different amounts of nanocellulose. The latter compound was isolated from agro-waste using chemo-mechanical treatments. The photocatalytic efficiency of 10 mg of nanocellulose doped zinc oxide (NC/ZnO) composite was investigated using RhB and was found to be 35% invisible, 76% in sunlight, respectively. But the paper could not explain anything about the influence of pH in the photodegradation process. In metal oxide doped composites, adsorption is the key to photodegradation, which will influence the crystallite size of the composite. Herein, the doping of ZnO with cellulose was in the ratio of about 2:1, this ratio must increases the crystallinity of the prepared composite and it may be one of the reasons for the low degradation rate.

Peng et al. (2019) fabricated a novel manganese dioxide cellulose (MnO_2/cellulose) composite film through a low-temperature physical sol–gel gelation. The composite exhibited high mechanical strength and could buckle flexibly. Composite had a three-dimensional porous frame structure where MnO_2 particles were uniformly dispersed. The MnO_2/cellulose film showed better Acid Orange (AO) dye removal efficiency than pure cellulose film (76.1% after 90 min) and is highly recyclable. The XPS analysis revealed the +3 and +4 valence of manganese at the peak of 641.2 eV and 642.6 eV, indicating the existence of different valencies of Mn due to the surface defects. This +3 valence of manganese will enhance the catalytic oxidation properties via the surface oxygen on the composite.

Somehow, the cellulose matrix shows some resistance with the semiconductors at the extreme environment (e.g., hardness, pH value, conductivity, and tensile strength) that does not meet some experimental requirements. So, the researchers were forced to imprint other substances in cellulose membrane (Singh et al. 2020). Oxygen cavities in inorganic semiconductors have a significant role in reducing electron–hole recombination, which may have vital implications in photocatalysis. The visible-light active p-type semiconductor cuprous oxide (Cu_2O) is one such catalyst. However, the synthesis of photostable Cu_2O enriched with oxygen defects remains a challenge. Tu et al. (2014) synthesized a portable visible-light photocatalyst through in-situ process for the degradation of Methyl Orange (MO) dye. The micropores of regenerated cellulose (RC)/graphene oxide (GO) matrix were used as a microreactor for the development of Cu_2O nanoparticles. Cu_2O nanoparticles were immobilized and evenly distributed in the cellulose matrix to excite and generate free photoelectrons and electron holes. It can be seen that, among various Cu_2O compositions are loaded on RC/GO, the lowest degradation rate was shown by the photocatalyst with highest Cu_2O loading. This is because higher dosage of Cu_2O nanoparticles formed in this case may aggregate inside the cellulose matrix and block the cavities of the cellulose composite films. Thus, the activity point of Cu_2O was diminished, and the exchange of MO solution was inhibited.

Rajeswari et al. (2017) fabricated cellulose acetate-polyurethane (CA-PU) membrane imprinted with nano-ZnO as a photocatalyst through solution dispersion blending method and utilized for anionic Reactive Red 11 (RR11) and cationic Reactive Orange 84 (RO84) dyes degradation. In this study, the maximum dye removal observed at neutral pH needed a more convincing explanation, since the natures of the two dyes are entirely different. The prepared membrane surface of this study confirmed the existence of micropores which were randomly distributed on the surface. This reveals the asymmetric nature of the membrane, and it may be the reason for the highest percentage of degradation of the dye molecules.

Linda et al. (2016b) explored the solution cast method for the synthesis of cellulose/ PVC/ZnO thin film and was effectively utilized for the degradation of Congo Red (CR) and Crystal Violet (CV) dyes. The photocatalytic activity studies of the PVC/ZnO composite reveals that the cellulose/PVC/ZnO composite has better photocatalytic activity than that of the PVC/ZnO composite. However, the aspects that activity of these types of composites is influenced by pH and crystallite size were not explained in this research. Fan et al. (2019) synthesized the cotton (C)-PDA–Ag–WO_3(c) composite photocatalytic material and investigated the photocatalytic performance with Reactive Blue 19 (RB 19) dye. A 98% degradation of the solution was accomplished within 180 min using visible-light irradiation. The (C)-PDA–Ag–WO_3(c) exhibited better photodegradation than C-PDA-Ag–WO_3(a), of 50 mg/L dye solution. The FE-SEM clearly showed the regular three-dimensional structure of Ag nanoparticles with a size of about 40 nm. The XPS of the composite, 36.31 and 34.21 eV, was related to W^{6+} atoms, and the peaks at 37.07 and 34.66 eV was well established with the W^{5+} atoms (Cao et al. 2018; Ding et al. 2017; Meng et al. 2015). The results confirmed that the tungsten element extensively existed in two oxidation states such as W^{6+} and W^{5+} (Zhang et al. 2017a) suggesting the presence of oxygen vacancies in WO_3, which will enhance the photodegradation capacity of the (C)-PDA–Ag–WO_3(c) composite matrix.

9.4.2.2 Metal Sulfide-Based Cellulose Composite

Metal sulfides are supposed to be the promising photocatalysts as they have narrow band gaps together with valence bands at the negative potential in comparison to oxides. Thus, it results in an enhanced visible-light photocatalytic activity (Melinte et al. 2019). But, due to their photo-corrosion phenomenon (Yue et al. 2016), they are the most unstable photocatalyst (Wang et al. 2017b; Kumar et al. 2016; Lim et al. 2018). Hence, organic–inorganic hybrid synthesis is found to be an efficient method to improve physicochemical properties. It will further enhance the photocatalytic activity metal sulfide structures.

Saeed et al. (2019) prepared 3D macroporous CuS-functionalized cellulose-based aerogel (CBA) structure through in-situ deposition of visible-light-responsive CuS photocatalyst and utilized it for the effective degradation of Methylene Blue (MB) dye molecule. CuS/CBA composite catalyst in visible light showed the degradation 94.1%, while, 67.4% for pure CuS at pH 6.8. Even after five consecutive cycles, it maintained the MB degradation rate above 80%. During the preparation, the CuS was dumped onto the outer and inner walls of CBA aerogel structure. In this study, there was formlessly explained that the samples with high concentration load of CuS on CBA demonstrate better degradation performance against MB. Since the increase in the ratio of metal sulfide and cellulose will cause the agglomeration, and it will retard the rate of photocatalysis in the absence of any organic capping agent.

Zhou et al. (2019a) prepared 2D porous CdS sheets held on bacterial cellulose (BC@CdS) via the biotemplate method. Compared with pure CdS nanoparticles, visible-light-driven H_2 production activity of BC@CdS composite was improved by ~3.5 times. In this preparation, CB acts as a scaffold for the proper development of CdS at its surface. The BC@CdS-2 and BC@CdS-4 TEM shows a consistent epitaxial growth of CdS nanocrystals. The diffraction peaks of BC were broadened due to the impure crystalline structure of natural material. Because of the capping nature of the CdS nanocrystals, the peak intensity of BC decreases, whereas broadening increases, which emphasizes the semi-crystalline nature of the composite.*Capping provide a large surface area to the composite, thereby improving its adsorption during photocatalysis. In addition, even after four cycles test of degradation of Methylene Blue (MB), the degradation ability of BC@CdS sheets did not show a significant decrease. This indicates the excellent stability of the composite as well as recycling efficiency.

Tavker et al. (2019) hydrothermally synthesized agro-waste-extracted, cellulose-supported $CuInS_2$ nanocomposite photocatalysts and studied its efficiency in dye degradation using Rhodamine B (RhB) dye. The paper put forth the crystalline nature of the cellulose. However, the XRD pattern emphasized the semi-crystallinity of the composite. The FE-SEM image of CIS shows hierarchic hollow structures made up of nanoflakes with the thickness of ~50–55 nm. These Cel/CIS have superior visible-light-driven photocatalytic activity, as the electrostatic interaction that operates between cellulose and CIS helps in delaying the charge carrier. The Photocatalytic efficiency of bare $CuInS_2$ is very low due to speedy recombination of the charge carriers, electron, and hole. However, the cellulose supported $CuInS_2$ exhibited tenfold higher photocatalytic efficiency than pristine-supported $CuInS_2$. The blending of $CuInS_2$ with biopolymeric cellulose delays the electron–hole recombination interaction and thus facilitate degradation.

Prashantha Kumar & Ashok Kumar (2019) fabricated new visible-light-active nanosized double sulfides, Ag_2S–ZnS loaded on cellulose (AZCE) by a precipitation method. The XRD results revealed the semi-crystalline nature of the composite. The TEM showed that Ag_2S–ZnS nanoparticles were evenly distributed and also encapsulated on the surface of CE fiber. The photodegradation ability of the prepared composite under visible light had been investigated using Rhodamine B (RhB) dye. 30 mg/L of the dye have been effectively oxidized using 30 mg of the AZCE dose in the pH 4–12 range within 90 min. The cellulose acts as catalyst surface, and the alkaline pre-treated cellulose provides an active surface hydroxyl group that will enhance the deposition capacities of Ag_2S and ZnS. Consequently, a large amount of visible light could be absorbed, and the photodegradation ability of the composite was increased.

9.4.2.3 Gold- and Silver-Based Cellulose Composite

Gold and silver are considered to have of promising photocatalytic activity to decompose organic pollutants because of the exceedingly dispersive band structure (Reddy and Yang 2015; Azetsu et al. 2011). Moreover, the size of Ag remains relatively large, which might have an influence on its activity. The higher surface area of smaller particle size is thought to be of benefit to the photocatalytic reaction, which mostly occurs on the catalyst surface (Linsebigler et al. 1995). Still, a tendency to self-aggregate due to their sizeable active surface energy causes a massive drop in catalytic activity (Azetsu et al. 2011). Construction of nanoparticles with the highly uniform size is also an effective method to improve the photocatalytic efficiency in eliminating the effluents from the solution. Conversely, the removal nanoparticles from solution are very difficult, which causes secondary environmental pollution. To resolve this difficulty, imprinting of nanoparticles with renewable biomass macromolecules such as cellulose, chitin, and chitosan, has been employed (Chang and Chang 2010; Madhusudhan and Reddy 2019).

Alle et al. (2020) exhibited well-dispersed synthesize gold nanoparticles (AuNPs) through cellulose nanocrystals (CNC), which acts mutually as reductant and sustaining agent. The prepared composite AuNPs were well-adhered on the CNC surface and had a homogeneous spherical shape with an average diameter of 8 ± 5.3 nm. The dye mineralization of the adsorbent molecule is investigated extensively using Allura Red (AR), Congo Red (CR), Rhodamine B (RhB), and Amaranth (AM) with $NaBH_4$, the latter compound acted as an electron donor, and the dye molecules act as an electron acceptor. The biosynthesized AuNPs/CNC nanocomposite showed exceptional degradation properties in the mineralization of dyes such as AR, CR, RhB, and AM. Most of the synthesized AuNPs have been successfully embedded on the surface of the cellulose nanocrystals.

Wang et al. (2014) introduced a method for the synthesis Ag_3PO_4/cellulose hydrogels nanocomposite by in-situ reductions and oxidations of Ag_3PO_4 nanoparticles on cellulose matrix and this, when used for the mineralization of RhB dye, showed 50% degradation. The Ag_3PO_4 were dispersed homogeneously in the regenerated cellulose hydrogels without having any agglomeration, with an average diameter of Ag_3PO_4 particles from 3.1 ± 2.7 to 11 ± 4.5 nm and enhanced Ag ion concentration. The SEM images of the Ag_3PO_4/cellulose composite hydrogels displayed incorporation of the Ag_3PO_4 nanoparticles on the cellulose matrix and were packed into the cavities of the cellulose hydrogels that emphasized the formation of uniformly distributed composite matrix. In order to improve the photo-mineralization (Zhou et al. 2016), cellulose acetate was used as a support for Ag/AgCl composite. The CA supported cubic Ag/AgCl catalyst exhibited high catalytic efficiency in the degradation of Methyl Orange (MO) in visible light. As per the authors' findings, the local surface plasmon resonance (LSPR) of silver nanoparticles can improve the local internal electromagnetic field and caused enhanced photocatalytic degradation. At the same time, Ag@AgCl structure has high stability, which can maintain the degradation efficiency of 73% after three cycles (Zhou et al. 2016). The XPS and XRD analysis emphasized that the samples in-situ developed on electrospun CA fibers were Ag@AgCl, of which silver existed as the metallic state.

Zhang et al. (2018) prepared the Ag@AgCl/PVP/cellulose composite film through one-step coagulation process. The PVP was added into the cellulose solution in order to regulate the structure and size of Ag@AgCl in the cellulose matrix. The synthesized composite was used to investigate the MO dye degradation and exhibited more than 90% mineralization. The PVP was used as the stabilizing agent of AgCl, which could adsorb the {100} facet of AgCl and inhibit its growth (Zhang et al. 2018). PVP was also able to decrease the size of Ag@AgCl, to improve the specific area, and made them partly embedded in cellulose matrix to enhance the stability of Ag@AgCl on the surface of cellulose film which will enhance the photo-mineralization.

Gopiraman et al. (2019) reported the first composite of bimetallic Ag–Au cellulose nanofiber (Ag@Au/CNCs) fabricated through a very simple green preparation process and used not for photocatalytic but catalytic degradation of organic compounds. An aqueous leaf extract of *Moringa oleifera* was used to gain the bimetallic Ag@Au/CNC nanocomposite. The formed bimetallic Ag@Au/CNC catalyst performed remarkably well in the reduction of nitrophenols catalytically. The Ag@Au/CNC catalyst proved to have about a onefold catalytic improvement compared with the Ag/CNCs and Au/CNCs. The Ag@Au/CNC catalyst could be reused in the aza-Michael reaction at least for ten cycles without a loss of its catalytic activity.

9.4.2.4 g-C_3N_4 Based Cellulose Composites

For visible-light photodegradation, graphitic carbon nitride (g-C_3N_4) is particularly attractive because of its visible-light response and excellent chemical stability (Wang et al. 2009; Yan et al. 2010). Even though g-C_3N_4 has been reported for CO_2 reduction (Mao et al. 2013), photocatalytic water splitting (Wang et al. 2009), and organic contaminants degradation (Yan et al. 2010), its fast recombination of photogenerated electrons weakens its photocatalytic efficiency. In recent times, researchers found that exfoliating bulk g-C_3N_4 into ultra-thin two-dimensional (2D) g-C_3N_4 nanosheet is an effective strategy to enhance its photocatalytic ability by improving the mobility and density of the photogenerated charges (Xu et al. 2013). However, the photocatalytic efficiency remains far from satisfaction on modifying membrane for water treatment. To overcome this complexity, doping of nanoparticles with renewable biological macromolecules such as cellulose, chitin, and chitosan has been much employed.

Zhao et al. (2016) synthesized a graphitic carbon nitride nanosheet/reduced graphene oxide/cellulose acetate composite photocatalytic membrane (g-C_3N_4 NS/RGO/CA) by assembling a g-C_3N_4 NS/RGO photocatalyst on the surface of commercial CA membrane. Owing to the attractive photocatalytic efficiency of g-C_3N_4 NS under visible-light irradiation and photogenerated charge separation resulting in the formation of the unique heterostructure between g-C_3N_4 NS and RGO, g-C_3N_4 NS/RGO/CA. The prepared matrix was used for the degradation of Rhodamine B (RhB) with the initial dye concentration of 10 mg L^{-1}, and 5 mg of photocatalyst was employed in the photodegradation process. The band gap of g-C_3N_4 NS was found to be 2.9 eV, indicating that the composite was a potential material in the visible-light irradiation.

Wang et al. (2020) prepared sulfuric acid treated graphitic carbon nitride (SA-g-C_3N_4) embedded within a porous cellulose network (denoted here as CN/CA films). The SA-g-C_3N_4 content in the films were varied from 0 to 50 wt. %. The H_2SO_4 treatment brings in sulfonyl and carboxyl groups on the surface of g-C_3N_4, results in stable hydrogen bonding interactions with the hydroxyl groups of cellulose acetate. The composite used to degrade the Rhodamine B (RhB), Crystal Violet (CV), Methylene Blue (MB), and Malachite Green (MG) dyes solutions showed reasonable dye degradation capacity. More importantly, the CN/CA composites exhibited recyclability and exceptional stability during the photocatalysts process, with no loss in activity even after five test cycles. Results encourage the broader development of porous composite films for photocatalytic applications. The composite also shows some exceptional ability to reduce Cr (VI), achieving a reduction efficiency of 95% for Cr (VI) (5 mg/L) in 100 min.

9.4.2.5 Bismuth-Based Cellulose Composites

Semiconductor photocatalysts are the most widely accepted options in the environmental purification system and that have often been researched due to their excellent performance in the degradation of organic pollutants, decomposition of water to produce hydrogen, and other processes (Chen et al. 2019; Li et al. 2017). Several metal oxide semiconductor such as TiO_2 (Fujishima et al. 2000), ZnO (Fujishima et al. 2000), Cu_2O, (Mateo et al. 2017), and Bi_2O_3 (Bian et al. 2008) were reported so far, but their activity has been limited in the ultraviolet light (UV). However, bismuth-based catalysts, especially bismuth oxyhalides (BiOX, X = F, Cl, Br, I), have got more attention due to their outstanding properties such as low toxicity, low cost, and reusability (Wu et al. 2016; Zhang et al. 2017b). More prominently, the suitable band gap, unique lamellar structure, and excellent fluorescence characteristics facilitated BiOX into a series of electro-optical and photoelectric, and photocatalytic properties, which can degrade organic pollutants in water under ultraviolet (UV) or visible-light irradiation (Wang et al. 2018a, b). Cellulose is an important biomaterial that comprised of macromolecule polysaccharide with high quantity of glucose and hydroxyl groups. The hydroxyl groups in cellulose can efficiently interact with metal cations. Therefore, metal oxides can be easily distributed uniformly on the surface of cellulose (Zhao et al. 2017). These binding properties of the cellulose make in use for the preparation of an organic–inorganic hybrid photocatalyst of bismuth compounds.

Zhou et al. (2019b) synthesized microcrystalline cellulose (MCC)-BiOBr composite through one-step in-situ wet-chemical method. In this preparation process, MCC acted as a matrix to provide template and support. Due to the reaction with cellulose hydroxyl group, the nanosheets of BiOBr grow into flower-like as shown by the ruling morphology. According to BJH adsorption study, the average pore diameter of the prepared photocatalyst was 43.72 nm, which suggests that the MCC can successfully afford support for the BiOBr nanosheets. The 20 mg of the prepared composite was dispersed into 100 mL of 30 mg/L Rhodamine B (RhB) dye solution and irradiated with a wavelength of 552 nm. The photocatalyst mineralized more than 90% of RhB in the aqueous solution within 70 min.

Tian et al. (2019) fabricated a high-performance flower-like BiOCl (BOC) nanomaterial on sustainable cellulose nanofibrils (CNFs) as regulator via a facile one-pot hydrothermal method. The flower-like morphology of CNFs doped BiOCl (CBOC) formed was depicted in Fig. 9.13.

The size of the particle was reduced during nanoparticle formation. Due to the reduced size, the enhanced photosensitization and photocatalysis were observed during the degradation of Rhodamine B (RhB). They were degraded by CBOC-5 (5% of CNFs) within 16 min upon visible-light irradiation. The schematic illustration of the dye degradation process was shown in Fig. 9.14. The radical oxygen ion and holes were very active during the RhB and CBOC-5 showed excellent stability after five times repeated use.

Tavker et al. (2020) fabricated cellulose/$BiVO_4$ nanocomposite by varying the amount of extracted cellulose and checked for the degradation of Methyl Orange dye under visible light (Tungsten 200 W) give 87% photodegradation efficiency.

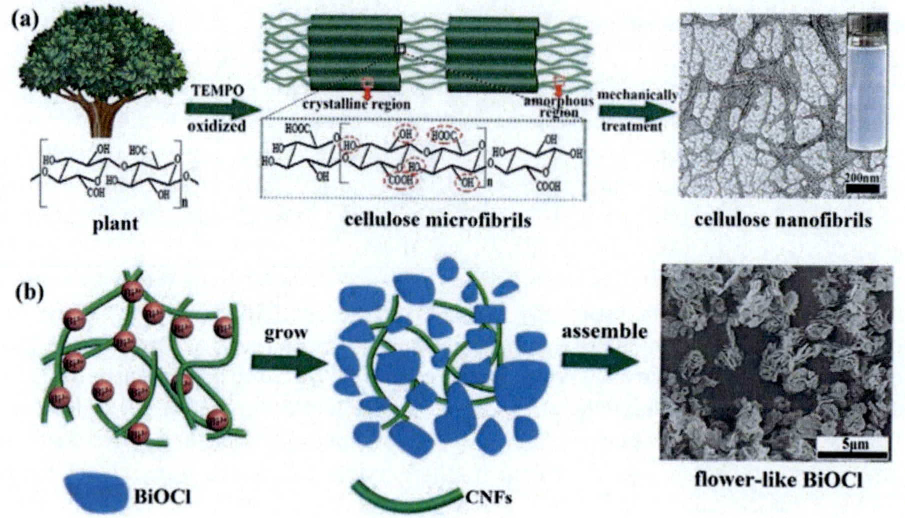

Fig. 9.13 Schematic illustration of the preparation of CNFs (**a**) and flower-like BOC (**b**) (Tian et al. 2019). *Source* Tian et al. (2019) Reproduced with permission from Copyright © 2019 Elsevier Ltd

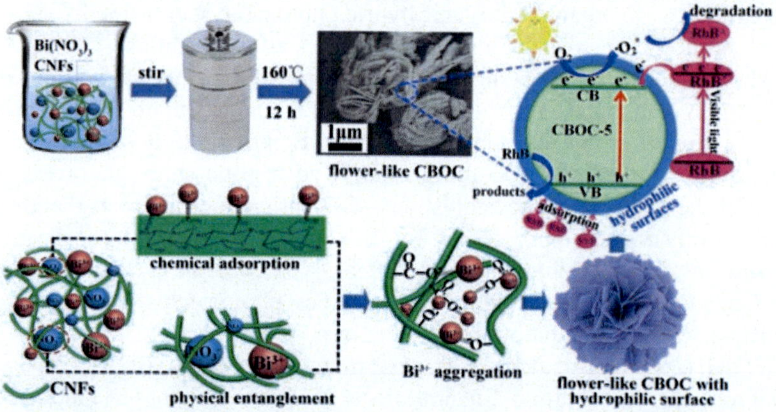

Fig. 9.14 Schematic illustration for the photocatalytic degradation of Rhodamine B (Tian et al. 2019). *Source* Tian et al. (2019) Reproduced with permission from Copyright © 2019 Elsevier Ltd

Even after five repetitive runs, it shows a photodegrading efficiency of 68%. The photocatalytic results accentuated that cellulose employed as a scaffold and thereby developed an interface with $BiVO_4$ that will reduce the recombination time of the electron–hole, which will enhance efficiency in photodegradation of the catalyst.

9.5 Future Perspective

This paper critically investigated the various types of metallic compound-imprinted chitosan and cellulose composites and their photocatalytic efficiency in degrading different dye molecules. The exceptional physical and chemical applicability of these biopolymers encouraged the researchers to develop advanced water purification technologies. It has also been recognized that the imprinting of various semiconductors like TiO_2, ZnO, CdS, BiOX, AgI, g-C_3N_4, etc., with chitosan, cellulose, and synthetic polymers composites will enhance the photocatalytic ability of the composite due to the increased specific surface area and the improved pore structures. Moreover, apparent hydrophilicity, optical transparency, adsorption, mechanical properties, and thermal stability also were attained in most cases. These studies will provide thoughtful insight into the basics of semiconductor/biopolymer composites photocatalyst, which still demands deep study on improving the stability of the photocatalysts.

The semiconductor-blended biopolymers were utilized for the photodegradation of a variety of organic pollutants due to their synergetic effect since the doping improves the adsorption capacity of the biopolymer and at the same time suppresses the recombination of photogenerated electrons and holes. It is possible to enhance these two effects by further imprinting these biopolymers with other suitable diverse non-metallic nanostructure materials and also by modifying the catalytic surface.

Some of the semiconductors used here are toxic, In this regard; their use is restricted only in the mineralization of the noxious pollutants, and their disposal causes a further hazard. Therefore, it has been recommended that future researches in this area should focus on developing sustainable materials and promoting the development of robust, smart, and cost-effective composites for possible application in medicinal field and commercialization.

There are some ambiguous reports against the photostability of the biopolymer composites and needs thorough examination toward the photostability of the biopolymer composites. Besides, the large scale use of these biopolymers composites with high photostability, high photocatalytic efficiency, and cost-effective production, is most wanted and demanding as well. Hence, this matter would seek the attention of the researchers to ensure its future commercialization.

9.6 Conclusion

This paper decisively overviewed the present scenario of biopolymer-based semiconductors with substantial prospective to be utilized as a potential photocatalyst in the field of the ecological purification process, particularly for water detoxification. The disadvantage of bare metallic, non-metallic composites and nanoparticles including their low absorption capacity of the photon in the visible region, small quantum yield, speedy recombination of electron and hole, poor stability, etc., were surpassed by imprinting the metallic semiconductors with the biopolymeric materials. Diverse approaches have been invented to enhance the photocatalytic ability of the metals such as coupling, doping, surface sensitization, etc., with the biopolymer materials, which helped synergistically to improve the electrical conductivity, photostability, and photocatalytic activity. The literature discussed in this study suggested that biopolymers like chitosan, cellulose, and their composites, containing binary, ternary, and complex metal oxides, sulfides, halides, etc., exhibited enhanced photodegradation of organic dyes under visible-light irradiations. Thus, along with other applications, chitosan and cellulose, macro and nanocomposites come into view as promising candidates for water detoxification and purification. This paper can be an input to understand the basics of photocatalytic and detoxification properties of biopolymer incorporated metal oxides, sulfides, halides, etc., for water treatment. However, considerable progress in the field of photocatalytic disinfection by metal biopolymer composites has not been achieved yet at large scale for environmental application. Thus, extensive possibilities exist in the area of research to expand the perspective of biopolymer composite as photocatalytic material.

References

Aboamera NM, Mohamed A, Salama A, Khattab A (2019) Characterization and mechanical properties of electrospun cellulose acetate/graphene oxide composite nanofibers. Mech Adv Mater Struct 26(9):765–769. https://doi.org/10.1080/15376494.2017.1410914

Alle M, Lee SH, Kim JC (2020) Ultrafast synthesis of gold nanoparticles on cellulose nanocrystals via microwave irradiation and their dyes-degradation catalytic activity. J Mater Sci Technol 41:168–177. https://doi.org/10.1016/j.jmst.2019.11.003

Alzahrani E (2018) Chitosan membrane embedded with ZnO/CuO nanocomposites for the photodegradation of fast green dye under artificial and solar irradiation. Anal Chem Insights 13:1–13. https://doi.org/10.1177/1177390118763361

An C, Ming X, Wang J, Wang S (2012) Construction of magnetic visible-light-driven plasmonic Fe_3O_4@SiO_2@AgCl: Ag nanophotocatalyst. J Mater Chem 22(11):5171–5176. https://doi.org/10.1039/c2jm16622d

Ashori A, Cordeiro N, Faria M, Hamzeh Y (2013) Effect of chitosan and cationic starch on the surface chemistry properties of bagasse paper. Int J Biol Macromol 58:343–348. https://doi.org/10.1016/j.ijbiomac.2013.04.056

Azetsu A, Koga H, Isogai A, Kitaoka T (2011) Synthesis and catalytic features of hybrid metal nanoparticles supported on cellulose nanofibers. Catalysts 1(1):83–96. https://doi.org/10.3390/catal1010083

Aziz A, Ali N, Khan A, Bilal M, Malik S, Ali N, Khan H (2020) Chitosan-zinc sulfide nanoparticles, characterization and their photocatalytic degradation efficiency for azo dyes. Int J Biol Macromol 153:502–512. https://doi.org/10.1016/j.ijbiomac.2020.02.310

Bahal M, Kaur N, Sharotri N Sud D (2019) Investigations on amphoteric chitosan/TiO_2 bionanocomposites for application in visible light induced photocatalytic degradation. Adv Polym Technol 2019:1–9. https://doi.org/10.1155/2019/2345631

Barry TI, Stone FS (1960) The reactions of oxygen at dark and irradiated zinc oxide surfaces. In: Proceedings of the Royal Society of London. Series A. Mat Phys Sci 255(1280):124–144. https://doi.org/10.1098/rspa.1960.0058

Bian Z, Zhu J, Wang S, Cao Y, Qian X, Li H (2008) Self-assembly of active Bi_2O_3/TiO_2 visible photocatalyst with ordered mesoporous structure and highly crystallized anatase. J Phys Chem C 112(16):6258–6262. https://doi.org/10.1021/jp800324t

Bickley RI, Stone FS (1973) Photoadsorption and photocatalysis at rutile surfaces: I. Photoadsorption of oxygen. J Catal 31(3):389–397. https://doi.org/10.1016/0021-9517(73)90310-2

Cao C, Xiao L, Liu L, Zhu H, Chen C, Gao L (2013) Visible-light photocatalytic decolorization of reactive brilliant red X-3B on Cu_2O/crosslinked-chitosan nanocomposites prepared via one step process. Appl Surf Sci 271:105–112. https://doi.org/10.1016/j.apsusc.2013.01.135

Chang CC, Chang CH (2010) Preparation and characterization of polyurethane-gold nanocomposites prepared using encapsulated gold nanoparticles. Polym Int 59(7):910–916. https://doi.org/10.1002/pi.2805

Chen C, Ma W, Zhao J (2010) Semiconductor-mediated photodegradation of pollutants under visible-light irradiation. Chem Soc Rev 39(11):4206–4219. https://doi.org/10.1039/b921692h

Chen F, Huang H, Guo L, Zhang Y, Ma T (2019) The role of polarization in photocatalysis. Angew Chem 131(30):10164–10176. https://doi.org/10.1002/ange.201901361

Dette C, Pérez-Osorio MA, Kley CS, Punke P, Patrick CE, Jacobson P, Giustino F, Jung SJ, Kern K (2014) TiO_2 anatase with a bandgap in the visible region. Nano Lett 14(11):6533–6538. https://doi.org/10.1021/nl503131s

Dhanya A, Aparna K (2016) Synthesis and evaluation of TiO_2/chitosan based hydrogel for the adsorptional photocatalytic degradation of Azo and Anthraquinone dye under UV light irradiation. Proc Technol 24:611–618. https://doi.org/10.1016/j.protcy.2016.05.141

Fan J, Yu D, Wang W, Liu B (2019) The self-assembly and formation mechanism of regenerated cellulose films for photocatalytic degradation of C.I. Reactive Blue 19. Cellulose 26(6): 3955–3972. https://doi.org/10.1007/s10570-019-02350-y

Farzana MH, Meenakshi S (2015) Exploitation of zinc oxide impregnated chitosan beads for the photocatalytic decolorization of an azo dye. Int J Biol Macromol 72:900–910. https://doi.org/10.1016/j.ijbiomac.2014.09.038

Fujishima A, Rao TN, Tryk DA (2000) Titanium dioxide photocatalysis. J Photochem Photobiol C 1:1–21. https://doi.org/10.1016/S1389-5567(00)00002-2

Gao ST, Liu WH, Shang NZ, Feng C, Wu QH, Wang Z, Wang C (2014) Integration of a plasmonic semiconductor with a metal-organic framework: a case of Ag/AgCl@ZIF-8 with enhanced visible light photocatalytic activity. RSC Adv 4(106):61736–61742. https://doi.org/10.1039/c4ra11364k

Gopiraman M, Saravanamoorthy S, Baskar R, Ilangovan A, Ill-Min C (2019) Green synthesis of Ag@Au bimetallic regenerated cellulose nanofibers for catalytic applications. New J Chem 43(43):17090–17103. https://doi.org/10.1039/c9nj04428k

Gupta VK, Saleh TA (2013) Sorption of pollutants by porous carbon, carbon nanotubes and fullerene—An overview. Environ Sci Pollut Res 20(5):2828–2843. https://doi.org/10.1007/s11356-013-1524-1

Han F, Kambala VSR, Srinivasan M, Rajarathnam D, Naidu R (2009) Tailored titanium dioxide photocatalysts for the degradation of organic dyes in wastewater treatment: a review. Appl Catal A 359(1–2):25–40. https://doi.org/10.1016/j.apcata.2009.02.043

He L, Tong Z, Wang Z, Chen M, Huang N, Zhang W (2018) Effects of calcination temperature and heating rate on the photocatalytic properties of ZnO prepared by pyrolysis. J Colloid Interface Sci 509:448–456. https://doi.org/10.1016/j.jcis.2017.09.021

Huang R, Liu Q, Zhang L, Yang B (2015) Utilization of cross-linked chitosan/bentonite composite in the removal of methyl orange from aqueous solution. Water Sci Technol 71(2):174–182. https://doi.org/10.2166/wst.2014.478

Jayalakshmi A, Kim IC, Kwon YN (2015) Cellulose acetate graft-(glycidylmethacrylate-g-PEG) for modification of AMC ultrafiltration membranes to mitigate organic fouling. RSC Adv 5(60):48290–48300. https://doi.org/10.1039/c5ra03499j

Jeevanantham V, Hemalatha KV, Satheeskumar S (2018) Photodegradation activity of pure, PVP capped and chitosan capped ZnO nanoparticles against Azo red dye under UV irradiation. J Ovonic Res 14(4):269–275

Jiang Y, Lawan I, Zhou W, Zhang M, Fernando GF, Wang L, Yuan Z (2020) Synthesis, properties and photocatalytic activity of a semiconductor/cellulose composite for dye degradation-a review. Cellulose 27(2):595–609. https://doi.org/10.1007/s10570-019-02851-w

Khan A, Shah SJ, Mehmood K, Awais NA, Khan H (2019) Synthesis of potent chitosan beads a suitable alternative for textile dye reduction in sunlight. J Mater Sci Mater Electron 30(1): 406–414. https://doi.org/10.1007/s10854-018-0305-5

Kim K, Ingole PG, Yun S, Choi W, Kim J, Lee H (2015) Water vapor removal using CA/PEG blending materials coated hollow fiber membrane. J Chem Technol Biotechnol 90(6):1117–1123. https://doi.org/10.1002/jctb.4421

Kudo A, Omori K, Kato H (1999) A novel aqueous process for preparation of crystal form-controlled and highly crystalline $BiVO_4$ powder from layered vanadates at room temperature and its photocatalytic and photophysical properties. J Am Chem Soc 121(49):11459–11467. https://doi.org/10.1021/ja992541y

Kumar DP, Hong S, Reddy DA, Kim TK (2016) Noble metal-free ultrathin MoS_2 nanosheet-decorated CdS nanorods as an efficient photocatalyst for spectacular hydrogen evolution under solar light irradiation. J Mater Chem A 4(47):18551–18558. https://doi.org/10.1039/c6ta08628d

Kuo WS, Ho PH (2006) Solar photocatalytic decolorization of dyes in solution with TiO_2 film. Dyes Pigm 71(3):212–217. https://doi.org/10.1016/j.dyepig.2005.07.003

Li FB, Li XZ, Hou MF (2004) Photocatalytic degradation of 2-mercaptobenzothiazole in aqueous La^{3+}-TiO_2 suspension for odor control. Appl Catal B 48(3):185–194. https://doi.org/10.1016/j.apcatb.2003.10.003

Li G, Nandgaonkar AG, Wang Q, Zhang J, Krause WE, Wei Q, Lucia LA (2017b) Laccase-immobilized bacterial cellulose/TiO_2 functionalized composite membranes: evaluation for photo- and bio-catalytic dye degradation. J Membr Sci 525:89–98. https://doi.org/10.1016/j.memsci.2016.10.033

Li S, Hao X, Dai X, Tao T (2018) Rapid photocatalytic degradation of pollutant from water under UV and sunlight via cellulose nanofiber aerogel wrapped by TiO_2. J Nanomater 2018:1-12. https://doi.org/10.1155/2018/8752015

Li W, Zou Y, Geng X, Xiao F, An G, Wang D (2017a) Constructing highly catalytic oxidation over BiOBr-based hierarchical microspheres: importance of redox potential of doped cations. Molecular Catal 438:19–29. https://doi.org/10.1016/j.mcat.2017.05.017

Liang H, Hu X (2016) A quick review of the applications of nano crystalline cellulose in wastewater treatment. J Bioresources Bioproducts 2016(4): 199–204. http://dx.doi.org/10.21967/jbb.v1i4.65

Liang S, Xiao K, Mo Y, Huang X (2012) A novel ZnO nanoparticle blended polyvinylidene fluoride membrane for anti-irreversible fouling. J Membr Sci 394–395:184–192. https://doi.org/10.1016/j.memsci.2011.12.040

Lightcap IV, Kosel TH, Kamat PV (2010) Anchoring semiconductor and metal nanoparticles on a two-dimensional catalyst mat. storing and shuttling electrons with reduced graphene oxide. Nano Letters 10(2): 577–583. https://doi.org/10.1021/nl9035109

Lim WY, Wu H, Lim YF, Ho GW (2018) Facilitating the charge transfer of $ZnMoS_4$/CuS p-n heterojunctions through ZnO intercalation for efficient photocatalytic hydrogen generation. J Mater Chem A 6(24):11416–11423. https://doi.org/10.1039/c8ta02763c

Linda T, Muthupoongodi S, Shajan XS, Balakumar S (2016a) Fabrication and characterization of chitosan templated CdO/NiO nano composite for dye degradation. Optik 127(20):8287–8293. https://doi.org/10.1016/j.ijleo.2016.06.025

Linda T, Muthupoongodi S, Sahaya Shajan X, Balakumar S (2016b) Photocatalytic degradation of congo red and crystal violet dyes on cellulose/PVC/ZnO composites under UV light irradiation. Mater Today Proc 3(6):2035–2041. https://doi.org/10.1016/j.matpr.2016.04.106

Linsebigler AL, Lu G, Yates JT (1995) Photocatalysis on TiO_2 surfaces: principles, mechanisms, and selected results. Chem Rev 95(3):735–758. https://doi.org/10.1021/cr00035a013

Ma X, Zhang B, Cong Q, He X, Gao M, Li G (2016) Organic/inorganic nanocomposites of ZnO/CuO/chitosan with improved properties. Mater Chem Phys 178:88–97. https://doi.org/10.1016/j.matchemphys.2016.04.074

Madhusudhan A, Reddy GB, Krishana IM (2019) Green synthesis of gold nanoparticles by using natural gums. Nanomaterials Plant Potential 111–134.https://doi.org/10.1007/978-3-030-05569-1_4

Mao J, Peng T, Zhang X, Li K, Ye L, Zan L (2013) Effect of graphitic carbon nitride microstructures on the activity and selectivity of photocatalytic CO2 reduction under visible light. Catal Sci Technol 3(5):1253–1260. https://doi.org/10.1039/c3cy20822b

Mateo D, Esteve-Adell I, Albero J, Primo A, García H (2017) Oriented 2.0.0 Cu_2O nanoplatelets supported on few-layers graphene as efficient visible light photocatalyst for overall water splitting. Appl Catal B 201:582–590. https://doi.org/10.1016/j.apcatb.2016.08.033

Meenakshi S, Farzana M (2013) Synergistic effect of Chitosan and Titanium dioxide on the removal of toxic dyes by photodegradation technique. Industrial Eng Chem Res 53: 55–63. http://pubs.acs.org/doi/abs/10.1021/ie402347g

Melinte V, Stroea L, Chibac-Scutaru AL (2019) Polymer nanocomposites for photocatalytic applications. Catalysts 9(12):986. https://doi.org/10.3390/catal9120986

Mittal A, Mittal J, Malviya A, Gupta VK (2010) Removal and recovery of Chrysoidine Y from aqueous solutions by waste materials. J Colloid Interface Sci 344(2):497–507. https://doi.org/10.1016/j.jcis.2010.01.007

Mohamed MA, Salleh WNW, Jaafar J, Ismail AF, Abd Mutalib M, Jamil SM (2015) Incorporation of N-doped TiO_2 nanorods in regenerated cellulose thin films fabricated from recycled newspaper as a green portable photocatalyst. Carbohyd Polym 133:429–437. https://doi.org/10.1016/j.carbpol.2015.07.057

Nasr C, Vinodgopal K, Fisher L, Hotchandani S, Chattopadhyay AK, Kamat PV (1996) Environmental photochemistry on semiconductor surfaces. Visible light induced degradation of a textile diazo dye, naphthol blue black, on TiO_2 nanoparticles. J Phys Chem 100(20): 8436–8442. https://doi.org/10.1021/jp953556v

Natarajan S, Bajaj HC, Tayade RJ (2018) Recent advances based on the synergetic effect of adsorption for removal of dyes from waste water using photocatalytic process. J Environ Sci (china) 65:201–222. https://doi.org/10.1016/j.jes.2017.03.011

Nešić J, Manojlović DD, Anđelković I, Dojčinović BP, Vulić PJ, Krstić J, Roglić GM (2013) Preparation, characterization and photocatalytic activity of lanthanum and vanadium co-doped mesoporous TiO_2 for azo-dye degradation. J Mol Catal A Chem 378:67–75. https://doi.org/10.1016/j.molcata.2013.05.018

Nithya A, Jothivenkatachalam K, Prabhu S, Jeganathan K (2014) Chitosan based nanocomposite materials as photocatalyst—a review. Mater Sci Forum 781:79–94. https://doi.org/10.4028/www.scientific.net/MSF.781.79

Peng R, Zhang H, Gui L, Wu Z, Yu P, Luo Y (2019) Facile Synthesis of MnO_2 @Cellulose composite film. Environ Eng Sci 36(5):583–588. https://doi.org/10.1089/ees.2018.0442

Plantard G, Goetz V, Sacco D (2011) TiO_2-coated foams as a medium for solar catalysis. Mater Res Bull 46(2):231–234. https://doi.org/10.1016/j.materresbull.2010.11.011

Prashantha Kumar TKM, Ashok Kumar SK (2019) Visible-light-induced degradation of rhodamine B by nanosized Ag_2S-ZnS loaded on cellulose. Photochem Photobiol Sci 18(1):148–154. https://doi.org/10.1039/c8pp00330k

Rajeswari A, Vismaiya S, Pius A (2017) Preparation, characterization of nano ZnO-blended cellulose acetate-polyurethane membrane for photocatalytic degradation of dyes from water. Chem Eng J 313:928–937. https://doi.org/10.1016/j.cej.2016.10.124

Reddy N, Yang Y (2014) Introduction to Natural Cellulose Fibers from Renewable Resources. Innovative Biofibers from Renewable Resources 3–4. https://doi.org/10.1007/978-3-662-45136-6_1

Saeed RMY, Bano Z, Sun J, Wang F, Ullah N, Wang Q (2019) CuS-functionalized cellulose based aerogel as biocatalyst for removal of organic dye. J Appl Polym Sci 136(15):1–10. https://doi.org/10.1002/app.47404

Saleh TA, Gupta VK (2011) Functionalization of tungsten oxide into MWCNT and its application for sunlight-induced degradation of rhodamine B. J Colloid Interface Sci 362(2):337–344. https://doi.org/10.1016/j.jcis.2011.06.081

Saravanan R, Gupta VK, Narayanan V, Stephen A (2014) Visible light degradation of textile effluent using novel catalyst $ZnO/\gamma\text{-}Mn_2O_3$. J Taiwan Inst Chem Eng 45(4):1910–1917. https://doi.org/10.1016/j.jtice.2013.12.021

Saravanan R, Karthikeyan S, Gupta VK, Sekaran G, Narayanan V, Stephen A (2013) Enhanced photocatalytic activity of ZnO/CuO nanocomposite for the degradation of textile dye on visible light illumination. Mater Sci Eng C 33(1):91–98. https://doi.org/10.1016/j.msec.2012.08.011

Selloni A (2008) Anatase shows its reactive side. Nat Mater 7(8):613–615. https://doi.org/10.1038/nmat2241

Senthil Kumar P, Selvakumar M, Babu SG, Jaganathan SK, Karuthapandian S, Chattopadhyay S (2015) Novel CuO/chitosan nanocomposite thin film: facile hand-picking recoverable, efficient and reusable heterogeneous photocatalyst. RSC Adv 5(71):57493–57501. https://doi.org/10.1039/c5ra08783j

Sheng J, Tong S, He Z, Yang R (2017) Recent developments of cellulose materials for lithium-ion battery separators. Cellulose 24(10):4103–4122. https://doi.org/10.1007/s10570-017-1421-8

Singh P, Shandilya P, Raizada P, Sudhaik A, Rahmani-Sani A, Hosseini-Bandegharaei A (2020) Review on various strategies for enhancing photocatalytic activity of graphene based nanocomposites for water purification. Arab J Chem 13(1):3498–3520. https://doi.org/10.1016/j.arabjc.2018.12.001

Sirajudheen P, Karthikeyan P, Basheer MC, Meenakshi S (2020) Adsorptive removal of anionic azo dyes from effluent water using Zr (IV) encapsulated carboxymethyl cellulose-montmorillonite composite. Environ Chem Ecotoxicol 2:73–82. https://doi.org/10.1016/j.enceco.2020.04.002

Sirajudheen P, Meenakshi S (2019a) Facile synthesis of chitosan-La^{3+}-graphite composite and its influence in photocatalytic degradation of methylene blue. Int J Biol Macromol 133:253–261. https://doi.org/10.1016/j.ijbiomac.2019.04.073

Sirajudheen P, Meenakshi S (2019b) Lanthanum (III) incorporated chitosan-montmorillonite composite as flexible material for adsorptive removal of azo dyes from water. Mater Today Proceedings 27(1):318-326. https://doi.org/10.1016/j.matpr.2019.11.040

Sirajudheen P, Meenakshi S (2020) Encapsulation of Zn–Fe layered double hydroxide on activated carbon and its litheness in tuning anionic and rhoda dyes through adsorption mechanism. Asia-Pacific J Chem Eng 15(5):1–13. https://doi.org/10.1002/apj.2479

Sirajudheen P, Sanoop KB, Rashid M (2016) Visible light induced $ZnTiO_3$ photocatalyst synthesized by co-precipitation process. In: Recent advances in chemical engineering, pp 227–234. https://doi.org/10.1007/978-981-10-1633-2

Sivakumar A, Murugesan B, Loganathan A, Sivakumar P (2014) A review on decolourisation of dyes by photodegradation using various bismuth catalysts. J Taiwan Inst Chem Eng 45(5):2300–2306. https://doi.org/10.1016/j.jtice.2014.07.003

Sohrabnezhad S, Zanjanchi MA, Razavi M (2014) Plasmon-assisted degradation of methylene blue with Ag/AgCl/montmorillonite nanocomposite under visible light. Spectrochimica Acta Part A Mol Biomolecular Spectroscopy 130:129–135. https://doi.org/10.1016/j.saa.2014.02.188

Taghizadeh MT, Siyahi V, Ashassi-Sorkhabi H, Zarrini G (2020) ZnO, AgCl and AgCl/ZnO nanocomposites incorporated chitosan in the form of hydrogel beads for photocatalytic degradation of MB, *E. coli* and *S. aureus*. Int J Biol Macromol 147:1018–1028. https://doi.org/10.1016/j.ijbiomac.2019.10.070

Tang J, Durrant JR, Klug DR (2008) Mechanism of photocatalytic water splitting in TiO_2. Reaction of water with photoholes, importance of charge carrier dynamics, and evidence for four-hole chemistry. J Am Chem Soc 130(42): 13885–13891. https://doi.org/10.1021/ja8034637

Tavker N, Gaur UK, Sharma M (2019) Highly active agro-waste-extracted cellulose-supported $CuInS_2$ nanocomposite for visible-light-induced photocatalysis. ACS Omega 4(7):11777–11784. https://doi.org/10.1021/acsomega.9b01054

Tavker N, Gaur U, Sharma M (2020) Cellulose supported bismuth vanadate nanocomposite for effective removal of organic pollutant. J Environ Chem Eng 8(4):104027. https://doi.org/10.1016/j.jece.2020.104027

Tavker N, Sharma M (2018) Enhanced photocatalytic activity of nanocellulose supported zinc oxide composite for RhB dye as well as ciprofloxacin drug under sunlight/visible light. AIP Conf Proc 1961:030013-8. https://doi.org/10.1063/1.5035215

Tian C, Zhang Q, Wu A, Jiang M, Jiang B, Fu H (2012) Cost-effective large-scale synthesis of ZnO photocatalyst with excellent performance for dye photodegradation. Chem Commun 48(23):2858–2860. https://doi.org/10.1039/c2cc16434e

Tian C, Luo S, She J, Qing Y, Yan N, Wu Y, Liu Z (2019). Cellulose nanofibrils enable flower-like BiOCl for high-performance photocatalysis under visible-light irradiation. Appl Surface Sci 464:606–615. https://doi.org/10.1016/j.apsusc.2018.09.126

Tu K, Wang Q, Lu A, Zhang L (2014) Portable Visible-Light Photocatalysts Constructed from Cu_2O Nanoparticles and Graphene Oxide in Cellulose Matrix. J Phys Chem C 118(13):7202–7210. https://doi.org/10.1021/jp412802h

Viswanathan B (2017) Photocatalytic degradation of dyes: an overview. Current Catal 7(2):99–121. https://doi.org/10.2174/2211544707666171219161846

Wan Ngah WS, Teong LC, Hanafiah MAKM (2011) Adsorption of dyes and heavy metal ions by chitosan composites: a review. Carbohyd Polym 83(4):1446–1456. https://doi.org/10.1016/j.carbpol.2010.11.004

Wang CY, Zhang YJ, Wang WK, Pei DN, Huang GX, Chen JJ, Zhang X, Yu HQ (2018a) Enhanced photocatalytic degradation of bisphenol A by Co-doped BiOCl nanosheets under visible light irradiation. Appl Catal B Environmental 221(July 2017):320–328. https://doi.org/10.1016/j.apcatb.2017.09.036

Wang H, Wu Y, Wu P, Chen S, Guo X, Meng G, Peng B, Wu J, Liu Z (2017a) Environmentally benign chitosan as reductant and supporter for synthesis of Ag/AgCl/chitosan composites by one-step and their photocatalytic degradation performance under visible-light irradiation. Front Mater Sci 11(2):130–138. https://doi.org/10.1007/s11706-017-0383-y

Wang H, Zhang W, Li X, Li J, Cen W, Li Q, Dong F (2018b) Highly enhanced visible light photocatalysis and in situ FT-IR studies on Bi metal@defective BiOCl hierarchical microspheres. Appl Catal B 225:218–227. https://doi.org/10.1016/j.apcatb.2017.11.079

Wang Q, Cai J, Zhang L (2014) In situ synthesis of Ag_3PO_4/cellulose nanocomposites with photocatalytic activities under sunlight. Cellulose 21(5):3371–3382. https://doi.org/10.1007/s10570-014-0340-1

Wang S, Li F, Dai X, Wang C, Lv X, Waterhouse GIN, Fan H, Ai S (2020) Highly flexible and stable carbon nitride/cellulose acetate porous films with enhanced photocatalytic activity for contaminants removal from wastewater. J Hazardous Mater 384:121417. https://doi.org/10.1016/j.jhazmat.2019.121417

Wang X, Maeda K, Thomas A, Takanabe K, Xin G, Carlsson JM, Domen K, Antonietti M (2009) A metal-free polymeric photocatalyst for hydrogen production from water under visible light. Nat Mater 8(1):76–80. https://doi.org/10.1038/nmat2317

Wang Z, Wang J, Li L, Zheng J, Jia S, Chen J, Liu B, Zhu Z (2017b) Fabricating efficient CdSe-CdS photocatalyst systems by spatially resetting water splitting sites. J Mater Chem A 5(38):20131–20135. https://doi.org/10.1039/c7ta06085h

Wittmar A, Thierfeld H, Köcher S, Ulbricht M (2015) Routes towards catalytically active TiO_2 doped porous cellulose. RSC Adv 5(45):35866–35873. https://doi.org/10.1039/c5ra03707g

Wu P, Peng H, Wu Y, Li L, Hao X, Peng B, Meng G, Wu J, Liu Z (2020) A green strategy to synthesize Ag/Ag_3PO_4/Chitosan composite photocatalyst and their photocatalytic degradation performance under visible light irradiation J Electronic Sci Technol 18(2):100019. https://doi.org/10.1016/j.jnlest.2020.100019

Wu T, Li X, Zhang D, Dong F, Chen S (2016) Efficient visible light photocatalytic oxidation of NO with hierarchical nanostructured 3D flower-like BiOClxBr 1–x solid solutions. J Alloy Compd 671:318–327. https://doi.org/10.1016/j.jallcom.2016.01.267

Xia SJ, Liu FX, Ni ZM, Xue JL, Qian PP (2013) Layered double hydroxides as efficient photocatalysts for visible-light degradation of Rhodamine B. J Colloid Interface Sci 405(3):195–200. https://doi.org/10.1016/j.jcis.2013.05.064

Xie J, Wang H, Duan M, Zhang L (2011) Synthesis and photocatalysis properties of ZnO structures with different morphologies via hydrothermal method. Appl Surf Sci 257(15):6358–6363. https://doi.org/10.1016/j.apsusc.2011.01.105

Xu J, Zhang L, Shi R, Zhu Y (2013) Chemical exfoliation of graphitic carbon nitride for efficient heterogeneous photocatalysis. J Mater Chem A 1(46):14766–14772. https://doi.org/10.1039/c3ta13188b

Xu T, Zhang L, Cheng H, Zhu Y (2011) Significantly enhanced photocatalytic performance of ZnO via graphene hybridization and the mechanism study. Appl Catal B 101(3–4):382–387. https://doi.org/10.1016/j.apcatb.2010.10.007

Yan SC, Li ZS, Zou ZG (2010) Photodegradation of rhodamine B and methyl orange over boron-doped g-C_3N_4 under visible light irradiation. Langmuir 26(6):3894–3901. https://doi.org/10.1021/la904023j

Yue M, Wang R, Ma B, Cong R, Gao W, Yang T (2016) Superior performance of $CuInS_2$ for photocatalytic water treatment: full conversion of highly stable nitrate ions into harmless N_2 under visible light. Catal Sci Technol 6(23):8300–8308. https://doi.org/10.1039/c6cy01858k

Zayed MF, Eisa WF, Hosam AEHM, Abou Zeid AM (2020) Spectroscopic investigation of chitosan-supported Cu_2O/CuO nanocomposite; a separable catalyst for water-pollutants degradation. J Alloy Compd 835:155306. https://doi.org/10.1016/j.jallcom.2020.155306

Zhang H, Xu P, Du G, Chen Z, Oh K, Pan D, Jiao Z (2011a) A facile one-step synthesis of TiO_2/graphene composites for photodegradation of methyl orange. Nano Res 4(3):274–283. https://doi.org/10.1007/s12274-010-0079-4

Zhang H, Luo X, Tang H, Zheng M, Huang F (2017a) A novel candidate for wound dressing: transparent porous maghemite/cellulose nanocomposite membranes with controlled release of doxorubicin from a simple approach. Mater Sci Eng C 79:84–92. https://doi.org/10.1016/j.msec.2017.05.019

Zhang P, Qiu Y, Yang S, Jiao Y, Ji C, Li Y, Chen B, Fan H (2017b) Oxygen-deficient bismuth oxychloride nanosheets: superior photocatalytic performance. Mater Res Bull 96:478–484. https://doi.org/10.1016/j.materresbull.2017.03.071

Zhang X, Shu Y, Su S, Zhu J (2018) One-step coagulation to construct durable anti-fouling and antibacterial cellulose film exploiting Ag@AgCl nanoparticle- triggered photo-catalytic degradation. Carbohydrate Polym 181: 499–505. https://doi.org/10.1016/j.carbpol.2017.10.041

Zhang X, Chen W, Lin Z, Yao J, Tan S (2011b) Preparation and photocatalysis properties of bacterial cellulose/TiO_2 composite membrane doped with rare earth elements. Synth React Inorg Met-Org Nano-Met Chem 41(8):997–1004. https://doi.org/10.1080/15533174.2011.591334

Zhao C, Yan Q, Wang S, Dong P, Zhang L (2018) Regenerable g-C_3N_4-chitosan beads with enhanced photocatalytic activity and stability. RSC Adv 8(48):27516–27524. https://doi.org/10.1039/c8ra04293d

Zhao H, Chen S, Quan X, Yu H, Zhao H (2016) Integration of microfiltration and visible-light-driven photocatalysis on g-C_3N_4 nanosheet/reduced graphene oxide membrane for enhanced water treatment. Appl Catal B 194:134–140. https://doi.org/10.1016/j.apcatb.2016.04.042

Zhao SW, Zheng M, Zou XH, Guo Y, Pan QJ (2017) Self-Assembly of hierarchically structured Cellulose@ZnO composite in solid-liquid homogeneous phase: synthesis, DFT calculations, and enhanced antibacterial activities. ACS Sustain Chem Eng 5(8):6585–6596. https://doi.org/10.1021/acssuschemeng.7b00842

Zhong Q, Tian J, Liu T, Guo Z, Ding S, Li H (2018) Preparation and antibacterial properties of carboxymethyl chitosan/ZnO nanocomposite microspheres with enhanced biocompatibility. Mater Lett 212:58–61. https://doi.org/10.1016/j.matlet.2017.10.062

Zhong R, Zhong Q, Huo M, Yang B, Li H (2020) Preparation of biocompatible nano-ZnO/chitosan microspheres with multi-functions of antibacterial, UV-shielding and dye photodegradation. Int J Biol Macromol 146:939–945. https://doi.org/10.1016/j.ijbiomac.2019.09.217

Zhou M, Chen J, Hou C, Liu Y, Xu S, Yao C, Li Z (2019) Organic-free synthesis of porous CdS sheets with controlled windows size on bacterial cellulose for photocatalytic degradation and H_2 production. Appl Surface Sci 470:908–916. https://doi.org/10.1016/j.apsusc.2018.11.207

Zhou W, Sun S, Jiang Y, Zhang M, Lawan I, Fernando GF, Wang L, Yuan Z (2019) Template in situ synthesis of flower-like BiOBr/microcrystalline cellulose composites with highly visible-light photocatalytic activity. Cellulose 26(18):9529–9541. https://doi.org/10.1007/s10570-019-02722-4

Zhou Z, Peng X, Zhong L, Wu L, Cao X, Sun RC (2016) Electrospun cellulose acetate supported Ag@AgCl composites with facet-dependent photocatalytic properties on degradation of organic dyes under visible-light irradiation. Carbohyd Polym 136:322–328. https://doi.org/10.1016/j.carbpol.2015.09.009

Zhu H, Jiang R, Xiao L, Chang Y, Guan Y, Li X, Zeng G (2009) Photocatalytic decolourization and degradation of Congo Red on innovative crosslinked chitosan/nano-CdS composite catalyst under visible light irradiation. J Hazard Mater 169(1–3):933–940. https://doi.org/10.1016/j.jhazmat.2009.04.037

Zou XH, Zhao SW, Zhang JG, Sun HL, Pan QJ, Guo YR (2019) Preparation of ternary ZnO/Ag/cellulose and its enhanced photocatalytic degradation property on phenol and benzene in VOCs. Open Chem 17(1):779–787. https://doi.org/10.1515/chem-2019-0088

Chapter 10
Metal Organic Frameworks for Removal of Heavy Metal Cations and Emerging Organic Pollutants

Kamlesh Kumar, Simant Kumar Srivastav, and Swatantra P. Singh

Abstract The toxicity of wastewater generated from industrial plants is one of the serious environmental issues. The wastewater often contains toxic compounds such as inorganic pollutants (e.g. oxyanions/cations and heavy metal ions) and organic pollutants (e.g. organic dyes, phenols, biphenyls, pesticides, fertilizers, hydrocarbons, plasticizers, detergents, oils, greases, pharmaceuticals, proteins, carbohydrates etc.) which cause severe environmental and health problems. Many organic pollutants are chemically stable and they are not very prone to biodegradable. Hence, their removal from wastewater only by biological processes is challenging. Many physical, chemical and biological techniques have been developed for wastewater treatment and physical adsorption method has been considered the most effective. Thus materials with adsorption properties have gained wide attention in the scientific community. In recent years, metal–organic frameworks (MOFs) have been employed in technologies to bring an inspiring breakthrough for wastewater treatment. The MOF is class of materials with the exceptionally high surface areas and larger porosity, easier in pore structure designing, and structural modifications. This chapter aims to give insight into the latest developments onto the use of MOFs in the removal of inorganic and emerging organic contaminants present in the wastewater.

Keywords Metal–organic framework · Synthesis · Metal ions · Emerging organic pollutants · Water purification

---

K. Kumar (✉)
Department of Chemistry, Institute of Science, Banaras Hindu University (BHU), Varanasi 221005, India
e-mail: kamlesh.kumar@bhu.ac.in

S. K. Srivastav
Department of Chemistry, Lalit Narayan Mithila University, Darbhanga 846008, Bihar, India

S. P. Singh
Department of Environmental Science and Engineering (ESE), Indian Institute of Technology Bombay, Mumbai 400076, Maharashtra, India

S. P. Singh et al. (eds.), *Nanomaterials and Nanocomposites for Environmental Remediation*, Energy, Environment, and Sustainability,
https://doi.org/10.1007/978-981-16-3256-3_11

10.1 Introduction

Water pollution has become a serious environmental problem in the last few decades due to the tremendous population growth and industrial revolution. Although, industrial growths are necessary for the great benefits of human being, these have also created a deep negative impact on the Earth's ecosystem (Xu 2015). The first consequences of this environmental pollution and misuse of natural resources are mostly observed in the Earth's water ecosystem. In present time, nearly half of the world's population is living in water stressed areas, and it is projected to become worse in the following decades with the increase in global temperatures as a consequence of global warming. Hence, providing clean and freshwater to every person is one of the main goals of national mission.

In this context, developing cost-effective water remediation technologies are currently in high demand. Many methods such as adsorption, oxidation–reduction, membrane filtration and coagulation-flocculation have extensively been used for the removal of toxic materials from contaminated water (Giusti 2009; Gupta et al. 2012). Among all these methods, adsorption has long been used to remove contaminants in the water treatment processes and varieties of materials such as zeolites, bio-adsorbents, polymeric resins, activated carbon materials as well as mesoporous clay materials have been utilized for this purpose (Blanchard et al. 1984; Bailey et al. 1999; Gupta and Saleh 2013; Malik et al. 2016). However, these materials have their own set of limitations. Their synthesis and also modifications of pore size and pore structures are difficult. To discover alternative pollutant adsorbents with precise pore structures and exceptional capacity to adsorb pollutants with high selectivity is always on high demand. This is main parameter for further developments of highly efficient water pollutant adsorbents. In this context, metal–organic frameworks (MOFs) have been considered as one of the most effective porous materials employed in the technologies to purify polluted water (Okoro et al. 2018; Yuan et al. 2018; Mon et al. 2018; Li et al. 2019). Table 10.1 contains list of inorganic and organic pollutants which are removed from wastewater using MOFs materials. MOFs are one of the emerging classes of crystalline porous materials constructed by self-assembly of bi/multi-dentate organic linkers which are held periodically with metal ions or clusters through coordination bonds (Xiang et al. 2010; Yuan et al. 2018; Slater and Cooper 2015). MOFs possess unique properties e.g. extremely high surface area, easy structure modifications, tunable porosity, and controlled functionality. These exceptional physical properties make them excellent platforms for potential applications in diverse fields such as gas storage and separation, heterogeneous catalysis, sensing and photonics, biomedical sciences, etc. The growth of MOF chemistry is a consequence of two noteworthy features of these materials (i) characterization of MOFs by X-ray crystallography which has resulted extensive knowledge about the structural characteristics including pore's size and structures, and (ii) an intriguing host–guest chemistry in different media tuned by a fine control over the size, shape and functionality of MOF channels.

Table 10.1 Inorganic and organic pollutants in wastewater can be removed using MOFs materials (Mon et al. 2018)

Inorganic Pollutants		**Organic Pollutants**	
Metal Cations		**Organic dyes**	
	Lead		Anionic
	Cadmium		Cationic
	Mercury	**Industrial products**	
	Other Transition Metal ions		Aromatic compounds
Oxyanions and cations			Aliphatic compounds
	Arsenic derivatives		Oils
	Chromium	**Agricultural Products**	
	Phosphate		Fertilizers
	Selenium derivatives		Herbicides
	Sulfate, perchlorates and nitrates		Pesticides
Nuclear wastes		**Pharmaceuticals and personal care products**	
	Uranium		Antibiotics
	Pertechnetate/perrhenate/permanganate		Anti-inflammatory
	Cesium(I)		Make-up products
	Thorium(IV)		

In this chapter, the latest developments performed on the use of MOFs as adsorbents will be demonstrated. The chapter will start with a discussion on synthetic and post-synthetic modification methods of MOFs followed by presenting various studies on the removal of heavy metal cations and emerging organic pollutants from contaminated water. In the end, conclusion and opinions on the future research in this field will be addressed.

10.2 Synthesis of Metal–organic Frameworks

The studies have shown that the synthesis of MOFs as well as their functionalization (*i.e.*, post-synthetic modification) are an effective and useful tool for the modification of their structure and also other physio-chemical properties. Therefore, their synthetic processes have been an extremely attractive research topic over the last three decades and different synthetic methods have been developed (Stock and Biswas 2012). Some commonly used synthetic procedures include conventional heating (hydrothermal and solvothermal synthesis), electrochemical synthesis, mechanochemical synthesis, microwave assisted synthesis and sonochemical synthesis. The extensive studies on synthetic protocols for MOFs gave better understanding of their mechanism of formation and also gives idea of factors which affect morphology and properties.

10.2.1 Conventional Methods of MOFs Synthesis

The term conventional synthesis of MOFs is generally applied to the reactions carried out by conventional electric heating without any parallelization of reactions and two different types of reactions such as solvothermal and non-solvothermal are classified in this category. Solvothermal synthesis usually refer to the reactions occurring in closed containers in aqueous or non-aqueous environments under autogenous pressure above the boiling point of the solvent whereas, non-solvothermal reactions take place below or at the boiling point of the solvent under ambient pressure. The solvothermal synthesis of MOFs is most common synthetic procedure and there are many examples of MOFs which are synthesized through this process. Some examples are HKUST-1 (HKUST = Hong Kong University of Science and Technology) (Min Wang et al. 2002), MOF-5 (MOF = metal–organic framework) (Rosi et al. 2003), MOF-74 (Rosi et al. 2005), ZIF-8 (ZIF = zeolitic imidazolate framework) (Park et al. 2006), MIL-53 (MIL = Material of Institute Lavoisier) (Serre et al. 2002), MIL-101 (Férey et al. 2005), UiO-66 (UiO = University of Oslo) (Cavka et al. 2008) and PCN series (PCN = porous coordination network) (Zhao et al. 2008). The synthetic parameters such as metal/ligand ratio, reaction medium, temperature etc. are not only affect the formation of phase-pure materials but they also influence the crystal morphology. Therefore, identifying an optimized synthetic condition for a specific MOF is challenging task. High throughput methods for solvothermal synthesis have also been demonstrated which present a systematic strategy to optimize the synthetic conditions (Stock and Biswas 2012).

10.2.2 Alternative MOFs Synthetic Methods

The input of energy is required for any chemical reaction and therefore, synthesis of MOFs is performed in a suitable solvent at temperatures ranging from room temperature to ~250 °C. The required energy for such type of reaction is supplied by conventional electric heating of reaction vessels in a hot oven. However, an alternatively energy can also be generated by an electric potential, electromagnetic radiation, ultrasound waves or mechanically. The energy source is closely related to the duration, pressure, and energy per molecule and also these parameters have pronounced effects on the product formation as well as crystal morphology. Therefore, material synthesized by adopting alternative routes may have different particle sizes and morphologies, consequently a change in material's properties is observed.

10.2.2.1 Microwave-Assisted Synthesis

Microwave irradiation has been used to provide energy in the synthesis of a variety of MOFs. Microwave-assisted synthesis is based on the interaction of electromagnetic waves with mobile electric charges (e.g. polar solvent molecules or ions in the solution) and therefore, this represents a very energy efficient method of heating. The advantages of this method include fast synthesis, high phase purity of materials, smaller crystal size as well as morphology control. This is due to the direct heating of the solvents and the higher nucleation rate during the reaction.

The first example of microwave-assisted MOF synthesis is MIL-100(Cr). This was synthesized with 44% yield in 4 h at 220 °C. This process is quick and result is comparable to the conventional method which requires 4 days reaction time. Since then, many metal(III) carboxylate-based MOFs (M = Fe, Al, Cr, V, Ce) have been prepared by microwave-assisted synthesis method e.g. IRMOF-1, HKUST-1, Fe-MIL-100, Ni-glutarates, $[Co_4O(BDPB)_3]$ etc.

10.2.2.2 Mechanochemical Synthesis

Mechanical force is applied in the mechanochemical synthesis which induces the breaking and formation of a chemical bond. The mechanochemical synthesis of MOFs is of great economic and environmental interest due to its multiple reasons: solvent free reaction, quantitative yield of product in short reaction time, use of metal oxide in place of metal ions etc. Therefore, this route has become a recent research hotspot as a green synthesis method. Cupper(II) containing MOF $[Cu(INA)_2]$ (INA = isonicotinate ion) was first example to synthesized by using solvent-free mechanochemical method. The grinding of $Cu(OAc)_2$ and bridging organic ligand (isonicotinic acid) for only 10 min resulted highly crystalline and single phase MOF product with acetic acid and water molecule in the pores. It has been shown that just grinding is necessary to initiate the reaction. Manganese-based MOF (Mn-MOFs) has been prepared using the mechanochemical method in 10 min whereas, liquid phase reaction took 24 h for same reaction.

10.2.2.3 Sonochemical Synthesis

Sonochemistry deals with the chemistry occurring upon application of high-energy ultrasound to a reaction mixture. The ultrasonic environment creates cyclic alternative areas of high and low pressures in the solvent which result into the formation of small bubbles and cavities. The bubbles grow under the alternating pressure conditions through the diffusion of solute vapor into the volume of the bubble and ultrasonic energy is accumulated. After reaching to a maximum size bubbles collapse and energy is released. This process of bubble formation, growth, and collapse is called cavitation and it leads to the rapid release of energy with heating and cooling rates of $>10^{10}$ Ks^{-1}, temperature ~5000 K and pressure ~1000 bar. The primary reason for using a sonochemical synthesis method is to achieve a fast, environmentally friendly, and energy-efficient approach of MOFs synthesis. The first sonochemical MOF synthesis was Zn-based MOFs e.g. Zn carboxylates, $Zn_3(BTC)_2{\bullet}12H_2O$ (BTC = benzene-1,3,5-tricarboxylate) in solution of 20% ethanol in water at room temperature and atmospheric pressure in an ultrasonic bath (Qiu et al. 2008).

10.2.2.4 Electrochemical Synthesis

Electrochemical synthesis of MOFs has also been performed with aim to exclude the anions (nitrate, perchlorate, or chloride) during the synthesis processes. In this synthesis, the metal cations are continuously diffused through anode into the reaction medium where they react with dissolved organic linkers and formation of MOFs takes place. The electrochemical route is applied in the synthesis of Zn, Cu and Al-based MOFs. This method is beneficial for large-scale production of MOFs and also suitable for the formation of MOF films and patterned coating (Ameloot et al. 2009).

10.3 Stability of Metal-organic Frameworks in Water

The stability of MOFs in aqueous medium is one of the most important parameters to determine their potential use in the wastewater treatment process. In the presence of water, MOFs may undergo a decomposition process in which metal coordinated linkers are replaced by H_2O molecules or OH^- ions (Yuan et al. 2018). This indicates the stability of MOFs can be enhanced fundamentally by increasing the strength of the coordination bonds between metal ions and ligands. The Pearson's HSAB theory can be apply to predict and design the stable MOFs. According to the HSAB theory, the interactions between hard Lewis acids and hard Lewis bases, or soft Lewis acids and soft Lewis bases are much stronger than those between hard Lewis acids and soft Lewis bases, or soft Lewis acids and hard Lewis bases. Thus, stable MOFs could be constructed from carboxylate-based linkers (hard Lewis bases) and high-valent metal ions (hard Lewis acids e.g. Ti^{4+}, Zr^{4+}, Hf^{4+}, Ce^{4+}, Fe^{3+}, Cr^{3+} and Al^{3+}), or azolate-based ligands (soft Lewis bases) and low-valent transition metal ions i.e. soft Lewis acids e.g. Zn^{2+}, Cu^{2+}, Co^{2+} and Mn^{2+} (Yuan et al. 2018).

Tetravalent metal ions (Ti^{4+}, Zr^{4+}, Hf^{4+} and Ce^{4+}) with carboxylate ligands based MOFs have attracted attention in recent years because of their high stability in aqueous medium. The exceptional chemical stability of M^{4+}–MOFs is attributed to the (i) high charge and charge to radius ratio (*Z*/*R*) of these metal ions make them hard acids and therefore, form stronger coordination bonds with relatively hard carboxylate ligands, (ii) M^{4+} metal ions require more ligands to balance their charge, therefore their secondary building units (SBUs) tend to have a high connection number. As a result of this, the chance of attack by guest molecules (e.g. H_2O) is insignificant.

The trivalent metal ions (Fe^{3+}, Al^{3+}, and Cr^{3+}) also form stable MOFs with carboxylate ligands. The MIL (Material Institut Lavoisier) series MOFs for example, MIL-53, MIL-88, MIL-100, and MIL-101 contain $[M_3(\mu_3\text{-}O)(COO)_6]$ or linear $[M(OH)(COO)_2]_n$ trivalent metal clusters. The chemical stability in water, acids, and bases usually decreases in the order of Cr-MOF > Al-MOF > Fe-MOF with decreasing inertness of the M–O bonds.

Another class of stable MOFs are synthesized from azolate ligands (e. g. pyrazole, imidazole, triazole, and tetrazole) and softer Lewis acids (Zn^{2+}, Co^{2+}, Cu^{2+}, and Mn^{2+}). The notable examples of this class are zeolitic imidazolate frameworks (ZIFs) with M^{2+} metal ion and imidazolate ligand which exhibited exceptional chemical and thermal stability. Their stability is attributed to the strong metal–imidazole bond and M-Im-M bond angle of 145° that coincides with the Si-O-Si angle in zeolites.

The length of ligand and rigidity, connectivity, and hydrophobicity of the framework also contribute to the MOFs stability. The length of the ligand can affect the rate of its replacement by attacking species, and therefore, MOFs with shorter ligands always show better stability compare to the ones containing longer organic linkers. Furthermore, short and rigid ligands would bend to a larger angle in order to leave the coordination and this would raise the activation energy of decomposition, thereby improving the robustness of the MOF. The higher connectivity of ligands or metal nodes also enhances the MOFs stability by increasing the rate of repair of structural defects. The inclusion of hydrophobic groups in the linker ligands increases the water stability of MOFs as well. For example, the presence of methyl group on the imidazole ligand in ZIF-8 assists in blocking water molecules from attacking the ZnN_4 unit (Park et al. 2006).

10.4 Modifications of Metal-organic Frameworks

MOFs can easily be modified by various chemical treatments without affecting the porous and crystalline nature of the materials for e,g, ZIFs, MIL and UiO series (Yuan et al. 2018). The specific modification is very advantageous to the removal of targeted and specific pollutants as well as improving performance of MOFs in water treatment. Incorporation of different functional groups in the MOFs which may change the polarity and pore's structures.

Conventional modifications involve pre-synthetic modification method, in which the organic linkers are pre-functionalized with certain number of specific functional groups and such functionalized organic linkers then used to prepare functionalized MOFs. Functional groups such as $-NH_2$, halides, –COOH and $-SO_3H$ are often grafted onto the organic ligands in the pre-synthetic modification of MOFs.

However, decomposition of less thermal stable functional groups is possible during MOF synthesis through solvothermal route.

Another modification of MOF materials called post-synthetic modification (PSM) is also used and this is performed after the preparation of MOFs (Cohen 2012). The resulting structures generally have a similar topology but have different metal unit or linker. Since, majority of MOFs are unstable in aqueous conditions, PSM are particularly needed to improve their stability. Also, PSM is a useful method in cases where direct synthesis of a MOF is not possible. Both metal ion and linker exchanges are performed immersing pre-synthesized MOF into the solution of modifier. The optimal conditions are determined based on metal type and charge, as well as organic linker type and connectivity. According to the breaking and formation of chemical bonds, PSM is often divided into (i) covalent post-modification (ii) coordinate covalent post-modification and (iii) post-synthesis protection and de-protection. The covalent post-synthesis method has been used to incorporate many functional groups into MOFs containing –OH, $-NH_2$, azide and halogen functionalities. Coordinate covalent post-modification usually modifies the unsaturated coordination site of the metal in MOFs. After coordination covalent modification, the new coordination bonds between metal ion and modifier organic ligands are formed. The post-synthesis protection and de-protection process involves use of protected functional groups which is followed by their de-protection. This is a powerful tool to incorporate reactive functional groups into MOFs to prevent the loss of catalytic performance of MOFs.

10.5 Applications of Metal–organic Frameworks in Pollutants Removal

10.5.1 Inorganic Contaminants

Various types of inorganic pollutants such as heavy metal ions, oxyanions/cations and inorganic acids etc. are present in the wastewater. The removal of these pollutants using MOFs as adsorbents is very active area of research in present time and recent advances made in this area have provided useful information on the possible mechanisms of adsorption as well as removal of some contaminants. Furthermore, these studies have established that through specific modifications in the MOF structures, the adsorption performance can greatly be enhanced. We will discuss herein the recent progress made for the removal of some heavy metal ions such as Pb^{2+}, Hg^{2+}, As^{5+} and Cr^{6+} ions.

Cationic inorganic pollutants in industrial wastewater include heavy metals and their oxidized products which are harmful to the environment and directly related to the health of humans as well as other forms of life along the food chain. The exposure of lead, cadmium, mercury, chromium and arsenic metal ions are main threats to human health which have been regularly reviewed and monitored by international bodies such as the World Health Organization (WHO). These are very toxic and may damage central nervous function, lungs, kidneys, liver, and bone and also increase the risk of cancers. Recently, MOFs have been found to be promising adsorbents for the removal of heavy metal ions from solution.

10.5.1.1 Removal of Pb^{2+} Metal Ions

Lead(II) (Pb^{2+}) is recognized as one of the most toxic heavy metal ions present in wastewater and affects almost all parts of the body. The relatively long biological half-life (10 to 35 years) and its bio-accumulative nature makes its removal from aquatic ecosystems a significant environmental objective. The ion exchange between metal ions of the SBUs constituting the MOF and Pb^{2+} from solution has been proposed as a potential way to clean contaminated water. Wang et al. have studied the ability of six MOFs—MIL-53(Fe), MIL-101(Fe), UIO-66(Zr), IRMOF-3(Zn), MOF-5(Zn) and ZIF-8(Zn) to adsorption Pb^{2+} metal ion (Wang et al. 2017) The Zn-based MOFs exhibited the highest adsorption, particularly ZIF-8 adsorbs 449 mg g^{-1} at pH ~ 6. This was attributed to the stronger binding ability of Pb than Zn and also less stability of Zn-based MOFs in water. However, such process is not a best way to solve environmental issues because introducing another metal ion into the water ecosystem is not a promising idea.

The installation of sulfur-containing functionalities e.g. thiourea, isothiocyanate, and isocyanate into UiO-66-NH_2(Zr) through PSM does not change their structure and thermal stability (Saleem et al. 2016). However, removal efficiency of Pb^{2+}, Cr^{3+}, Cd^{2+} and Hg^{2+} metal ions from wastewater has been enhanced with compare to original MOF. The removal efficiency could reach 99% with maximum adsorption of 49, 117, 232, and 769 mg g^{-1} for Cd^{2+}, Cr^{3+}, Pb^{2+} and Hg^{2+} respectively with modified MOF UiO-66-NHC(S)NHMe.

A Zn-based MOF $\{[Zn_3L_3(BPE)1.5]\cdot4.5DMF\}_n$ (L = 4,4'-azoxydibenzoate, BPE = bis(4-pyridyl)ethylene) was recently synthesized (Yu et al. 2017). This MOF is decorated with O^- groups from an azoxy-type ligand and shows very high Pb^{2+} uptake capacity up to 616 mg g^{-1} and a high selectivity (>99%) even in the presence of competing metal ions e.g. Mg^{2+} and Ca^{2+}. The excellent removal capacity attributed to the strong electrostatic interaction between the highly accessible O^- groups and Pb^{2+}. Nanotube-like Tb-based MOFs also exhibit excellent adsorption capacity of Pb^{2+} (547 mg g^{-1}) (Zhu et al. 2019). The XPS spectral analysis revealed that the innersphere complexation between nitrogen functional groups of Tb-MOFs and Pb^{2+} is the reason for the high adsorption efficiency.

10.5.1.2 Removal of Hg^{2+} Metal Ions

Contamination of mercury at ppb levels causes a severe environmental and health problems. A Zn-based MOF; [Zn(hip)(L)(DMF)(H_2O)] (hip

= 5-hydroxyisophthalate and L = N^4,$N^{4'}$-di(pyridine-4-yl)biphenyl-4,4'-dicarboxamide) functionalized with acylamide and hydroxy groups is synthesized to remove Hg^{2+} at ppb level (Luo et al. 2015). It was demonstrated that this MOF has 66.5% Hg^{2+} removal capacity from aqueous solution containing 2 ppb Hg^{2+} concentration and maximum uptake capacity of 278 mg g^{-1} from a solution of Hg^{2+} with 20 ppm initial concentration. The interaction of free hydroxyl-groups and acylamide groups with Hg^{2+} have resulted high removal efficiency. Cation exchange processes have also been suggested for mercury removal. A water stable anionic MOF (AMOF-1) with formula $\{[(NH_2Me_2)_2][Zn_3(L)_2]\cdot 9H_2O\}$ (L = 5,5'-(1,4-phenylenebis(methylene))bis(oxy)diisophthalic acid) is reported (Chakraborty et al. 2016). Although, this MOF exhibited exchange of Me_2NH_2 cations by heavy metal ions (Hg^{2+}, Pb^{2+} and Cd^{2+}), but shows low efficiency of adsorption as well as poor selectivity with 98.7% Hg^{2+} removal efficiency from 20 mL solution (1 ppm Hg^{2+} concentration) in 24 h (Fig. 10.1).

Mercury cations are soft Lewis acids, so another strategy to capture of mercury species would be the installation of soft Lewis base into the MOFs. Xu and co-workers incorporated – SCH_3 groups into the MOFs structure for the removal of Hg^{2+} cations. The thiol functionalized Cu-BTC (BTC = benzene-1,3,5-tricarboxylate) exhibited Hg^{2+} adsorption efficiency of 714 mg g^{-1} (Ke et al. 2011). This remarkably high adsorption capacity is attributed to the binding of Hg^{2+} cations with the large numbers of exposed thiol groups. However highest adsorption capacity of 900 mg g^{-1} have been achieved in aqueous medium using a 3D bio-MOF (Mon et al. 2016) but shows slow adsorption kinetics and more than 3 h is required to reach equilibrium. Another sulfur-functionalized MOF FJI-H12 has proven to be excellent adsorbents for mercury species with relative satisfactory adsorption efficiency of 439.8 mg g^{-1}, and short equilibrium time (<50 min) but highly selective for Hg^{2+} over Mn^{2+}, Ba^{2+}, Ni^{2+}, and Cd^{2+} (Liang et al. 2016). Other important characteristics of FJI-H12 are

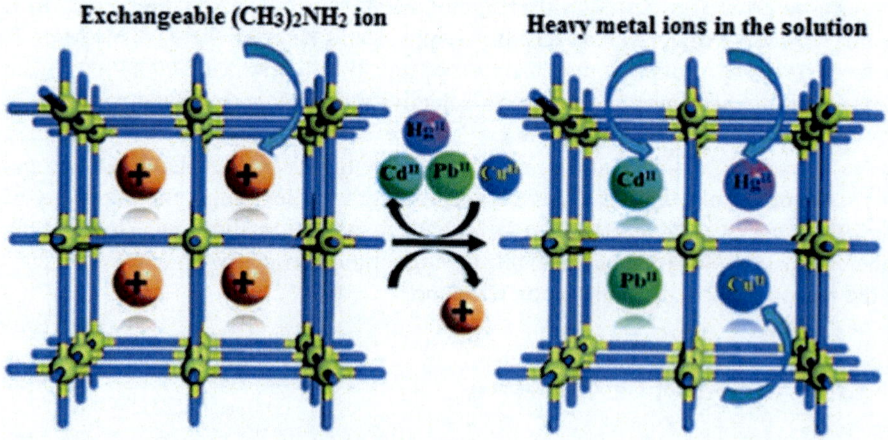

Fig. 10.1 $Me_2NH_2^+$ cations exchange with Hg^{2+}, Pb^{2+}, Cu^{2+} or Cd^{2+} using AMOF-1. Adopted and modified from (Chakraborty et al. 2016)

the easy and cheaper synthesis as well as recycling (1 day immersion of FJI-H12-Hg in the KSCN solution at ambient temperature). All these features makes FJI-H12 excellent Hg^{2+} adsorbent for the practical applications. This study also demonstrated that continuous and fast removal of Hg^{2+} using a column which implies its usefulness in wastewater treatment.

10.5.1.3 Removal of As^{3+}/As^{5+} Metal Ions

The acute toxicity and carcinogenic nature of arsenic compounds have triggered intensive studies for the removal of this pollutant from contaminated water bodies. According to WHO 10 ppb as the maximum amount of arsenic compounds in safe drinking water. However, current technologies face some difficulties to achieve this level. In this regard, MOF materials have shown high performance towards arsenic compound removal. Arsenic is generally present as arsenite ($H_xAsO_3{}^{3-x}$) and arsenate ($H_xAsO_4{}^{3-x}$) in the water ecosystems.

A remarkable arsenate adsorption capacity of UiO-66(Zr) in a wide range of pH (1–10) with maximum arsenate uptake capacity of 303 mg g^{-1} at pH = 2 has been reported (Wang et al. 2015). This excellent performance was even maintained in the presence of competing anions e.g. chloride, nitrate, carbonate and sulfates. This value is much higher than those of commercial adsorbents, but at pH = 7 this value is drastically decreases to 147 mg g^{-1}. Spectroscopic studies have indicated the excellent As^{5+} removal capacity of this MOF was related to hydroxyl and 1,4-benzenedicarboxylic acid (BDC) ligand exchange processes that lead to the formation of arsenic complexes on the Zr6 SBUs.

The MIL series MOFs are also effective absorbents to remove arsenic ions. MIL-100(Fe) removes arsenic (As^{5+}) from wastewater with sixfold higher capacity compared to commercial iron oxide powders (50 nm Fe_2O_3 nanoparticle) (Zhu et al. 2012). A study has indicated arsenic ions are absorbed onto the interior of the MOF instead of on the outer surface and formation of Fe—O—As bonds resulted high As^{5+} adsorption. A core-shell and mesoporous $CoFe_2O_4$@MIL-100(Fe) hybrid magnetic nanoparticles (MNPs) show best performance for adsorptive removal of As^{3+} and As^{5+} and maximum adsorption capacities are 143.6 and 114.8 mg g^{-1} for As^{3+} and As^{5+} ions, respectively (Yang and Yin 2017).

This material also exhibits rapid adsorption as 0.1 mg L^{-1} arsenic ions could be adsorbed in 2 min. Furthermore, this hybrid material works over a wide range of pH 4—10 and could selectively capture arsenic ions over various oxyanions. The formation of Fe—O—As inner-sphere complex has been proposed on the surface of MOF through hydroxyl substitution. Figure 10.2 represents the adsorption of arsenic species on $CoFe_2O_4$@MIL-100(Fe) and formation of multilayer structures through hydrogen bonding.

Similar composite material Fe_3O_4-ZIF-8 has also been reported for the removal of As^{3+} from polluted water with satisfactory adsorption with maximum capacity of 100 mg g^{-1} (Huo et al. 2018). The adsorption in this material is attributed to the strong surface–complex interactions. Study has indicated that As^{3+} removal with this

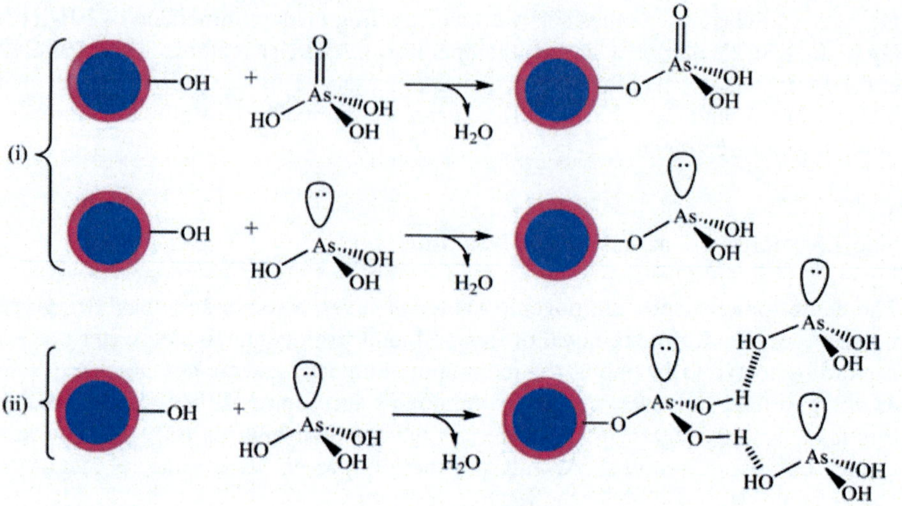

Fig. 10.2 Schematic representation of arsenic ions adsorption. **i** Indicates substitution of –OH groups on $CoFe_2O_4$@MIL-100(Fe) by the deprotonated arsenic species; **ii** indicates adsorption of H_3AsO_3 through hydrogen bonding to form a multilayer structure for the high adsorption capacity of As^{3+} species. Adopted from (Yang and Yin 2017)

composite material is hindered by the presence of carbonate and phosphate anions but the presence of sulphate, chloride and nitrate do not affect adsorption efficiency. Additionally, the usefulness of this composite material arsenic is the easy separation by using external magnetic field after finishing the water treatment process.

10.5.1.4 Removal of Cr^{6+} Metal Ions

Water bodies contaminated with Cr^{6+} shows a severe human health threat due to its teratogenicity, toxicity and mutagenicity. The conventional methods for Cr^{6+} removal technologies suffer from poor adsorption uptake and low selectivity. The Cr^{6+} metal ion could exist in different forms—$HCrO_4^-$, CrO_4^{2-} and $Cr_2O_7^{2-}$ under environmental conditions and formation of these species depend on the pH and concentration.

The efficient adsorption of Cr^{6+} species over a wide pH range (2–9) and Cr^{6+} concentration (5–100 mg L^{-1}), where $HCrO_4^-$ and CrO_4^{2-} are the predominant species has been achieved using TMU-30 MOF synthesized from lead(II) and isonicotinate N-oxide (Aboutorabi et al. 2016). TMU-30 exhibits rapid Cr^{6+} adsorption uptake (90% adsorption in 5 min) and maximum adsorption capacity of 145 mg g^{-1}. It has been suggested that the electrostatic interaction between the positively charged nitrogen from N-oxide groups and oxygen atoms from chromate leads to remarkable performance of TMU-30. Consequently, this material differentiates metal cations and oxyanions. However, it is difficult to distinguish different oxyanions. Niu et al.

reported 1D Fe-gallic acid MOF which shows exceptional adsorption capacity for $Cr_2O_7^{2-}$ ions up to 1709.2 mg g^{-1} (Niu et al. 2017). This ultrahigh adsorption capacity is attributed to the large number of exposed active sites of the Fe-gallic acid MOF to bind with $Cr_2O_7^{2-}$ anions. This MOF also demonstrated excellent selectivity of $Cr_2O_7^{2-}$ over several types of anions and metal cations and exhibited high stability in strongly acidic (pH = 2) and alkaline (pH = 11) conditions, but it takes about 72 h to reach equilibrium which limits its use in the water treatment process. On the other hand MOR-1-HA (Rapti et al. 2016) and MOR-2 (Rapti et al. 2017) showed fast uptake which could reach an adsorption capacity of 240 − 280 and 194 mg g^{-1} within 3 and 1 min, respectively. Furthermore, these two MOFs could also be used in a wide range of pH conditions and also be regenerated.

10.5.2 Organic Contaminants

In recent years, removal of organic contaminants from water using MOFs and/or their degradation into non-harmful products have received a great deal of attention. Organic pollutants such as antibiotics, pesticides, plasticizers, organic dyes, and other substances whose usage recently escalated are on the priority list. Here we present discussion on the removal of pharmaceuticals and personal care products (PPCPs) emerging pollutants.

There is huge increase in the consumption of pharmaceutical drugs, cosmetics and household chemicals in modern societies. The shortage of established efficient decontamination procedures have led to the accumulation of these pollutants in aqueous ecosystem. Also, poor efficiency of their removal by available technologies have resulted the search of more efficient and cost-effective methodologies.

Wang et al. reported two isostructural Zr_6^{IV}-based MOFs, BUT-12 and BUT-13 as best performing luminescent MOF-based sensing materials. They showed remarkable uptake capacity as well as sensing properties for some antibiotics and nitroaromatic compounds at ppb levels (Wang et al. 2016). It has been shown that BUT-12 exhibited a remarkable and selective uptake capacity towards nitrofurazone and nitrofurantoin, whereas BUT-13 showed a fast and remarkable uptake of nitrofurazone, nitrofurantoin, ornidazole and chloramphenicol. The excellent performances are attributed to the hydrophobicity and suitability of pore size to adsorb these organic pollutants. These materials can be recycled consecutively for up to six cycles which makes them potentially useful for the wastewater treatment applications. Fluoroquinolones (FQs) are another largest consumed antibiotics. The high performance of Cu-based MOF PCN-124-stu(Cu) toward the adsorption of three FQs—norfoxacin (NOR), ofoxacin (OFL) and enrofoxacin (ENR) has been reported with adsorption capacity 354, 292 and 198 mg g^{-1} respectively (Jin et al. 2017). Studies have indicated that open metal sites and amide groups are preferential adsorption sites and hydrogen bonding interactions main interaction between MOF and the antibiotics. However, the differences between adsorbed amounts of FQs are due to their different molecular sizes. Also, this material exhibited higher FQ adsorption as in comparison to some common MOFs such as PCN-124(Cu), MIL-100(Cr), HKUST-1(Cu), MOF-74(Mg) and UiO-66(Zr) and traditional adsorbents such as zeolite 13X and activated carbon, except for mesoporous MOF-MIL-100(Cr).

The adsorptive removal of a toxic sulfonamide antibiotic sulfachloropyradazine (SCP) from aqueous solution using UiO-66(Zr) and ZIF-67(Co) has been studied (Azhar et al. 2017). UiO-66 shows an adsorption capacity of 417 mg g^{-1} with a fast uptake kinetics. It has been revealed that π-π interaction and electrostatic interaction are cause of noticeable performance of UiO-66(Zr).

The MOFs of MIL series exhibit exceptional high stability in water which makes use of these materials in the removal of PPCPs. The post-synthetic modification in MIL-101(Cr) structure with hydrogen donor functionalities has been successful strategy for the efficient removal of PPCPs from contaminated water.

Recently, Seo et al. studied the adsorption of nonsteroidal anti-infammatory drugs (naproxen and ibuprofen) and a component of many sunscreen lotions (oxybenzone) from aqueous solutions using the functionalized MIL-101 (Seo et al. 2016). The study has demonstrated that MIL-101 functionalized with H-donor functional groups (−OH, $-(OH)_2$, $-NH_2$) are very effective in the adsorption of naproxen. The H-bonding interaction between O atoms on naproxen and H atoms on the adsorbent is behind the noticeable adsorption affinity. Additionally, the MIL-101-OH could be recycled several times by simple washing with ethanol, hence shows potential applicability in the removal of PPCPs from contaminated water. Amino functionalized MOF, MIL-101(Cr)-NH_2 have also been used to capture nitroimidazole antibiotics (dimetridazole, menidazole and metronidazole) from water. Hydrogen bonding again plays role in efficient adsorption of these PPCPs (Seo et al. 2017). It has also been observed that MIL-101(Cr) with larger number of hydroxyl groups always shows high uptake for PPCPs because of greater number of hydrogen bond formation. The maximum adsorption capacity of MIL-101$(OH)_3$(Cr) for p-chloro-m-xylenol, triclosan, ketoprofen and naproxen are 79, 112, 80 and 156 mg g^{-1}, respectively in 12 h (Song and Jhung 2017). MIL-101(Cr) has also been used to remove bisphenol A (BPA) from polluted water and average pore size and specific surface area of MIL-101(Cr) are the most important parameters determining the adsorption kinetics and capacity of BPA (Qin et al. 2014). The π-π interaction and hydrogen bonding are the key adsorption mechanisms of BPA over MIL-101(Cr).

A widely used nonsteroidal anti-inflammatory drug is diclofenac and advance oxidative processes (AOPs) have shown to be effective method for its removal from contaminated water. However, this process is associated with the generation of more toxic secondary pollutants and technical difficulty to perform on a large-scale (Xu et al. 2017). Therefore, adsorptive methods are alternative routes for the removal of diclofenac. The ZIF-67(Co) together with different amounts of cetyltrimethylammonium bromide (CTAB) surfactant exhibited noticeable diclofenac removal efficiency and maximum adsorption uptake is increases with the amount of surfactant (Andrew Lin et al. 2015).

10.6 Conclusions and Future Perspectives

The safety of water resources and providing clean water to everyone are now becoming top challenges worldwide. A large number of inorganic and organic toxic materials are present in water ecosystem and removal of these hazardous materials utilizing cost-effective water remediation technologies is always on high demand. In this context, chemistry of MOFs have received significant attention in twenty-first century and which has already shown great potentials in inorganic and organic pollutants removal in the wastewater treatment process. In this chapter, we have shown that MOFs can work as efficient adsorbents for the removal of toxic metal ions as well as emerging organic pollutants (pharmaceuticals and personal care products). Since, majority of MOFs are not very stable in the aqueous media but taking advantage of different synthetic routes, it has now been possible to synthesize many water stable MOFs for water remediation. Although remarkable advances achieved in the MOFs research, there are limited number of MOFs which are suitable to use in water treatment. These are mainly MIL, ZIF and UiO derivatives. Therefore, the development of novel families of water-stable MOFs, with potential application in the water treatment should be a research area of continued interest. We also need to make effort to design new organic ligands beyond carboxylate and imidazolate ligands and construct stable MOFs for water research. Additionally, it is highly desirable to develop methods for low-cost and large-scale green synthesis of MOFs using cheaper starting materials. Functionalization of MOFs to perform multiple tasks in a single material would also be interesting area in the water treatment process.

References

Aboutorabi L, Morsali A, Tahmasebi E, Buyukgungor O (2016) Metal-organic framework based on isonicotinate N-oxide for fast and highly efficient aqueous phase Cr(VI) adsorption. Inorg Chem 55(11):5507–5513. https://doi.org/10.1021/acs.inorgchem.6b00522

Ameloot R, Stappers L, Fransaer J, Alaerts L, Sels BF, De Vos DE (2009) Patterned growth of metal-organic framework coatings by electrochemical synthesis. Chem Mater 21(13):2580–2582. https://doi.org/10.1021/cm900069f

Andrew Lin K-Y, Yang H, Lee W-D (2015) Enhanced removal of diclofenac from water using a zeolitic imidazole framework functionalized with cetyltrimethylammonium bromide (CTAB). RSC Adv 5(99):81330–81340. https://doi.org/10.1039/c5ra08189k

Azhar MR, Abid HR, Periasamy V, Sun H, Tade MO, Wang S (2017) Adsorptive removal of antibiotic sulfonamide by UiO-66 and ZIF-67 for wastewater treatment. J Colloid Interface Sci 500:88–95. https://doi.org/10.1016/j.jcis.2017.04.001

Bailey SE, Olin TJ, Bricka RM, Adrian DD (1999) A review of potentially low-cost sorbents for heavy metals. Water Res 33(11):2469–2479. https://doi.org/10.1016/S0043-1354(98)00475-8

Blanchard G, Maunaye M, Martin G (1984) Removal of heavy metals from waters by means of natural zeolites. Water Res 18(12):1501–1507. https://doi.org/10.1016/0043-1354(84)90124-6

Cavka JH, Jakobsen S, Olsbye U, Guillou N, Lamberti C, Bordiga S, Lillerud KP (2008) A new zirconium inorganic building brick forming metal organic frameworks with exceptional stability. J Am Chem Soc 130(42):13850–13851. https://doi.org/10.1021/ja8057953

Chakraborty A, Bhattacharyya S, Hazra A, Ghosh AC, Maji TK (2016) Post-synthetic metalation in an anionic MOF for efficient catalytic activity and removal of heavy metal ions from aqueous solution. Chem Commun (Camb) 52(13):2831–2834. https://doi.org/10.1039/c5cc09814a

Cohen SM (2012) Postsynthetic methods for the functionalization of metal-organic frameworks. Chem Rev 112(2):970–1000. https://doi.org/10.1021/cr200179u

Férey G, Mellot-Draznieks C, Serre C, Millange F, Dutour J, Surblé S, Margiolaki I (2005) A chromium terephthalate-based solid with unusually large pore volumes and surface area. Science 309(5743):2040–2042. https://doi.org/10.1126/science.1116275

Giusti L (2009) A review of waste management practices and their impact on human health. Waste Manag 29(8):2227–2239. https://doi.org/10.1016/j.wasman.2009.03.028

Gupta VK, Ali I, Saleh TA, Nayak A, Agarwal S (2012) Chemical treatment technologies for waste-water recycling—an overview. RSC Adv 2(16):6380. https://doi.org/10.1039/c2ra20340e

Gupta VK, Saleh TA (2013) Sorption of pollutants by porous carbon, carbon nanotubes and fullerene—an overview. Environ Sci Pollut Res Int 20(5):2828–2843. https://doi.org/10.1007/s11356-013-1524-1

Huo J-B, Xu L, Yang J-CE, Cui H-J, Yuan B, Fu M-L (2018) Magnetic responsive Fe_3O_4-ZIF-8 core-shell composites for efficient removal of As(III) from water. Colloids Surf A 539:59–68. https://doi.org/10.1016/j.colsurfa.2017.12.010

Jin W-G, Chen W, Xu P-H, Lin X-W, Huang X-C, Chen G-H, Lu F, Chen X-M (2017) An exceptionally water stable metal-organic framework with amide-functionalized cages: selective CO_2/CH_4 uptake and removal of antibiotics and dyes from water. Chem Eur J 23(53):13058–13066. https://doi.org/10.1002/chem.201701884

Ke F, Qiu LG, Yuan YP, Peng FM, Jiang X, Xie AJ, Shen YH, Zhu JF (2011) Thiol-functionalization of metal-organic framework by a facile coordination-based postsynthetic strategy and enhanced removal of Hg^{2+} from water. J Hazard Mater 196:36–43. https://doi.org/10.1016/j.jhazmat.2011.08.069

Li X, Wang B, Cao Y, Zhao S, Wang H, Feng X, Zhou J, Ma X (2019) Water contaminant elimination based on metal-organic frameworks and perspective on their industrial applications. ACS Sustain Chem Eng 7(5):4548–4563. https://doi.org/10.1021/acssuschemeng.8b05751

Liang L, Chen Q, Jiang F, Yuan D, Qian J, Lv G, Xue H, Liu L, Jiang H-L, Hong M (2016) In situ large-scale construction of sulfur-functionalized metal–organic framework and its efficient removal of Hg(ii) from water. J Mater Chem A 4(40):15370–15374. https://doi.org/10.1039/c6ta04927c

Luo F, Chen JL, Dang LL, Zhou WN, Lin HL, Li JQ, Liu SJ, Luo MB (2015) High-performance Hg^{2+} removal from ultra-low-concentration aqueous solution using both acylamide- and hydroxyl-functionalized metal–organic framework. J Mater Chem A 3(18):9616–9620. https://doi.org/10.1039/c5ta01669j

Malik DS, Jain CK, Yadav AK (2016) Removal of heavy metals from emerging cellulosic low-cost adsorbents: a review. Appl Water Sci 7(5):2113–2136. https://doi.org/10.1007/s13201-016-0401-8

Min Wang Q, Shen D, Bülow M, Ling Lau M, Deng S, Fitch FR, Lemcoff NO, Semanscin J (2002) Metallo-organic molecular sieve for gas separation and purification. Micropor Mesopor Mater 55(2):217–230. https://doi.org/10.1016/S1387-1811(02)00405-5

Mon M, Bruno R, Ferrando-Soria J, Armentano D, Pardo E (2018) Metal–organic framework technologies for water remediation: towards a sustainable ecosystem. J Mater Chem A 6(12):4912–4947. https://doi.org/10.1039/c8ta00264a

Mon M, Lloret F, Ferrando-Soria J, Martí-Gastaldo C, Armentano D, Pardo E (2016) Selective and efficient removal of mercury from aqueous media with the highly flexible arms of a BioMOF. Angew Chem Int Ed 55(37):11167–11172. https://doi.org/10.1002/anie.201606015

Niu H, Zheng Y, Wang S, He S, Cai Y (2017) Stable hierarchical microspheres of 1D Fe–gallic acid MOFs for fast and efficient Cr(vi) elimination by a combination of reduction, metal substitution and coprecipitation. J Mater Chem A 5(32):16600–16604. https://doi.org/10.1039/c7ta04006g

Okoro HK, Ayika SO, Ngila JC, Tella AC (2018) Rising profile on the use of metal–organic frameworks (MOFs) for the removal of heavy metals from the environment: an overview. Appl Water Sci 8(6). https://doi.org/10.1007/s13201-018-0818-3

Park KS, Ni Z, Cote AP, Choi JY, Huang R, Uribe-Romo FJ, Chae HK, O'Keeffe M, Yaghi OM (2006) Exceptional chemical and thermal stability of zeolitic imidazolate frameworks. Proc Natl Acad Sci USA 103(27):10186–10191. https://doi.org/10.1073/pnas.0602439103

Qin F-X, Jia S-Y, Liu Y, Li H-Y, Wu S-H (2014) Adsorptive removal of bisphenol A from aqueous solution using metal-organic frameworks. Desalin Water Treat 54(1):93–102. https://doi.org/10.1080/19443994.2014.883331

Qiu LG, Li ZQ, Wu Y, Wang W, Xu T, Jiang X (2008) Facile synthesis of nanocrystals of a microporous metal-organic framework by an ultrasonic method and selective sensing of organoamines. Chem Commun (camb) 31:3642–3644. https://doi.org/10.1039/b804126a

Rapti S, Pournara A, Sarma D, Papadas IT, Armatas GS, Hassan YS, Alkordi MH, Kanatzidis MG, Manos MJ (2016) Rapid, green and inexpensive synthesis of high quality UiO-66 amino-functionalized materials with exceptional capability for removal of hexavalent chromium from industrial waste. Inorg Chem Frontiers 3(5):635–644. https://doi.org/10.1039/c5qi00303b

Rapti S, Sarma D, Diamantis SA, Skliri E, Armatas GS, Tsipis AC, Hassan YS, Alkordi M, Malliakas CD, Kanatzidis MG, Lazarides T, Plakatouras JC, Manos MJ (2017) All in one porous material: exceptional sorption and selective sensing of hexavalent chromium by using a Zr^{4+} MOF. J Mater Chem A 5(28):14707–14719. https://doi.org/10.1039/c7ta04496h

Rosi NL, Eckert J, Eddaoudi M, Vodak DT, Kim J, Keeffe M, Yaghi OM (2003) Hydrogen storage in microporous metal-organic frameworks. Science 300(5622):1127. https://doi.org/10.1126/science.1083440

Rosi NL, Kim J, Eddaoudi M, Chen B, O'Keeffe M, Yaghi OM (2005) Rod packings and metal−organic frameworks constructed from rod-shaped secondary building units. J Am Chem Soc 127(5):1504–1518. https://doi.org/10.1021/ja045123o

Saleem H, Rafique U, Davies RP (2016) Investigations on post-synthetically modified UiO-66-NH_2 for the adsorptive removal of heavy metal ions from aqueous solution. Micropor Mesopor Mater 221:238–244. https://doi.org/10.1016/j.micromeso.2015.09.043

Seo PW, Bhadra BN, Ahmed I, Khan NA, Jhung SH (2016) Adsorptive removal of pharmaceuticals and personal care products from water with functionalized metal-organic frameworks: remarkable adsorbents with hydrogen-bonding abilities. Sci Rep 6:34462. https://doi.org/10.1038/srep34462

Seo PW, Khan NA, Jhung SH (2017) Removal of nitroimidazole antibiotics from water by adsorption over metal–organic frameworks modified with urea or melamine. Chem Eng J 315:92–100. https://doi.org/10.1016/j.cej.2017.01.021

Serre C, Millange F, Thouvenot C, Noguès M, Marsolier G, Louër D, Férey G (2002) Very large breathing effect in the first nanoporous Chromium(III)-based solids: MIL-53 or $CrIII(OH)\cdot\{O_2C-C_6H_4-CO_2\}\cdot\{HO_2C-C_6H_4-CO_2H\}_x\cdot H_2O_y$. J Am Chem Soc 124(45):13519–13526. https://doi.org/10.1021/ja0276974

Slater AG, Cooper AI (2015) Porous materials. Function-led design of new porous materials. Science 348 (6238):aaa8075. https://doi.org/10.1126/science.aaa8075

Song JY, Jhung SH (2017) Adsorption of pharmaceuticals and personal care products over metal-organic frameworks functionalized with hydroxyl groups: quantitative analyses of H-bonding in adsorption. Chem Eng J 322:366–374. https://doi.org/10.1016/j.cej.2017.04.036

Stock N, Biswas S (2012) Synthesis of metal-organic frameworks (MOFs): routes to various MOF topologies, morphologies, and composites. Chem Rev 112(2):933–969. https://doi.org/10.1021/cr200304e

Wang B, Lv XL, Feng D, Xie LH, Zhang J, Li M, Xie Y, Li JR, Zhou HC (2016) Highly stable Zr(IV)-based metal-organic frameworks for the detection and removal of antibiotics and organic explosives in water. J Am Chem Soc 138(19):6204–6216. https://doi.org/10.1021/jacs.6b01663

Wang C, Liu X, Chen JP, Li K (2015) Superior removal of arsenic from water with zirconium metal-organic framework UiO-66. Sci Rep 5:16613. https://doi.org/10.1038/srep16613

Wang L, Zhao X, Zhang J, Xiong Z (2017) Selective adsorption of Pb (II) over the zinc-based MOFs in aqueous solution-kinetics, isotherms, and the ion exchange mechanism. Environ Sci Pollut Res Int 24(16):14198–14206. https://doi.org/10.1007/s11356-017-9002-9

Xiang Z, Cao D, Lan J, Wang W, Broom DP (2010) Multiscale simulation and modelling of adsorptive processes for energy gas storage and carbon dioxide capture in porous coordination frameworks. Energ Environ Sci 3(10):1469. https://doi.org/10.1039/c0ee00049c

Xu L (2015) Impact of climate change and human activity on the eco-environment. Springer, Berlin Heidelberg, Berlin, Heidelberg

Xu Y, Liu T, Zhang Y, Ge F, Steel RM, Sun L (2017) Advances in technologies for pharmaceuticals and personal care products removal. J Mater Chem A 5(24):12001–12014. https://doi.org/10.1039/c7ta03698a

Yang JC, Yin XB (2017) $CoFe_2O_4$@MIL-100(Fe) hybrid magnetic nanoparticles exhibit fast and selective adsorption of arsenic with high adsorption capacity. Sci Rep 7:40955. https://doi.org/10.1038/srep40955

Yu C, Shao Z, Hou H (2017) A functionalized metal-organic framework decorated with O(−) groups showing excellent performance for lead(ii) removal from aqueous solution. Chem Sci 8(11):7611–7619. https://doi.org/10.1039/c7sc03308g

Yuan S, Feng L, Wang K, Pang J, Bosch M, Lollar C, Sun Y, Qin J, Yang X, Zhang P, Wang Q, Zou L, Zhang Y, Zhang L, Fang Y, Li J, Zhou HC (2018) Stable Metal-organic frameworks: design, synthesis, and applications. Adv Mater 30(37):e1704303. https://doi.org/10.1002/adma.201704303

Zhao D, Yuan D, Zhou H-C (2008) The current status of hydrogen storage in metal–organic frameworks. Energ Environ Sci 1(2):222–235. https://doi.org/10.1039/B808322N

Zhu B-J, Yu X-Y, Jia Y, Peng F-M, Sun B, Zhang M-Y, Luo T, Liu J-H, Huang X-J (2012) Iron and 1,3,5-Benzenetricarboxylic metal-organic coordination polymers prepared by solvothermal method and their application in efficient As(V) removal from aqueous solutions. J Phys Chem C 116(15):8601–8607. https://doi.org/10.1021/jp212514a

Zhu H, Yuan J, Tan X, Zhang W, Fang M, Wang X (2019) Efficient removal of Pb^{2+} by Tb-MOFs: identifying the adsorption mechanism through experimental and theoretical investigations. Environ Sci Nano 6(1):261–272. https://doi.org/10.1039/c8en01066h

Chapter 11
Advanced Oxidation Processes: A Promising Route for Abatement of Emerging Contaminants in Water

Tadimeti Divya Kusuma, M. S. V. Naga Jyothi, Chebrolu Pulla Rao, and Shihabudheen M. Maliyekkal

Abstract The importance of clean water and the link between water and health are well-documented. Ailment and disease from the consumption of polluted water is a significant cause of human misery in the world. With the advancements in chemistry in general and in analytical chemistry in particular, our ability to detect and analyze pollutants in trace quantities has increased significantly over the years. Currently, many freshwater bodies are contaminated not only with conventional pollutants but also with several organic molecules, such as pharmaceutically active compounds, endocrine-disrupting compounds, surfactants, personal care products, and several others. These emerging contaminants (ECs) require special attention due to their toxicity to all forms of life. Besides, some of these pollutants are bioresistant and can even sustain after primary and secondary treatments of wastewater. Advanced oxidation is a promising technique and has gained immense importance in recent years due to its ability to degrade and mineralize complex organic molecules, including ECs. Advanced oxidation processes (AOPs) rely on the in-situ generation of reactive chemical species such as hydroxyl radicals for degradation. This chapter discusses the principles and working mechanisms of various AOPs, including ozonation, Fenton, photolysis, plasma, sonolysis, and photocatalysis, and their combinations. The application of AOPs in degrading ECs such as pesticides, pharmaceutically active compounds (PACs), personal care products (PCPs), and endocrine-disrupting compounds (EDCs) are discussed. This chapter also elucidates the origin, fate, and human and ecological health impacts of ECs in water bodies.

Keywords AOPs · Oxidative degradation · Emerging contaminants · Water pollution · Wastewater

---

T. D. Kusuma · M. S. V. Naga Jyothi · S. M. Maliyekkal (✉)
Department of Civil and Environmental Engineering, Indian Institute of Technology Tirupati, Tirupati 517506, India
e-mail: shihab@iittp.ac.in

C. P. Rao
Department of Chemistry, Indian Institute of Technology Tirupati, Tirupati 517506, India

S. P. Singh et al. (eds.), *Nanomaterials and Nanocomposites for Environmental Remediation*, Energy, Environment, and Sustainability,
https://doi.org/10.1007/978-981-16-3256-3_12

11.1 Introduction

Adequate access to clean water is essential for the survival of flora and fauna. Though our blue planet is an abundant source of water, the freshwater is limited and unevenly distributed. The mismanagement and pollution of freshwater resources have further restricted access to safe water. As a result, a large population across the globe is now facing severe freshwater scarcity. According to the recent report by the World Health Organization (WHO), 1 in 3 does not have access to safe drinking water (WHO 2019). Though the Indian subcontinent is rich in water resources, about 80% of the surface water and a similar percentage of the groundwater bodies are contaminated and hence is not safe for drinking (Araya et al. 2019). The presence of conventional organic, inorganic, and biological pollutants is a common problem in freshwater resources. However, our ability to characterize water has gone beyond traditional pollutants with the emergence of advanced analytical tools such as ultra-high-performance liquid chromatography (UHPLC), liquid chromatography coupled with mass spectrometry (LC–MS), gas chromatography with mass spectrometry (GC–MS), and tandem mass spectrometry (MS2). The excessive use of pesticides, pharmaceutically active compounds (PACs), personal care products (PCPs), veterinary medicines, endocrine-disrupting compounds (EDCs), surfactants, nanomaterials, and other types of industrial discharges have resulted in their way to water bodies (Söderström et al. 2009). Though these compounds are not regulated in drinking water, some are potentially toxic to humans and other aquatic organisms. These compounds, now identified as emerging pollutants, cause concern in the drinking water sector due to their toxicity and persistence nature at trace amounts (Daughton and Ternes 1999).

The first report mentioning the presence of emerging contaminants in water could probably link to Rachel Carson's book "silent spring." The book deals with the widespread use of dichloro diphenyl trichloroethane (DDT) to control pests in agricultural lands (Drury 1963). United States Environmental Protection Agency (USEPA) defines ECs as chemicals and other substances with no regulatory standard and are recently discovered in natural waters at detectable levels (Ankley et al. 2008). The complete scientific data regarding environment fate and ecotoxicological studies are not yet available (Sauvé and Desrosiers 2014). PACs are one of the widely reported emerging pollutants in water. As per global statistics, the average drug consumption per capita per annum is around 15 g (Ternes and Joss 2006). India is the third-largest producer and exporter of PACs. The primary production hubs are located in Bangalore, Mumbai, Hyderabad, and Ahmedabad. Numerous studies on the Yamuna river reveal the presence of antibiotics and other emerging contaminants (Bhargava 2006; Mutiyar et al. 2018). The widespread occurrence of trace concentrations of EDCs and PCPs in water is also a significant cause of concern. The use of PCPs in toothpaste, sunscreens, soaps, cosmetics, detergents, fragrances, lotions, and insect repellents is increasing. The increasing trend in consumption can lead to the accumulation of these compounds in water with time. Indiscriminate use of pesticides found their entry to the water streams through agricultural runoff and industrial effluents. Numerous works reported the presence of toxic pesticidal residues in the water and sediments of Ganges and Yamuna rivers in India (Singh et al. 2005; Esmaeil and Somashekar 2013). Most widely used nanomaterials are also known to be ECs due to their accumulation in the environment. Personal care products, paints, varnishes, food packaging, surface coatings, textiles, electronics, and cleaning products are the sources of different nano-metals and nano-metal oxides (McGillicuddy et al. 2017; Wang and Nowack 2018). Stable nanoscale materials have the potential to produce reactive oxygen species (ROS), damage cell membranes, and cause impaired nutrient uptake by cells (Klaine et al. 2008). In 1916, the Endocrine Society further classified ECs as EDCs based on their effect on the endocrine glands. More than 3000 compounds are designated as EDCs, which interferes with the hormonal action. The studies also revealed that these are suspects for hormonal disorders, heart diseases, circulatory problems, and diabetes (Zoeller et al. 2012). However, the mechanism of interaction and health impacts associated with ECs in the environment is not well established.

This chapter deals with various aspects of ECs, including their origin and fate in the aqueous medium, environmental and human health effects, and treatment by various AOPs. The chapter presents a detailed discussion on treatment processes such as ozonation, Fenton-type reaction, sonolysis, plasma, photolysis, and photocatalysis.

11.2 Origin and Fate of ECs in the Aqueous Medium

The rapid growth of industrial and agriculture sectors, and lifestyle changes has resulted in the accumulation of ECs in water as hazardous and non-biodegradable wastes. The possible sources of ECs include effluents of domestic wastewater treatment plants (WWTPs) and industries, waste discharges from laboratories and hospitals, agricultural runoff, and landfill leachate (Gogoi et al. 2018). Several factors like hydrogeological parameters of a region, type of industries, the extent of prescribed and illegal use of drugs, quality of sewage system in the area, access to advanced treatment technologies, waste disposal systems, and cultural and behavioural practices, decide the magnitude of the contamination. For instance, 90% of the ingested drugs enter the sewer system as waste through urine and faeces. Approximately 3000 different chemicals such as lipid regulators, antibiotics, anti-inflammatories, analgesics, contraceptives, neuro-active medicines, and beta-blockers are reported in water streams (Carson 2002; Yang et al. 2011). Veterinary antibiotics employed for animal food production are not absorbed in animals' gut completely. As a result, more than 30% of these compounds are excreted (Kim et al. 2011). These drugs can undergo biological transitions in human bodies and WWTPs. These contaminants have complex chemical structures making it difficult to understand their fate in the environment. These compounds undergo adsorption, biological, and photochemical degradation in solution and solid medium. Microorganisms like bacteria, fungi, microalgae, and protozoa can degrade ECs in biological processes through different mechanisms, including contaminant uptake, diffusion, enzyme complexation, and release of transformed products (Wilkinson et al. 2017). USEPA estimates about 50% of the biosolids produced in the USA are used in <1% of the national farmlands for nutrition (USEPA 2012). The retained ECs on biosolids can re-enter the environment through plants or groundwater. The residual concentrations of ECs in the effluents of WWTPs can further reduce by adsorption onto soils, sediments, or suspended matter according to the effluent receiving medium. The physicochemical properties of solids and chemicals, pH, temperature of the soil, rate of percolation, and degree of soil water-saturation affect the adsorption process (Barron et al. 2010). Martín et al. (2010) observed sediments of the Guadiamar river in southern Spain have pharmaceuticals like naproxen, salicylic acid, propranolol, caffeine, and 17α-ethinylestradiol at a loading of 11.2, 9.49, 3.37, 7.21, and 48.1 μg/kg, respectively. Compounds that are amphipathic with sizeable linear carbon chains and polar moieties have a higher affinity towards sediments (Wilkinson et al. 2017). Daliao river in China, which receives nearly 2074 million tons of industrial and domestic wastes, was investigated for organochlorine pesticides (OCPs). Nineteen OCPs were detected with a concentration of 3.7–30.1 ng/L in the aqueous phase, 157–493 ng/L in bore water, and 2.1–21.3 ng/g in sediments (Tan et al. 2009). Some of the chemicals like ibuprofen and triclosan derives more toxic products like 4-isobutylacetophenone and 2,8-dichlorodibenzo-p-dioxin, respectively when subjected to solar radiation or reacted by photosensitized species (Latch et al. 2003; Ruggeri et al. 2013). Direct photochemical transformation involves non-reversible cleavage of bonds or rearrangement of organic compound structure. Molecules with functional groups like $-CH{=}CH_2$, $-NO_2$, $-NH_2$, $-N{=}N-$, and $-OH$ are prone to photochemical transformation by donating π-electrons, which determines the adsorptive capacity of sunlight (Wilkinson et al. 2017). On the other side, chromophoric compounds produce reactive photo-oxidant radicals under solar light and promote indirect photochemical transformation. Factors such as the mechanism of degradation (direct or indirect), chemical nature of the compound, pH, temperature, time of day, water depth, cloud coverage, and topography will vary the process efficiency (Remucal 2014). Moreover, the fate of ECs and its transformation mechanism in various matrices are relatively unresolved. They require a detailed study to suggest necessary remediation measures.

11.3 Environmental and Human Health Effects of ECs

Even at minuscule concentrations, several ECs pose a severe threat to the environment and human life. However, adequate epidemic information, human health exposure, and threats to ecology and health are not well-documented. Pharmaceuticals and personal care products (PPCP), endocrine-disrupting compounds (EDC), and pesticides became inevitable in human life, making them ubiquitous in water bodies. Constant exposure to antibiotics gives rise to antibiotic-resistant bacteria making infection control troublesome. It is reported that streptomycin, penicillin, and tetracycline are responsible for antibiotic-resistant bacteria (Chudobova et al. 2014). Numerous studies have focused on the acute toxicity of human drugs on aquatic life. For instance, propranolol shows substantial toxicity towards zooplankton, silvestre, and benthos, whose EC_{50} is 3.17 mg/L (Damasceno de Oliveira et al. 2018). The most widely used 17α-ethinylestradiol (EE2) contraceptive arrested male gonad development in zebrafish at an environmental concentration of 2 ng/L (Larsen et al. 2008). Current research on carcinogenic perfluorooctanoic acid (PFOA) and perfluorooctane sulfonate (PFOS) proved their presence in soil, sea, wildlife, human serum, and even in breast milk. Intake of perfluorinated compounds by animals induces benign liver adenomas, pancreatic adenocarcinoma, and Leydig cell adenoma (Bonefeld-Jorgensen et al. 2011; Saikat et al. 2013). Coperchini et al. (2017) observed that the PFOS of 2.72–12.81 ng/mL was detected in patients who underwent autopsy thyroid operation. Another study says that the presence of PFC in lipid and cholesterol layers resulted in chronic kidney disease and liver malfunctions (Shankar et al. 2011; Gleason et al. 2015). Pesticides like ECs enter into the water through agricultural runoff or spillages. A lake in Florida contaminated with an excessive spill of dicofol and DDT from surrounding agriculture fields, and sewage systems exhibited abnormal ovarian morphology in female alligators (Lind et al. 2004). DDT is lipophobic and is quickly absorbed into fish and animal tissues. Humans who occupy the top place in the food chain would consume more pesticides. The significant effects include DNA damage, endocrine system disruption, reduced immune, and reproduction failures. EDCs can interfere with hormonal actions, abnormal thyroid functioning, and decreased fertility in birds, fishes, and mammals (Petrović et al. 2001; Balabanič et al. 2011; Careghini et al. 2015). In order to get clearer views of health impacts on the environment that is contaminated with ECs, extensive research is required.

11.4 The Need for Advanced Oxidation Processes

A conventional municipal wastewater treatment plant typically consists of primary and secondary treatment units, followed by a disinfection unit to remove organic and inorganic solids, dissolved organic compounds, and microorganisms. The secondary treatment works based on the biological process to remove carbonaceous and nitrogenous compounds (Metcalf et al. 2004). Since most of the ECs are recalcitrant, the traditional suspended, membrane, and attached growth biological reactors are not reliable for the complete removal of ECs at normal operating conditions (Gabet-Giraud et al. 2010; Alvarino et al. 2014, 2016). In membrane bioreactor, 60% removal of sulfamethoxazole compounds is obtained after the solid retention time of 33 days (Göbel et al. 2007). The biological degradation of ECs in wastewater treatment plants may be enhanced by combining anaerobic and aerobic treatment processes. Falås and his co-workers' extensive study on 15 diverse biological reactors summarized that many of the ECs still shown stability towards biological degradation (Falås et al. 2016). Hence the tertiary treatment is sometimes required to improve the quality of treated effluent and to meet the discharge standards. Table 11.1 shows the performance of various conventional treatment techniques tested for the removal of ECs in wastewater. It infers that the secondary treatment is not enough to remove ECs.

Though reverse osmosis and nanofiltration remove some ECs, the operating cost is high compared to the traditional process (Yangali-Quintanilla et al. 2011; Lin et al. 2016). In this context, AOPs seems to be a reliable choice for completely mineralize organic ECs from the aqueous medium. The process is also beneficial for inhibiting the disease-causing germs and enhancing the biodegradability of recalcitrant organic compounds (Kanakaraju et al. 2018). Figure 11.1 shows the significant functions of AOPs (Pham et al. 2020). The process produces reactive oxygen species (ROS) such as hydroxyl radical (OH·), superoxide ($O_2{\cdot}^-$), peroxide (OOH·), alkoxyl (OR·), peroxyl (OOR·) and nitrogen species (e.g., $N_3{\cdot}$, NH_4^+, NO_3^-) to oxidize OMs (An et al. 2010; Rivera-Utrilla et al. 2013; Lyu et al. 2017). Among ROS, OH· has the highest oxidizing capacity nearer to that of fluorine. Figure 11.2 shows the redox potential of various electronegative elements and ROS. Due to the radical nature of ROS, they rapidly interact with organic compounds in a non-selective manner (Sharma et al. 2011).

The interaction of OH· with organic compound (RH) follows hydrogen abstraction, hydroxyl reaction, and electron transfer, as shown in Eqs. (11.1) and (11.2) (Legrini et al. 1993).

$$RH + OH^{\cdot} \rightarrow H_2O + R^{\cdot} \tag{11.1}$$

$$R^{\cdot} + O_2 \rightarrow ROO^{\cdot} \tag{11.2}$$

If the target compound is aromatic, then the ring structure will be dissociated, resulting in a non-cyclic compound, which makes further degradation easier. The complete degradation by ROS can be achieved through radical–radical interactions, electron transfers, and hydrogen abstractions depending on the compound affinity towards radicals (Nakata and Fujishima 2012). The dependency of the generation of ROS varies with the energy supplied, the chemicals used, and the pollutant nature. The brief descriptions of various AOPs, including the Fenton process, ozonation, photocatalysis, plasma treatment, and sonolysis are discussed below.

11.5 Various Advanced Oxidation Processes

11.5.1 Fenton and Photo-Fenton Processes

In the nineteenth century, H. J. H. Fenton discovered that the oxidation of organic acids is enhanced in iron salts' presence. He observed that the ferrous ion could activate hydrogen peroxide (H_2O_2) and improve the degradation of tartaric, malic, glycolic, lactic, succinic, benzoic, and glyceric acids (Fenton and Jones 1900). Since then, the Fenton process is employed widely for the degradation of complex organic compounds (Chamarro et al. 2001). In this process, oxidation of ferrous ion by H_2O_2 yields OH radicals, as shown in Fig. 11.3 (Lipczynska-Kochany 1991). The excess

Table 11.1 Removal of ECs by conventional treatment techniques

EC class	Compound	Treatment technique	Removal (%)	Reference
Pesticides	Diuron	Activated sludge process	65–90	(Stasinakis et al. 2009)
	Chlorpyrifos	Membrane bioreactor	83–90	(Ghoshdastidar et al. 2012)
	Malathion			
	Lindane	Microalgae treatment	50	(Matamoros et al. 2016)
	Chlorpyrifos			
	Triazole	Wetlands	55–80	(Vymazal and Březinová 2015)
	Urea			
	Organophosphates			
	2,4-Dichlorophenoxyaceti acid	Adsorption	70–80	(Gani and Kazmi 2017)
Pharmaceuticals	Penicillin	Activated sludge process	70–90	(Gadipelly et al. 2014)
	Amoxicillin	Membrane bioreactor	90	Gadipelly et al. 2014)
	Acetaminophen	Wetlands	20–50	(Li et al. 2014)
	Diclofenac			
	Ibuprofen			
	Carbamazepine	Stabilization ponds	6–20	(Gruchlik et al. 2018)
	Propranolol			
	Atenolol	Adsorption	89–95	(Beijer et al. 2017)
	Diclofenac			
EDCs	Oestrone	Activated sludge process	40–90	(Ifelebuegu 2011)
	17-beta estradiol			
	17 alfa-ethinylestradiol			
	Bisphenol A(BPA)	Membrane bioreactor	93	(Zhu and Li 2013)
	BPA	Wetlands	62–73	(Toro-Vélez et al. 2016)
	Nonylphenol			
	BPA	Stabilization ponds	24–70	(Gruchlik et al. 2018)
	Octylphenol			
	Triclosan	Adsorption	48–93	(Solak et al. 2014)
	Estrogens			
PPCPs	Homosalate	Activated sludge process	20–90	(Yang et al. 2017)

(continued)

Table 11.1 (continued)

EC class	Compound	Treatment technique	Removal (%)	Reference
	DEET (N, N-Diethyl-meta-toluamide)			
	Ethylhexylsalicilate			
	Caffeine	Membrane bioreactor	34–99	(Kim et al. 2014)
	Triclosan			
	Caffeine	Wetlands	40–80	(Reyes-Contreras et al. 2012)
	Ketoprofen			
	Naproxen			
	Triclosan	Stabilization ponds	85–90	(Gruchlik et al. 2018)
	Oxybenzone			
	Chlorophene	Adsorption	13–37	(Reyes-Contreras et al. 2012)
	3-Methyllindole			
	1-H-Benzotriazole			

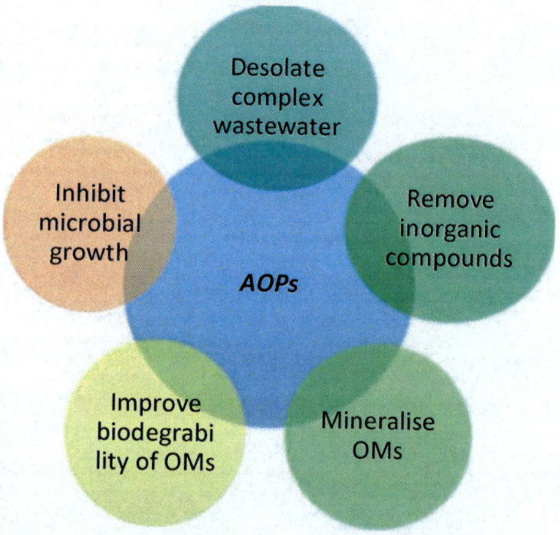

Fig. 11.1 Different roles of the AOPs

H_2O_2 reacts with ferric ion and forms peroxide radical. These radicals are responsible for the degradation of organic molecules. The Eqs. (11.3)–(11.11) show possible reactions that can occur upon H_2O_2 consumption (Huang 2013; Babuponnusami and Muthukumar 2014).

$$Fe^{2+} + H_2O_2 \rightarrow Fe^{3+} + OH^{\cdot} + OH^{-} \tag{11.3}$$

$$Fe^{3+} + H_2O_2 \rightarrow Fe^{2+} + OOH^{\cdot} + H^{+} \tag{11.4}$$

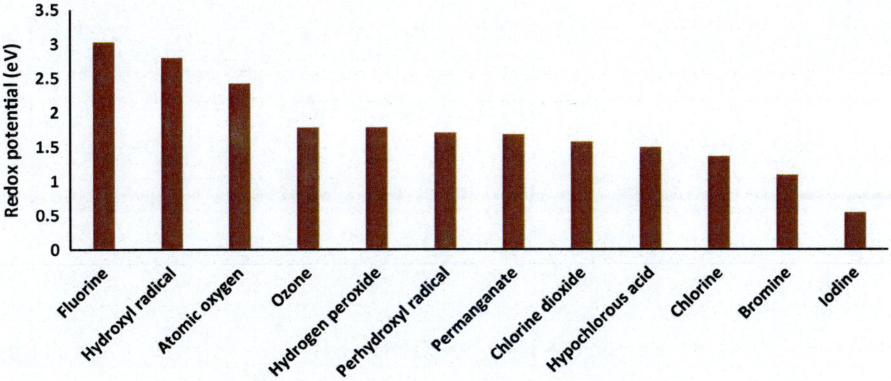

Fig. 11.2 The redox potentials of various chemical species. Adapted from Litter (2005)

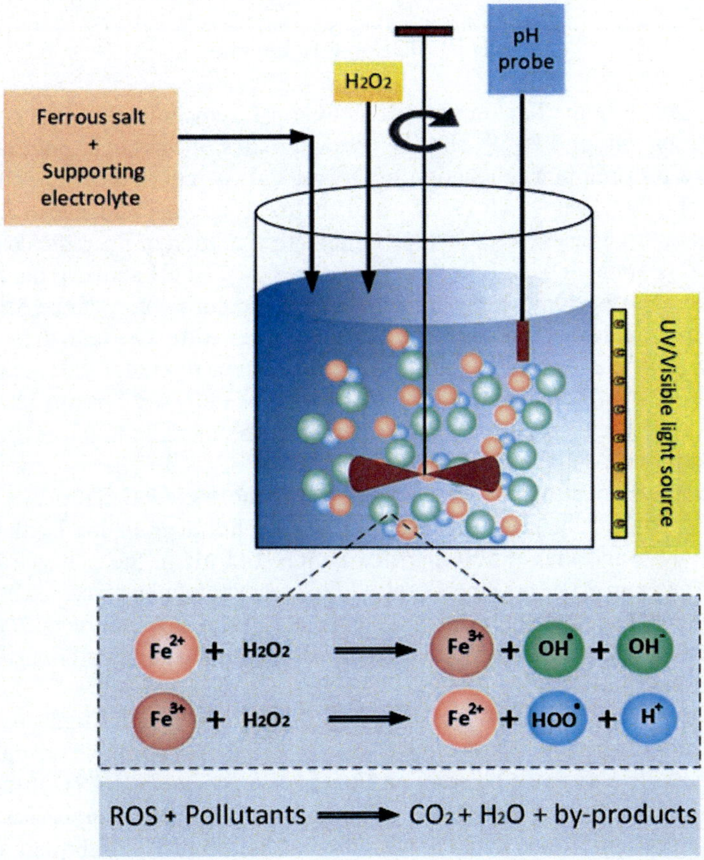

Fig. 11.3 Diagrammatic representation of Fenton and photo-Fenton processes for the removal of pollutants in water

$$Fe^{2+} + OH^{\cdot} \rightarrow Fe^{3+} + OH^{-} \tag{11.5}$$

$$Fe^{2+} + OOH^{\cdot} \rightarrow Fe^{3+} + HOO^{-} \tag{11.6}$$

$$Fe^{3+} + OOH^{\cdot} \rightarrow Fe^{2+} + O_2 + H^{+} \tag{11.7}$$

$$OH^{\cdot} + OH^{\cdot} \rightarrow H_2O \tag{11.8}$$

$$OH^{\cdot} + H_2O_2 \rightarrow OOH^{\cdot} + H_2O \tag{11.9}$$

$$OOH^{\cdot} + OOH^{\cdot} \rightarrow H_2O_2 + O_2 \tag{11.10}$$

$$OH^{\cdot} + OOH^{\cdot} \rightarrow H_2O + O_2 \tag{11.11}$$

Factors affecting the Fenton process include pH, temperature, initial concentration of ferrous ions and H_2O_2, and the characteristics of the target pollutants. The pH required for optimum production of OH radicals is reported to be around pH 3 (Leung et al. 1992; Zepp et al. 1992). At pH below 2, there is a chance of formation of stable oxonium ions $(H_3O_2)^+$, which inhibits the creation of the radicals by stabilizing H_2O_2 (Babuponnusami and Muthukumar 2014). At pH above 4, the hydroxyl radical formation is significantly reduced due to the conversion of the ferrous ion into ferric hydroxide precipitates in the presence of peroxide. The rate of degradation also depends on the loading of the ferrous ion. Chamarro et al. (2001) studied the degradation of 4-chlorophenol (4-CP) as function of Fe^{2+}/ 4-CP ratios (1, 0.1, and 0.01) and initial 4-CP concentration of 300 mg/L. The increase in the concentration of Fe^{2+} has significantly increased the degradation of 4-CP. The amount of H_2O_2 present in the system also plays a significant role in the degradation process. Lofrano et al. (2009) studied the effect of H_2O_2 in the catechol degradation by the Fenton process. They found that at 500 mg/L of $FeSO_4$ and pH 3, the increase in H_2O_2 concentration from 75 to 600 mg/L, raised the removal efficiency of catechol from nearly 75–93%. Complete removal is achieved at H_2O_2 concentration of 2000 mg/L for the same conditions (Lofrano et al. 2009). However, the addition of excess H_2O_2 may increase capital costs.

Fenton process is modified by combining it with UV or solar light to enhance the degradation. The process is shown in Fig. 11.3. The principle of the photo-Fenton process is elucidated by reaction Eq. (11.12). Saien et al. (2011) studied the degradation of direct red dye (DR16) by Fenton and photo-Fenton processes. Fenton process has been improved by 14% with the irradiation of UV light and achieved aromatic ring degradation of 68%. It is summarized that 85–99% of discoloration of DR16 of 30 mg/L initial concentration is obtained at optimum conditions after 30 min of reaction time.

$$FeOH^{2+} + h\upsilon \rightarrow Fe^{2+} + OH^{\cdot} \quad (11.12)$$

The formation of ferrous hydroxide precipitate at higher pH is still a worry in the photo-Fenton process. The photo-Fenton process is modified by adding organic ligands such as ethylenediaminetetraacetic acid (EDTA), ethylenediaminediacetic acid (EDDA), and citric acid in water to overcome the limitation. The organic ligand increases the radical production rate by stabilizing iron and decreases precipitate formation even at a high pH range (Litter 2005; Clarizia et al. 2017). Pereira et al. (2014) used oxalic acid as a ligand source to form ferricarboxylate complexes. They have shown more than 98% degradation of oxytetracycline and 32% TOC removal at pH 6. Similar studies are done by Trovó and Nogueira (2011) by replacing synthetic ligands like EDTA with safe and natural compounds such as citrates. They conducted the reaction in natural light to degrade diclofenac (DCF). At pH 7, 62% of DCF oxidation is achieved with an initial concentration of 33.4 mg/L in 10 min of light irradiation. Though the modified process is superior to the conventional photo-Fenton process, the high chemical requirement and sludge generation need attention. Heterogeneous catalysts like iron oxides prevent the sludge generation in Fenton processes by stable iron species within the structure. Huang (2013) reported 80.8% degradation of bisphenol A (BPA) with goethite, H_2O_2, and ethylenediamine-N, N'-disuccinic acid (EDDS) under UV light for 9 h at pH 6.2. At the end of 9 h, the leached iron is nearly 5 μmol/L.

11.5.2 Ozonation

Ozonation is a unit process that involves the use of triatomic oxygen (ozone, O_3) for the degradation of pollutants. On an industrial scale, it is generally produced by the excitation of the oxygen molecule in a high electric field (Miichi et al. 2002; Chen et al. 2015). The ozone gas can also be produced by other methods, including UV irradiation (Zoschke et al. 2014). The nascent oxygen generated in the process combines with an O_2 and form the O_3, a powerful oxidizing agent. It is prepared on-site and pumped directly to the treatment plant due to its poor stability. The typical ozone treatment plant consists of an ozone generator and a reactor setup to bubble ozone into wastewater (Miichi et al. 2002). Two reactor columns in series are operated in down-flow and up-flow modes, as shown in Fig. 11.4 (Huber et al. 2005).

The process is widely used in water and wastewater treatment plants for disinfection and also as for the pre-treatment and post-treatment of recalcitrant compounds. The method is superior to chlorination in terms of disinfection power and does not generate any halothanes (Fiessinger et al. 1981). The interaction of ozone with ECs is direct or through the formation of secondary oxidants. Equations (11.13) and (11.14) shows the formation of hydroxyl radicals when ozone interacts with water. It also forms hydrogen peroxides, as demonstrated in Eq. (11.15).

$$O_3 \rightarrow O^{\cdot} + O_2 \quad (11.13)$$

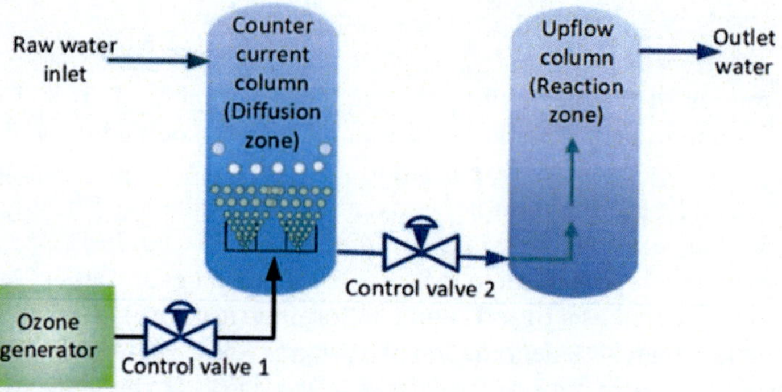

Fig. 11.4 Process flow diagram of ozonation in water treatment plant

$$O^{\cdot} + H_2O_{(v)} \rightarrow 2OH^{\cdot} \tag{11.14}$$

$$O^{\cdot} + H_2O_{(l)} \rightarrow H_2O_2 \tag{11.15}$$

Factors affecting the efficacy of the ozonation are pH, ozone loading, and the presence of natural organic matter (NOM). The formation of OH radicals is favoured in alkaline pH. Molecular ozone selectively reacts with organic molecules having nucleophilic moieties such as carbon–carbon double bonds, aromatic rings, and the functional groups bearing phosphorus, sulphur, nitrogen, and oxygen atoms (Legrini et al. 1993). Whereas OH radicals are non-selective and the degradation mechanism involves hydrogen abstraction, radical–radical reactions, electrophilic addition, and the electron transfer reactions (Legrini et al. 1993). Presence of unsaturated and aromatic hydrocarbons with $-OH$, $-CH_2$ and $-NH_2$ groups react with OH radicals to form organic radicals, which upon addition of O_2 converts to superoxide anion (O_2^-). The formed O_2^- undergoes electron transition with ozone, decomposes, and forms again OH radicals. On the other side, the alkyl group, eg., t-butyl group, and bicarbonates scavenge OH radicals and terminate the radical formation by forming their respective radicals (Staehelin and Hoigne 1985).

It is reported that ozonation is effective against compounds like bisphenol A, nonylphenol, 17α-ethinylestradiol, 17β-estradiol, and estriol. The process showed 100% removal efficiency at an initial concentration of 50 μg/L (Ahmed et al. 2017). Zhao et al. (2017) reported complete degradation of indomethacin in 7 min with 2 mg/L ozone dosage. By increasing ozone loading to 35 mg/L, 50% of TOC is removed. Studies are also reported that the complete mineralization of ECs is not achieved through the ozonation process alone (Cai and Lin 2016). Tay and Madehi (2015) observed incomplete mineralization of ofloxacin with ozone and the formation of by-products. The irradiation of UV light or the presence of H_2O_2 shows the complete mineralization of the compound. Equation (11.16) shows the formation of ozone molecules upon the UV radiation (Beltrán et al. 1999).

$$3O_2 + h\vartheta \leftrightarrow 2O_3 \tag{11.16}$$

The ozonation is superior to the Fenton process in terms of chemical requirements and sludge production. For instance, a study on the degradation of phenolic compounds shows that the estimated cost of the ozonation process is around 0.3–0.61 EUR/kg of pollutants. The cost of the Fenton process is double the value reported for ozonation. Also, it requires skilled labour for operation and maintenance. However, for faster degradation, the addition of Fenton's reagent is preferable (Krichevskaya et al. 2011). The possible formation of carcinogenic by-products (in the case of brominated compounds present in wastewater), potential fire hazards, and ozone toxicity are of concerns (Deng 2020). Zeng et al. (2013) studied the combination of ozonation and Fenton processes for the degradation of phenol in a rotating packed bed reactor. The degradation rate in ozonised-Fenton is 20% higher than that of the ozone process and reached 98.3% upon the addition of H_2O_2 (1.6 mM). Parameters like rotational speed, inlet ozone, and the H_2O_2 concentration increases the degradation rate, and the optimum range of pH is 5–6.8 with Fe (II) concentration of 0.1 mM.

11.5.3 Photolysis

Several ECs are photosensitive and undergo degradation upon exposure to light (Minella et al. 2016). Photolysis is the natural process of degradation of photoactive ECs by light, as shown in Fig. 11.5. The addition of ozone and hydrogen peroxide can enhance the process performance through the generation of hydroxyl radicals, as shown in chemical reactions Eqs. (11.17)–(11.19).

$$O_3 + h\vartheta + H_2O \rightarrow H_2O_2 + O_2 \tag{11.17}$$

$$H_2O_2 + h\vartheta \rightarrow 2OH^{\cdot} \tag{11.18}$$

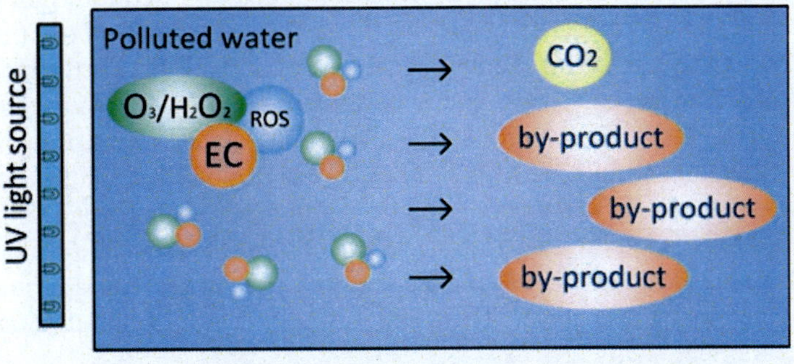

Fig. 11.5 Schematic representation of photolytic degradation of the pollutants in water

$$OH^{\cdot} + EC \rightarrow \text{Degradation products} \tag{11.19}$$

Water quality parameters, including the concentration of pollutants, organic matter, pH, and the intensity of light, influence the process. The photoreactivity shows higher at a lower concentration of contaminants and near-neutral pH (Jacobs et al. 2011; Babić et al. 2013). The photodegradation experiments of ECs also showed that that the half-lives of most of the compounds depend on the light intensity and their absorbance in the sunlight radiation. The direct photolysis of photolabile compounds such as ketoprofen, diclofenac, triclosan, and naproxen can lead to the generation of toxic oxidative compounds (Schweitzer and Noblet 2018). This process typically generates by-products like carbazole derivatives and benzophenones, depending on the pollutants degraded even after longer irradiation times (Brigante et al. 2005; Koumaki et al. 2015).

Trapido et al. (1997) compared ozonation and photolysis in treating chlorophenols. Ozonation showed fewer degradation by-products when compared to photolysis. However, ozonation shows higher degradation at higher pH, and UV treatment shows higher degradation at lower pH. It is also inferred that toxic products are generated in photolysis, even with the addition of H_2O_2 and ozone.

11.5.4 Plasma Treatment

In Greek, plasma means "mouldable substance" or "jelly" and is one of the four fundamental states of matter. In 1928, Irving Langmuir, a chemist, coined this word to describe an ionized gas (Tonks and Langmuir 1929). By definition, plasma is a state of matter that exists in the form of electrons and ions. The ionized gas is obtained from the exposure of the material to a high electric field, magnetic field, or thermal energy. Depending on the temperature and electron density, the plasmas are classified as thermal or equilibrium plasma and non-thermal plasma or non-equilibrium plasma (Conrads and Schmidt 2000). In thermal plasma, the temperature of the electron and the bulk gas molecule are the same. They are in thermodynamic equilibrium with each other. Examples of thermal plasma are arc discharges, torches, and radiofrequency plasma. The high flux of heat generation enables thermal plasma for the incineration of solid waste (Tang et al. 2013). In the case of non-thermal plasma, the electron temperature is much higher than that of bulk gas molecules. Thus, they are not in thermodynamic equilibrium with each other. Types of non-thermal plasma include corona discharge, dielectric barrier discharge, gliding arc discharge, glow discharge, and spark discharge (refer to Fig. 11.6). Thermal plasma is associated with sufficient energy to be in thermal equilibrium, hence requires high power. Non-thermal plasma can be obtained using less power. Therefore, the application of non-thermal plasma is more economical (Jiang et al. 2014).

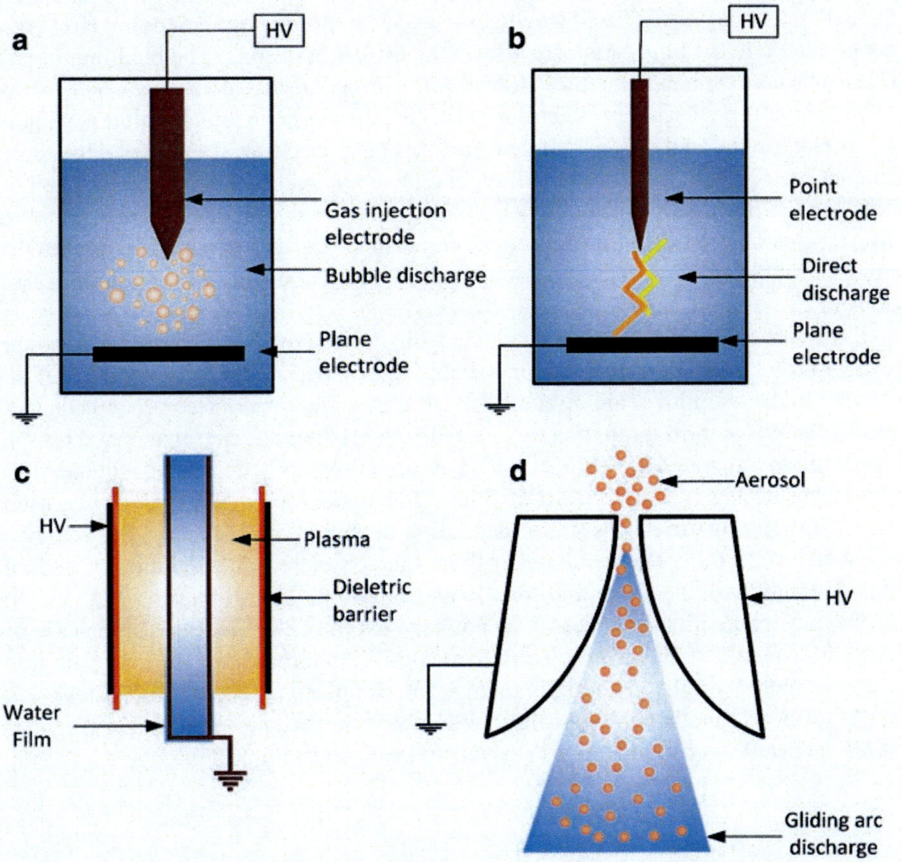

Fig. 11.6 Schematic representation of non-thermal plasma reactors like A) Bubble discharge, B) Direct discharge, C) Dielectric discharge, and D) Gliding arc discharge reported for the water treatment. Adapted with permission from Perinban et al. (2019)

In non-thermal plasma, the energetic electrons can collide with background molecules (N_2, O_2, H_2O, Ar) to produce secondary electrons, photons, and chemically reactive species such as OH·, H_2O_2, O_3, OOH·, O atom, and ions (O_2^-, O_2^+, H_3O^+, O_3^-) (Foster et al. 2012). Glow discharge, dielectric barrier discharge, and gliding arc plasma come under the non-thermal surface plasma discharge methods. In surface discharge, the liquid surface acts as one of the electrodes to generate plasma. It involves the breakdown of gas molecules at the liquid surface and diffuses the formed ions into liquid. Hence the plasma produced is the source of charged particles, excited species, shock waves, ultrasound radicals, and UV radiation. The plasma-produced ions and radicals have a short life span and are harmful. The produced species and their diffusion influence the chemical reactivity of plasma.

In direct electric discharge, plasma is produced by direct injection of high pulse voltage across the submerged electrodes. Due to the gas–liquid density variations, the generated ions are diffused slowly into the surrounding liquid medium. However, the conductivity of liquid majorly affects the diffusion of ions. The bubbling of gas is introduced to enhance the mass transfer in a direct electric discharge.

A bubble-type discharge is similar to surface discharge in which liquid itself acts as an electrolyte, and gas is bubbled into it. These discharged bubbles have more contact surface with the liquid medium and hence increases the efficiency. The UV light formed during the discharge helps in degrading compounds via photolysis, and also creates OH radicals through dissociation of H_2O_2 and ozone. This process is superior to the surface discharge due to less energy consumption and more mass transfer due to bubbling (Jiang et al. 2014).

Extensive studies are being conducted on the removal of ECs in wastewater using non-thermal plasma treatment (Jiang et al. 2012, 2014). Tichonovas et al. (2013) conducted pilot-scale research for the application of semi-continuous dielectric discharge for the degradation of 13 textile dyes. However, energy demand is high compared to other AOPs. Ten out of 13 dyes are completely degraded, and the toxicity of wastewater is near zero within 5 min. In another study, Yan et al. (2005) used gas–liquid gliding arc discharge for degrading phenol and achieved 91.8% removal efficiency in 38 min. The results also show the complete mineralization of phenol obtained only with the air supply for a longer duration.

The plasma treatment process does not need external chemicals, as in the case of conventional AOPs. However, electrode erosion, less throughput capacity, and less process volume are the constraints involved in the plasma generation (Foster et al. 2012). Though this technology offers high removal efficiency, coupling with other AOPs reduces the cost and allows scalability (Vanraes et al. 2015).

11.5.5 Sonolysis

Sonolysis is a process that describes the degradation of pollutants using ultrasound waves. Interaction of a liquid medium with the ultrasound waves generates kinetic and thermal energy, which induces fine bubbles and temperature changes in the liquid. Ultrasound waves propagate in water either as continuous or pulsed waves to form cavitation bubbles. The formed bubbles will oscillate with compression and rarefaction until they reach a critical size, as shown in Fig. 11.7. Localized hotspots induce the critical-sized bubbles to implode, releasing enormous energy into the solution. These hotspots provide sufficient heat for dissociation of chemical bonds in pollutants through thermal decomposition and hydrolyze water to form radicals (OH·) to degrade the pollutant (Hua and Hoffmann 1996; Xiao et al. 2014; Sasi et al. 2015).

Degradation rate in sonolysis is dependent on the initial concentration of pollutants, sonication frequency, and the temperature (Güyer and Ince 2011; Golash and Gogate 2012; Nejumal et al. 2014). This process requires excessive chemical input for the enhancement of degradation. For example, the addition of sodium chloride, ruthenium, and platinum catalysts increased the acetic acid degradation percentage

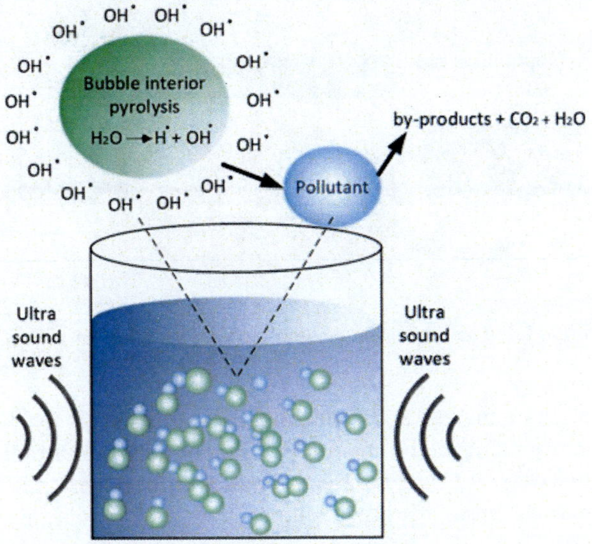

Fig. 11.7 Schematic representation of sonolysis process for the degradation of pollutants. Adapted with modifications from Mishra et al. (2017)

from 10 to 100 at 473 K with 0.6–1 g of catalyst in 200 mL of solution in a pressurized reactor. (Findik and Gündüz 2007). The presence of bromide ion increased radical production during the degradation of 4-cumyl phenol. Degradation of 100% is achieved in 20 min at a temperature of 293 K, power of 80 W, and pH 7.6. The presence of bicarbonates in natural water improved the degradation at a low concentration of 0.05 mg/L of 4-cumyl phenol (Chiha et al. 2011). The process shall be hyphenated with other methods to achieve complete degradation. Peller et al. (2003) conducted a study on the degradation of 2,4,6 trichlorophenol by coupling of sonolysis and TiO_2 photocatalysis. The study revealed that the total degradation time with sonolysis alone is 12 min, and TiO_2 photocatalysis alone is 18 min. However, combined treatment degraded the compound in 8 min. Similar studies are reported with AOPs like ozonation (Weavers et al. 1998), membrane reactor, or concurrent systems (Secondes et al. 2014).

11.5.6 Photocatalysis

According to the IUPAC definition, photocatalysis is a catalytic reaction involving light absorption by a catalyst or substrate (Agustina et al. 2005). A photocatalyst (PC) in the presence of oxygen/air, and a light source, as shown in Fig. 11.8, will generate ROS, and ROS will degrade the organic compounds. Typically, semiconducting metal oxides are used as PCs due to their distinct low bandgap. When the incident light with energy higher than that of bandgap energy is applied, electrons excite from valency band (VB) to conduction band (CB). Positive holes are formed in VB in place of excited electrons, as given in Eq. (11.20). The formed electron–hole pairs

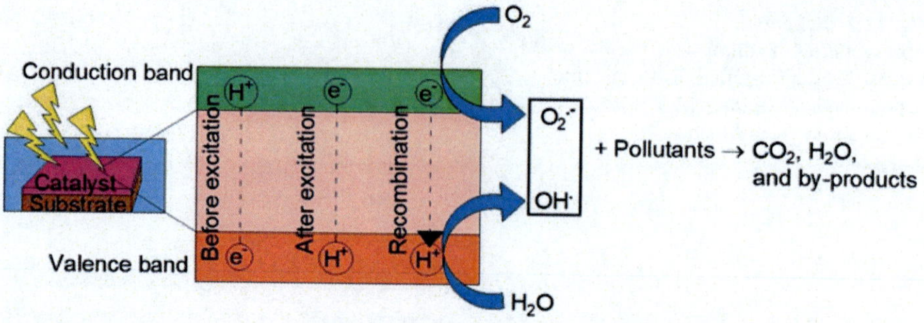

Fig. 11.8 The schematic diagram showing the mechanism of photocatalysis

act as reducing and oxidizing agents, respectively. Redox reactions happen with surrounding liquid medium to generate hydroxyl radical and superoxide, as given in Eqs. (11.21) and (11.22). Thus, formed superoxide further converts into peroxyl radical through protonation, as given in Eqs. (11.23) and (11.24) Fujishima et al. (2017).

$$PC_s + h\vartheta \rightarrow e_s^- + h_s^+ \tag{11.20}$$

$$OH^- + h^+ \rightarrow OH^{\cdot} \tag{11.21}$$

$$O_2 + e^- \rightarrow O_{2.}^- \tag{11.22}$$

$$H_2O \rightarrow H^+ + OH^- \tag{11.23}$$

$$O_2^- + H^+ \rightarrow HOO^{\cdot} \tag{11.24}$$

The radical production can be decreased by 'recombination', which is a phenomenon of electrons re-occupying the VB from CB. The peroxyl radicals, which act as electron acceptors, prevent the recombination rate by scavenging photo-induced electrons and generates H_2O_2, as shown in Eqs. (11.25) and (11.26). Thus the photocatalytic reaction's efficiency is enhanced by the decrease of recombination and the production of hydrogen peroxide (Alfano et al. 1997).

$$HOO^{\cdot} + e^- \rightarrow HOO^- \tag{11.25}$$

$$HOO^- + H^+ \rightarrow H_2O_2 \tag{11.26}$$

The degradation pathways are based on the formation of the radicals and the nature of pollutants. The presence of functional groups like alkyl (Huang et al. 2020), nitro

(Di Paola et al. 2003), or sulfate (Ahmed et al. 2010) can form radicals other than ROS, which may influence the mineralization of the target analyte.

Titanium dioxide (TiO_2) is a well-known photocatalyst because of its stability, longevity, high UV light absorption capacity, abundant availability, and ease of large-scale production (Reck and Richards 1999). In nature, TiO_2 exists in three crystalline forms as anatase, rutile, and brookite. Brookite is a rare and unstable form. Commercially, TiO_2 is available as Degussa P25 with 25% rutile and 75% anatase crystalline phases, which is a benchmark form of TiO_2. However, TiO_2 is not active in the visible region of light. Sivagami (2011) studied the degradation of monocrotophos (MCP) at an initial concentration range of 10–23.4 mg/L using TiO_2 in batch reactors and the pH range of 3.6–10.4. At optimized conditions (15 mg/L MCP, 4 g/L TiO_2), 78% of degradation is achieved. The authors inferred that degradation is influenced by the initial concentration of pesticides and the pH of the solution. Zinc oxide (ZnO) is another example of UV light active photocatalyst with strong oxidizing capacity and good photocatalytic properties. It is cheaper than TiO_2 and also known for its application in the removal of ECs in water. Daneshvar et al. (2007) reported the degradation of diazinon using ZnO. They considered the initial concentration of 20 mg/L diazinon solution and loading of ZnO as 150 mg/L. After UV irradiation of 80 min, 80% of degradation is achieved.

The visible light photoactivity of TiO_2 and ZnO can be achieved by bandgap modification. The bandgap can be tuned by doping/co-doping with metals and non-metals. Besides enhancing the activity in visible light, the doping technique rectifies constraints like recombination and deactivation of photocatalysts, and accomplishes higher process productivity. Generally, doping of metal ion helps in decreasing the bandgap by creating additional energy levels between the VB and CB of a primary photocatalyst. Numerous studies are reported using metallic doping agents such as iron, cobalt, nickel, and copper in ZnO (Sharma and Jha 2017; Basri et al. 2018; El Haimeur et al. 2018; Shakil et al. 2018), and molybdenum, chromium, lanthanum, and cerium in TiO_2 for visible light activity (Shayegan et al. 2018; Vieira et al. 2019; Zaki and Lee 2019). Non-metal doping agents like carbon (C), nitrogen (N), sulphur (S), and fluorine (F) are also used to modify the bandgap and improve the visible light activity of the PC. The dopant will replace the oxygen atom in the lattice of semiconductor, forming crystal defects (Wang et al. 2017). The crystal defects act as localized centres for electrons, and reduces the recombination rate. Table 11.2 gives band energies of different pure and doped photocatalysts and their excitation wavelengths (Papoulis 2019; Shayegan et al. 2019; Ahmad et al. 2020).

However, these modifications are dependent on the material synthesis route, which significantly affects the production cost. Among various approaches, sol-gel is the commonly used synthesis technique. The process involves the conversion of a sol, which is a solid particle dispersion in a liquid medium to gel, a three-dimensional solid with liquid inside. Metal alkoxides are typically used, precursors. Precursor solution with or without doping agents undergoes hydrolysis, condensation, drying, and thermal decomposition during the synthesis. The final compound has strong covalent bonds between the dopant element and the precursor of the semiconductor. Numerous studies used the sol-gel method for doping of metals and non-metals to

Table 11.2 Different photocatalysts and their excitation wavelength and band energy

Photocatalyst	Amount of dopant (wt%)	Band energy (eV)	Excitation wavelength (nm)	Reference
TiO_2	–	3.2	388	(Fujishima et al. 2017)
ZnO	–	3.3	376	(Hong et al. 2006)
CdS	–	2.4	517	(Gangu et al. 2019)
ZrO_2	–	5.0	248	(Liu et al. 2010)
ZnS	–	3.6	345	(Lee and Wu 2017)
Ru—doped TiO_2	1	2.9	428	(Barrocas et al. 2019)
Co—doped TiO_2	1	2.43	511	(Osawa et al. 2020)
Fe—doped TiO_2	1	3.42	363	(Osawa et al. 2020)
Ir—doped TiO_2	0.5	3.14	395	(Shayegan et al. 2018)
Pt—doped TiO_2	0.3–1.6	2.98–2.82	416–440	(Nishiyama and Yamazaki 2017)
C—doped TiO_2	2.8	2.74	453	(Huang et al. 2008)
S—doped TiO_2	1.5	2.65	468	(Wang et al. 2017)
N—doped TiO_2	4.64	3.01	412	(Liu et al. 2016)
B—doped TiO_2	5	3.11	399	(Shayegan et al. 2018)
F—doped TiO_2	0.3	3.09	402	(Khalilzadeh and Fatemi 2016)

prepare visible light active TiO_2 and ZnO photocatalysts (Samat and Nor 2013). Li and Li (2001) studied Au/Au^{3+} doped TiO_2 synthesized through a sol-gel process for the degradation of methylene blue (MB). They reported that the optimum molar percentage of gold doping/deposition on TiO_2 is 0.5. Abatement studies were done in a cylindrical Pyrex photoreactor, and 110 W sodium lamp was taken as a visible light source. Degradation studies were done taking 12 mg/L MB and 0.2 g of the photocatalyst in the pH range of 2.4–8.4. At a pH of 5.98, Au^{3+}–TiO_2 achieved high total organic carbon (TOC) removal of 73.6% as compared to that of 43.7% by simple TiO_2. Vaiano et al. (2015) studied MB degradation by continuous catalytic fixed-bed reactor under artificial and solar light. N-doped TiO_2 catalyst with N/Ti molar ratio of 18.6 is immobilized on glass spheres by dip-coating technique and followed by calcination. About 75 and 50% of 10 mg/L of MB is degraded under UV and visible light, respectively, in 4.4 h of reaction time.

Another well-known synthesis technique is a hydrothermal process that requires high temperature and water pressure. If the solvent used is non-water-based, the process is termed as solvothermal technique (Shen et al. 2015). Compared to the sol-gel technique, this technique requires controlled conditions that allow the generation of different morphologies like hollow TiO_2 particles and flower-like F—doped TiO_2 hollow microspheres (Pastrana-Martínez et al. 2013; Khalilzadeh and Fatemi 2016).

Even though doping enhances the photocatalyst performance, the major issue that remains to be addressed is photocatalyst fouling. The catalyst fouling can happen due to surface blockage, bulk blockage of photocatalyst, and various ions in the water (Katz et al. 2015). Some compounds like phosphate ions, chromates ions, and humic acid target the sites on the PC's surface, resulting in the blockage of active sites. The blocked sites will neither adsorb pollutants nor produce radicals. Certain divalent ions will form a bridge between the PC particles and agglomerates particles. This agglomeration can be addressed by the immobilization of PC (Pozzo et al. 2002). The bulk blockage occurs due to the interfering compounds like chlorides and carbonates, which scavenges the hydroxyl radicals. Sometimes the ionic strength also results in the fouling or activation of catalyst depending on the target compound. Aguedach et al. (2008) investigated the discoloration of reactive black dye with TiO_2/SiO_2 composite. The order of initial discoloration was reported as $Ca^{2+} > K^+ > Na^+ > Li^+$. The control strategies to address catalyst fouling include self-cleaning by photooxidation, feed pre-treatment, aeration, applying electric field, and desorption through heating (Zhang et al. 2016).

11.6 Conclusion

With the advancement in the state-of-the-art analytical equipment, our ability to detect and analyze pollutants in trace quantities has improved significantly. Recent findings have established varying concentrations of organic, inorganic, and biological contaminants in water bodies worldwide. The frequent occurrence of unconventional pollutants in water bodies has brought a paradigm shift in the criteria for characterizing wastewater effluent and drinking water. A new class of contaminants named 'emerging pollutants' is now the focus of several research studies across the globe. Though these pollutants are not commonly monitored in water, they possess known or suspected toxicity to human health and aquatic organisms. Most of the conventional wastewater treatment techniques are not competent enough to remove emerging contaminants in an aqueous medium due to their recalcitrant nature. In this context, advanced treatment technologies hold the key to the management of these pollutants. Studies on the risk assessment of ECs and their metabolites and the investigations on the degradation mechanism also need to be addressed to design and develop an efficient treatment process.

Advanced oxidation processes such as ozonation, Fenton oxidation, photolysis, plasma, sonolysis, photocatalysis, and their combination are known for their ability to detoxify or completely mineralize emerging contaminants in water. They are suitable for deactivating a broad spectrum of pathogenic organism as well. The AOP process primarily relies on the in-situ generation of reactive chemical species with high redox potential. Among the AOPs, plasma, ozonation, and sonolysis are efficient in degrading the pollutants. However, they are energy-intensive and expensive. Fenton, photo-Fenton, and photocatalysis processes are relatively affordable. Photo-Fenton and photocatalysis can be made more economically viable by using sunlight as the source of energy. However, catalyst fouling, post-separation of chemical compounds, and the effect of water matrix need to be addressed adequately to improve their prospects in the field-level applications.

In some cases, the intermediate products of degradation are biologically active and toxic than the parent compounds. The complete removal of pollutants in such a situation adds to the cost of the treatment. The hybrid AOPs or AOPs, in combination with traditional techniques, seems to be a viable and practical option in terms of achieving high degradation efficacy, operation feasibility, and economic viability.

Acknowledgements Authors gratefully acknowledge the Department of Science and Technology (DST), Technology Mission Division, Government of India [Grant no: DST/TM/WTI/WIC/2K17/82(C)]. The authors also acknowledge the support of IIT Tirupati in the preparation of the chapter.

References

Aguedach A, Brosillon S, Morvan J, Lhadi EK (2008) Influence of ionic strength in the adsorption and during photocatalysis of reactive black 5 azo dye on TiO2 coated on non woven paper with SiO2 as a binder. J Hazard Mater 150:250–256. https://doi.org/10.1016/j.jhazmat.2007.04.086

Agustina TE, Ang HM, Vareek VK (2005) A review of synergistic effect of photocatalysis and ozonation on wastewater treatment. J Photochem Photobiol C Photochem Rev 6:264–273. https://doi.org/10.1016/j.jphotochemrev.2005.12.003

Ahmad I, Shoaib Akhtar M, Ahmed E, Ahmad M, Keller V, Qamar Khan W, Khalid NR (2020) Rare earth co-doped ZnO photocatalysts: solution combustion synthesis and environmental applications. Sep Purif Technol 237:116328. https://doi.org/10.1016/j.seppur.2019.116328

Ahmed MB, Zhou JL, Ngo HH, Guo W, Thomaidis NS, Xu J (2017) Progress in the biological and chemical treatment technologies for emerging contaminant removal from wastewater: a critical review. J Hazard Mater 323:274–298. https://doi.org/10.1016/j.jhazmat.2016.04.045

Ahmed S, Rasul MG, Martens WN, Brown R, Hashib MA (2010) Heterogeneous photocatalytic degradation of phenols in wastewater: a review on current status and developments. Desalination 261:3–18. https://doi.org/10.1016/j.desal.2010.04.062

Ain Samat N, Md Nor R (2013) Sol-gel synthesis of zinc oxide nanoparticles using Citrus aurantifolia extracts. In: Ceramics International

Alfano OM, Cabrera MI, Cassano AE (1997) Photocatalytic reactions involving hydroxyl radical attack: I. Reaction kinetics formulation with explicit photon absorption effects. J Catal 172:370–379. https://doi.org/10.1006/jcat.1997.1858

Alvarino T, Suárez S, Garrido M, Lema JM, Omil F (2016) A UASB reactor coupled to a hybrid aerobic MBR as innovative plant configuration to enhance the removal of organic micropollutants. Chemosphere 144:452–458. https://doi.org/10.1016/j.chemosphere.2015.09.016

Alvarino T, Suarez S, Lema JM, Omil F (2014) Understanding the removal mechanisms of PPCPs and the influence of main technological parameters in anaerobic UASB and aerobic CAS reactors. J Hazard Mater 278:506–513. https://doi.org/10.1016/j.jhazmat.2014.06.031

An T, Yang H, Li G, Song W, Cooper WJ, Nie X (2010) Kinetics and mechanism of advanced oxidation processes (AOPs) in degradation of ciprofloxacin in water. Appl Catal B Environ 94:288–294. https://doi.org/10.1016/j.apcatb.2009.12.002

Ankley GT, Hoff DJ, Mount DR, Lazorchak RJ, Beaman J, Linton TK, Erickson RJ (2008) Aquatic life criteria for contaminants of emerging concern

Araya V, Maliyekkal SM, Philip L (2019) Water pollution and treatment technologies—Indian perspective. In: Mjumdar PP, Tiwari VM (eds) Water futures of India, Status of science and technology. Indian National Science Academy, pp 215–252

Babić S, Periša M, Škorić I (2013) Photolytic degradation of norfloxacin, enrofloxacin and ciprofloxacin in various aqueous media. Chemosphere 91:1635–1642. https://doi.org/10.1016/j.chemosphere.2012.12.072

Babuponnusami A, Muthukumar K (2014) A review on Fenton and improvements to the Fenton process for wastewater treatment. J Environ Chem Eng 2:557–572. https://doi.org/10.1016/j.jece.2013.10.011

Balabanič D, Rupnik M, Klemenčič AK (2011) Negative impact of endocrine-disrupting compounds on human reproductive health. Reprod Fertil Dev 23:403–416. https://doi.org/10.1071/RD09300

Barrocas BT, Oliveira MC, Nogueira HIS, Fateixa S, Monteiro OC (2019) Ruthenium-modified titanate nanowires for the photocatalytic oxidative removal of organic pollutants from water. ACS Appl Nano Mater 2:1341–1349. https://doi.org/10.1021/acsanm.8b02215

Barron L, Nesterenko E, Hart K, Power E, Quinn B, Kelleher B, Paull B (2010) Holistic visualisation of the multimodal transport and fate of twelve pharmaceuticals in biosolid enriched topsoils. Anal Bioanal Chem 397:287–296. https://doi.org/10.1007/s00216-010-3494-1

Basri SH, Majid WHA, Talik NA, Sarjidan MAM (2018) Tailoring electronics structure, electrical and magnetic properties of synthesized transition metal (Ni)-doped ZnO thin film. J Alloys Compd 769:640–648. https://doi.org/10.1016/j.jallcom.2018.08.056

Beijer K, Björlenius B, Shaik S, Lindberg RH, Brunström B, Brandt I (2017) Removal of pharmaceuticals and unspecified contaminants in sewage treatment effluents by activated carbon filtration and ozonation: evaluation using biomarker responses and chemical analysis. Chemosphere 176:342–351. https://doi.org/10.1016/j.chemosphere.2017.02.127

Beltrán FJ, Rivas J, Acedo B (1999) Atrazine removal by ozonation processes in surface waters. J Environ Sci Heal—Part B Pestic Food Contam Agric Wastes 34:449–468. https://doi.org/10.1080/03601239909373208

Bhargava DS (2006) Revival of Mathura's ailing Yamuna river. Environmentalist 26:111–122. https://doi.org/10.1007/s10669-006-7481-1

Bonefeld-Jorgensen EC, Long M, Bossi R, Ayotte P, Asmund G, Krüger T, Ghisari M, Mulvad G, Kern P, Nzulumiki P, Dewailly E (2011) Perfluorinated compounds are related to breast cancer risk in Greenlandic Inuit: a case control study. Environ Heal A Glob Access Sci Source 10:1–16. https://doi.org/10.1186/1476-069X-10-88

Brigante M, DellaGreca M, Previtera L, Rubino M, Temussi F (2005) Degradation of hydrochlorothiazide in water. Environ Chem Lett 2:195–198. https://doi.org/10.1007/s10311-004-0096-1

Cai MJ, Lin YP (2016) Effects of effluent organic matter (EfOM) on the removal of emerging contaminants by ozonation. Chemosphere 151:332–338. https://doi.org/10.1016/j.chemosphere.2016.02.094

Careghini A, Mastorgio AF, Saponaro S, Sezenna E (2015) Bisphenol A, nonylphenols, benzophenones, and benzotriazoles in soils, groundwater, surface water, sediments, and food: a review. Environ Sci Pollut Res 22:5711–5741. https://doi.org/10.1007/s11356-014-3974-5

Carson R (2002) Silent spring. Houghton Mifflin Harcourt, Newyork

Chamarro E, Marco A, Esplugas S (2001) Use of Fenton reagent to improve organic chemical biodegradability. Water Res 35:1047–1051. https://doi.org/10.1016/S0043-1354(00)00342-0

Chen Y, Sun L, Yu Z, Wang L, Xiang G, Wan S (2015) Synergistic degradation performance and mechanism of 17β-estradiol by dielectric barrier discharge non-thermal plasma combined with Pt-TiO2. Sep Pur Technol 152:46–54. https://doi.org/10.1016/j.seppur.2015.07.061

Chiha M, Hamdaoui O, Baup S, Gondrexon N (2011) Sonolytic degradation of endocrine disrupting chemical 4-cumylphenol in water. Ultrason Sonochem 18:943–950. https://doi.org/10.1016/j.ultsonch.2010.12.014

Chudobova D, Dostalova S, Blazkova I, Michalek P, Ruttkay-Nedecky B, Sklenar M, Nejdl L, Kudr J, Gumulec J, Tmejova K, Konecna M, Vaculovicova M, Hynek D, Masarik M, Kynicky J, Kizek R, Adam V (2014) Effect of ampicillin, streptomycin, penicillin and tetracycline on metal resistant and non-resistant Staphylococcus aureus. Int J Environ Res Public Health 11:3233–3255. https://doi.org/10.3390/ijerph110303233

Clarizia L, Russo D, Di Somma I, Marotta R, Andreozzi R (2017) Homogeneous photo-Fenton processes at near neutral pH: a review. Appl Catal B Environ 209:358–371. https://doi.org/10.1016/j.apcatb.2017.03.011

Conrads H, Schmidt M (2000) Plasma generation and plasma sources. Plasma Sources Sci Technol 9:441–454. https://doi.org/10.1088/0963-0252/9/4/301

Coperchini F, Awwad O, Rotondi M, Santini F, Imbriani M, Chiovato L (2017) Thyroid disruption by perfluorooctane sulfonate (PFOS) and perfluorooctanoate (PFOA). J Endocrinol Invest 40:105–121. https://doi.org/10.1007/s40618-016-0572-z

Damasceno de Oliveira LL, Nunes B, Antunes SC, Campitelli-Ramos R, Rocha O (2018) Acute and chronic effects of three pharmaceutical drugs on the tropical freshwater Cladoceran Ceriodaphnia silvestrii. Water Air Soil Pollut 229:116. https://doi.org/10.1007/s11270-018-3765-6

Daneshvar N, Aber S, Seyed Dorraji MS, Khataee AR, Rasoulifard MH (2007) Photocatalytic degradation of the insecticide diazinon in the presence of prepared nanocrystalline ZnO powders under irradiation of UV-C light. Sep Purif Technol 58:91–98. https://doi.org/10.1016/j.seppur.2007.07.016

Daughton CG, Ternes TA (1999) Pharmaceuticals and personal care products in the environment: agents of subtle change? Environ Health Perspect 107:907–938. https://doi.org/10.1289/ehp.99107s6907

Deng H (2020) A review on the application of ozonation to NF/RO concentrate for municipal wastewater reclamation. J Hazard Mater 391:122071. https://doi.org/10.1016/j.jhazmat.2020.122071

Di Paola A, Augugliaro V, Palmisano L, Pantaleo G, Savinov E (2003) Heterogeneous photocatalytic degradation of nitrophenols. J Photochem Photobiol A Chem 155:207–214. https://doi.org/10.1016/S1010-6030(02)00390-8

Drury WH (1963) Silent spring rachel carson. Auk vol 80. pp 209–213. https://doi.org/10.2307/4082572

El Haimeur A, El Gana L, Addou M, El Kenz A (2018) Effect of tuning the structure on the optical and magnetic properties by various transition metal doping in ZnO/TM (TM = Fe, FeCo, Cr, and Mn) thin films. J Supercond Nov Magn 31:569–576. https://doi.org/10.1007/s10948-017-4237-8

Esmaeil MS, Somashekar R (2013) Orgnochlorine and synthetic pyrethroid pesticides in agricultural soil and water from Chamaranagar district, Karnataka, India. J Environ Sci Water Resour 2:221–225

Falås P, Wick A, Castronovo S, Habermacher J, Ternes TA, Joss A (2016) Tracing the limits of organic micropollutant removal in biological wastewater treatment. Water Res 95:240–249. https://doi.org/10.1016/j.watres.2016.03.009

Fenton HJH, Jones HO (1900) VII—the oxidation of organic acids in presence of ferrous iron. Part I. J Chem Soc Trans 77:69–76. https://doi.org/10.1039/CT9007700069

Fiessinger F, Richard Y, Montiel A, Musquere P (1981) Advantages and disadvantages of chemical oxidation and disinfection by ozone and chlorine dioxide. Sci Total Environ 18:245–261. https://doi.org/10.1016/S0048-9697(81)80062-9

Findik S, Gündüz G (2007) Sonolytic degradation of acetic acid in aqueous solutions. Ultrason Sonochem 14:157–162. https://doi.org/10.1016/j.ultsonch.2006.03.009

Foster J, Sommers BS, Gucker SN, Blankson IM, Adamovsky G (2012) Perspectives on the interaction of plasmas with liquid water for water purification. IEEE Trans Plasma Sci 40:1311–1323. https://doi.org/10.1109/TPS.2011.2180028

Fujishima A, Rao TN, Tryk DA (2017) Titanium dioxide photocatalysis. J Photochem Photobiol C: Photochem Rev 1:5525

Gabet-Giraud V, Miège C, Choubert JM, Ruel SM, Coquery M (2010) Occurrence and removal of estrogens and beta blockers by various processes in wastewater treatment plants. Sci Total Environ 408:4257–4269. https://doi.org/10.1016/j.scitotenv.2010.05.023

Gadipelly C, Pérez-González A, Yadav GD, Ortiz I, Ibáñez R, Rathod VK, Marathe KV (2014) Pharmaceutical industry wastewater: review of the technologies for water treatment and reuse. Ind Eng Chem Res 53:11571–11592. https://doi.org/10.1021/ie501210j

Gangu KK, Maddila S, Jonnalagadda SB (2019) A review on novel composites of MWCNTs mediated semiconducting materials as photocatalysts in water treatment. Sci Total Environ 646:1398–1412. https://doi.org/10.1016/j.scitotenv.2018.07.375

Gani KM, Kazmi AA (2017) Contamination of emerging contaminants in Indian aquatic sources: first overview of the situation. J Hazardous, Toxic, Radioact Waste 21:1–12. https://doi.org/10.1061/(ASCE)HZ.2153-5515.0000348

Ghoshdastidar AJ, Saunders JE, Brown KH, Tong AZ (2012) Membrane bioreactor treatment of commonly used organophosphate pesticides. J Environ Sci Heal Part B Pestic Food Contam Agric Wastes 47:742–750. https://doi.org/10.1080/03601234.2012.669334

Gleason JA, Post GB, Fagliano JA (2015) Associations of perfluorinated chemical serum concentrations and biomarkers of liver function and uric acid in the US population (NHANES), 2007–2010. Environ Res 136:8–14. https://doi.org/10.1016/j.envres.2014.10.004

Göbel A, McArdell CS, Joss A, Siegrist H, Giger W (2007) Fate of sulfonamides, macrolides, and trimethoprim in different wastewater treatment technologies. Sci Total Environ 372:361–371. https://doi.org/10.1016/j.scitotenv.2006.07.039

Gogoi A, Mazumder P, Tyagi VK, Tushara Chaminda GG, An AK, Kumar M (2018) Occurrence and fate of emerging contaminants in water environment: a review. Groundw Sustain Dev 6:169–180. https://doi.org/10.1016/j.gsd.2017.12.009

Golash N, Gogate PR (2012) Degradation of dichlorvos containing wastewaters using sonochemical reactors. Ultrason Sonochem 19:1051–1060. https://doi.org/10.1016/j.ultsonch.2012.02.011

Gruchlik Y, Linge K, Joll C (2018) Removal of organic micropollutants in waste stabilisation ponds: a review. J Environ Manage 206:202–214. https://doi.org/10.1016/j.jenvman.2017.10.020

Güyer GT, Ince NH (2011) Degradation of diclofenac in water by homogeneous and heterogeneous sonolysis. Ultrason Sonochem 18:114–119. https://doi.org/10.1016/j.ultsonch.2010.03.008

Hong R, Pan T, Qian J, Li H (2006) Synthesis and surface modification of ZnO nanoparticles. Chem Eng J 119:71–81. https://doi.org/10.1016/j.cej.2006.03.003

Hua I, Hoffmann MR (1996) Kinetics and mechanism of the sonolytic degradation of CCL4: intermediates and by products. Environ Sci Technol 30:864–871. https://doi.org/10.1021/es9502942

Huang N, Wang W-L, Xu Z-B, Lee M-Y, Wu Q-Y, Hu H-Y (2020) A study of synergistic oxidation between ozone and chlorine on benzalkonium chloride degradation: reactive species and degradation pathway. Chem Eng J 382:122856. https://doi.org/10.1016/j.cej.2019.122856

Huang W (2013) Homogeneous and heterogeneous Fenton and photo-Fenton processes : impact of iron complexing agent ethylenediamine-N, N'-disuccinic acid (EDDS) Wenyu Huang to cite this version : diplômée de Master homogeneous and heterogeneous Fenton and photo-Fento

Huang Y, Ho W, Lee S, Zhang L, Li G, Yu JC (2008) Effect of carbon doping on the mesoporous structure of nanocrystalline titanium dioxide and its solar-light-driven photocatalytic degradation of NOx. Langmuir 24:3510–3516. https://doi.org/10.1021/la703333z

Huber MM, GÖbel A, Joss A, Hermann N, LÖffler D, McArdell CS, Ried A, Siegrist H, Ternes TA, von Gunten U (2005) Oxidation of pharmaceuticals during ozonation of municipal wastewater effluents: a pilot study. Environ Sci Technol 39:4290–4299. https://doi.org/10.1021/es048396s

Ifelebuegu AO (2011) The fate and behavior of selected endocrine disrupting chemicals in full scale wastewater and sludge treatment unit processes. Int J Environ Sci Technol 8:245–254. https://doi.org/10.1007/BF03326213

Jacobs LE, Fimmen RL, Chin YP, Mash HE, Weavers LK (2011) Fulvic acid mediated photolysis of ibuprofen in water. Water Res 45:4449–4458. https://doi.org/10.1016/j.watres.2011.05.041

Jiang B, Zheng J, Liu Q, Wu M (2012) Degradation of azo dye using non-thermal plasma advanced oxidation process in a circulatory airtight reactor system. Chem Eng J 204–205:32–39. https://doi.org/10.1016/j.cej.2012.07.088

Jiang B, Zheng J, Qiu S, Wu M, Zhang Q, Yan Z, Xue Q (2014) Review on electrical discharge plasma technology for wastewater remediation. Chem Eng J 236:348–368. https://doi.org/10.1016/j.cej.2013.09.090

Kanakaraju D, Glass BD, Oelgemöller M (2018) Advanced oxidation process-mediated removal of pharmaceuticals from water: a review. J Environ Manage 219:189–207. https://doi.org/10.1016/j.jenvman.2018.04.103

Katz A, McDonagh A, Tijing L, Shon HK (2015) Fouling and inactivation of titanium dioxide-based photocatalytic systems. Crit Rev Environ Sci Technol 45:1880–1915. https://doi.org/10.1080/10643389.2014.1000763

Khalilzadeh A, Fatemi S (2016) Spouted bed reactor for VOC removal by modified nano-TiO2 photocatalytic particles. Chem Eng Res Des 115:241–250. https://doi.org/10.1016/J.CHERD.2016.10.004

Kim KR, Owens G, Kwon SI, So KH, Lee DB, Ok YS (2011) Occurrence and environmental fate of veterinary antibiotics in the terrestrial environment. Water Air Soil Pollut 214:163–174. https://doi.org/10.1007/s11270-010-0412-2

Kim M, Guerra P, Shah A, Parsa M, Alaee M, Smyth SA (2014) Removal of pharmaceuticals and personal care products in a membrane bioreactor wastewater treatment plant. Water Sci Technol 69:2221–2229. https://doi.org/10.2166/wst.2014.145

Klaine SJ, Alvarez PJJ, Batley GE, Fernandes TF, Handy RD, Lyon DY, Mahendra S, McLaughlin MJ, Lead JR (2008) Nanomaterials in the environment: behavior, fate, bioavailability, and effects. Environ Toxicol Chem 27:1825–1851

Koumaki E, Mamais D, Noutsopoulos C, Nika MC, Bletsou AA, Thomaidis NS, Eftaxias A, Stratogianni G (2015) Degradation of emerging contaminants from water under natural sunlight: the effect of season, pH, humic acids and nitrate and identification of photodegradation by-products. Chemosphere 138:675–681. https://doi.org/10.1016/j.chemosphere.2015.07.033

Krichevskaya M, Klauson D, Portjanskaja E, Preis S (2011) The cost evaluation of advanced oxidation processes in laboratory and pilot-scale experiments. Ozone Sci Eng 33:211–223. https://doi.org/10.1080/01919512.2011.554141

Larsen MG, Hansen KB, Henriksen PG, Baatrup E (2008) Male zebrafish (Danio rerio) courtship behaviour resists the feminising effects of 17α-ethinyloestradiol-morphological sexual characteristics do not. Aquat Toxicol 87:234–244. https://doi.org/10.1016/j.aquatox.2008.02.003

Latch DE, Packer JL, Arnold WA, McNeill K (2003) Photochemical conversion of triclosan to 2,8-dichlorodibenzo-p-dioxin in aqueous solution. J Photochem Photobiol A Chem 158:63–66. https://doi.org/10.1016/S1010-6030(03)00103-5

Lee GJ, Wu JJ (2017) Recent developments in ZnS photocatalysts from synthesis to photocatalytic applications—a review. Powder Technol 318:8–22

Legrini O, Oliveros E, Braun AM (1993) Photochemical processes for water treatment. Chem Rev 93:671–698. https://doi.org/10.1021/cr00018a003

Leung SW, Watts RJ, Miller GC (1992) Degradation of perchloroethylene by Fenton's reagent: speciation and pathway. J Environ Qual 21:377–381. https://doi.org/10.2134/jeq1992.00472425002100030012x

Li XZ, Li FB (2001) Study of Au/Au3+-TiO2 photocatalysts toward visible photooxidation for water and wastewater treatment. Environ Sci Technol 35:2381–2387. https://doi.org/10.1021/es001752w

Li Y, Zhu G, Ng WJ, Tan SK (2014) A review on removing pharmaceutical contaminants from wastewater by constructed wetlands: design, performance and mechanism. Sci Total Environ 468–469:908–932. https://doi.org/10.1016/j.scitotenv.2013.09.018

Lin T, Yu S, Chen W (2016) Occurrence, removal and risk assessment of pharmaceutical and personal care products (PPCPs) in an advanced drinking water treatment plant (ADWTP) around Taihu Lake in China. Chemosphere 152:1–9. https://doi.org/10.1016/j.chemosphere.2016.02.109

Lind PM, Milnes MR, Lundberg R, Bermudez D, Örberg J, Guillette LJ (2004) Abnormal bone composition female juvenile American alligators from a pesticide-polluted lake (Lake Apopka, Florida). Environ Health Perspect 112:359–362. https://doi.org/10.1289/ehp.6524

Lipczynska-Kochany E (1991) Degradation of aqueous nitrophenols and nitrobenzene by means of the Fenton reaction. Chemosphere 22:529–536. https://doi.org/10.1016/0045-6535(91)90064-K

Litter MI (2005) Introduction to photochemical advanced oxidation processes for water treatment. Environ Photochem Part II:325–366. https://doi.org/10.1007/b138188

Liu C, Zhang L, Liu R, Gao Z, Yang X, Tu Z, Yang F, Ye Z, Cui L, Xu C, Li Y (2016) Hydrothermal synthesis of N-doped TiO2 nanowires and N-doped graphene heterostructures with enhanced photocatalytic properties. J Alloys Compd 656:24–32. https://doi.org/10.1016/j.jallcom.2015.09.211

Liu XL, Pappas I, Fitzgerald M, Zhu YJ, Eibling M, Pan L (2010) Solvothermal synthesis and characterization of ZrO2 nanostructures using zirconium precursor. Mater Lett 64:1591–1594. https://doi.org/10.1016/j.matlet.2010.04.044

Lofrano G, Rizzo L, Grassi M, Belgiorno V (2009) Advanced oxidation of catechol: a comparison among photocatalysis, Fenton and photo-Fenton processes. Desalination 249:878–883. https://doi.org/10.1016/j.desal.2009.02.068

Lyu G, Shi G, Tang L, Fang H, Wu M (2017) Mechanism of degradation of a nitrogenous heterocycle induced by a reductive radical: decomposition of a: sym-triazine ring. Phys Chem Chem Phys 19:9354–9357. https://doi.org/10.1039/c7cp00004a

Martín J, Santos JL, Aparicio I, Alonso E (2010) Multi-residue method for the analysis of pharmaceutical compounds in sewage sludge, compost and sediments by sonication-assisted extraction and LC determination. J Sep Sci 33:1760–1766. https://doi.org/10.1002/jssc.200900873

Matamoros V, Rodríguez Y, Albaigés J (2016) A comparative assessment of intensive and extensive wastewater treatment technologies for removing emerging contaminants in small communities. Water Res 88:777–785. https://doi.org/10.1016/j.watres.2015.10.058

McGillicuddy E, Murray I, Kavanagh S, Morrison L, Fogarty A, Cormican M, Dockery P, Prendergast M, Rowan N, Morris D (2017) Silver nanoparticles in the environment: sources, detection and ecotoxicology. Sci Total Environ 575:231–246. https://doi.org/10.1016/j.scitotenv.2016.10.041

Metcalf L, Eddy HP, Tchobanoglous G (2004) Wastewater engineering: treatment, disposal, and reuse, 4th edn. McGraw-Hill, Newyork

Miichi T, Hayashi N, Ihara S, Satoh S, Yamabe C (2002) Generation of radicals using discharge inside bubbles in water for water treatment. Ozone Sci Eng 24:471–477. https://doi.org/10.1080/01919510208901636

Minella M, Maurino V, Minero C, Vione D (2016) A model assessment of the ability of lake water in Terra Nova Bay, Antarctica, to induce the photochemical degradation of emerging contaminants. Chemosphere 162:91–98. https://doi.org/10.1016/j.chemosphere.2016.07.049

Mishra NS, Reddy R, Kuila A, Rani A, Mukherjee P, Nawaz A, Pichiah S (2017) A review on advanced oxidation processes for effective water treatment. Curr World Environ 12:470–490. https://doi.org/10.12944/CWE.12.3.02

Mutiyar PK, Gupta SK, Mittal AK (2018) Fate of pharmaceutical active compounds (PhACs) from River Yamuna, India: an ecotoxicological risk assessment approach. Ecotoxicol Environ Saf 150:297–304. https://doi.org/10.1016/j.ecoenv.2017.12.041

Nakata K, Fujishima A (2012) TiO 2 photocatalysis: design and applications. J Photochem Photobiol C Photochem Rev 13:169–189. https://doi.org/10.1016/j.jphotochemrev.2012.06.001

Nejumal KK, Manoj PR, Aravind UK, Aravindakumar CT (2014) Sonochemical degradation of a pharmaceutical waste, atenolol, in aqueous medium. Environ Sci Pollut Res 21:4297–4308. https://doi.org/10.1007/s11356-013-2301-x

Nishiyama N, Yamazaki S (2017) Effect of mixed valence states of platinum ion dopants on the photocatalytic activity of titanium dioxide under visible light irradiation. ACS Omega 2:9033–9039. https://doi.org/10.1021/acsomega.7b01393

Osawa RA, Monteiro OC, Oliveira MC, Florêncio MH (2020) Comparative study on photocatalytic degradation of the antidepressant trazodone using (Co, Fe and Ru) doped titanate nanowires: kinetics, transformation products and in silico toxicity assessment. Chemosphere 259:127486. https://doi.org/10.1016/j.chemosphere.2020.127486

Papoulis D (2019) Halloysite based nanocomposites and photocatalysis: a review. Appl Clay Sci 168:164–174. https://doi.org/10.1016/j.clay.2018.11.009

Pastrana-Martínez LM, Morales-Torres S, Papageorgiou SK, Katsaros FK, Romanos GE, Figueiredo JL, Faria JL, Falaras P, Silva AMT (2013) Photocatalytic behaviour of nanocarbon-TiO2 composites and immobilization into hollow fibres. Appl Catal B Environ 142–143:101–111. https://doi.org/10.1016/j.apcatb.2013.04.074

Peller J, Wiest O, Kamat PV (2003) Synergy of combining sonolysis and photocatalysis in the degradation and mineralization of chlorinated aromatic compounds. Environ Sci Technol 37:1926–1932. https://doi.org/10.1021/es0261630

Pereira JHOS, Queirós DB, Reis AC, Nunes OC, Borges MT, Boaventura RAR, Vilar VJP (2014) Process enhancement at near neutral pH of a homogeneous photo-Fenton reaction using ferricarboxylate complexes: application to oxytetracycline degradation. Chem Eng J 253:217–228. https://doi.org/10.1016/j.cej.2014.05.037

Perinban S, Orsat V, Raghavan V (2019) Nonthermal plasma-liquid interactions in food processing: a review. Compr Rev Food Sci Food Saf 18:1985–2008. https://doi.org/10.1111/1541-4337.12503

Petrović M, Eljarrat E, López de Alda MJ, Barceló D (2001) Analysis and environmental levels of endocrine-disrupting compounds in freshwater sediments. TrAC—Trends Anal Chem 20:637–648. https://doi.org/10.1016/S0165-9936(01)00118-2

Pham TH, Bui HM, Bui TX (2020) Advanced oxidation processes for the removal of pesticides. Elsevier B.V.

Pozzo RL, Giombi JL, Baltanás MA, Cassano AE (2002) Radiation extinction of slurried TiO2 as a function of mechanical action and ionic composition of the suspending media: a key factor in the photocatalytic efficiency. Appl Catal B Environ 38:61–69. https://doi.org/10.1016/S0926-3373(02)00027-9

Reck E, Richards M (1999) Articles TiO2 manufacture and life cycle analysis. Pigment Resin Technol 28:149–157

Remucal CK (2014) The role of indirect photochemical degradation in the environmental fate of pesticides: a review. Environ Sci Process Impacts 16:628–653. https://doi.org/10.1039/c3em00549f

Reyes-Contreras C, Hijosa-Valsero M, Sidrach-Cardona R, Bayona JM, Bécares E (2012) Temporal evolution in PPCP removal from urban wastewater by constructed wetlands of different configuration: a medium-term study. Chemosphere 88:161–167. https://doi.org/10.1016/j.chemosphere.2012.02.064

Rivera-Utrilla J, Sánchez-Polo M, Ferro-García MÁ, Prados-Joya G, Ocampo-Pérez R (2013) Pharmaceuticals as emerging contaminants and their removal from water. a review. Chemosphere 93:1268–1287. https://doi.org/10.1016/j.chemosphere.2013.07.059

Ruggeri G, Ghigo G, Maurino V, Minero C, Vione D (2013) Photochemical transformation of ibuprofen into harmful 4-isobutylacetophenone: pathways, kinetics, and significance for surface waters. Water Res 47:6109–6121. https://doi.org/10.1016/j.watres.2013.07.031

Saien J, Soleymani AR, Sun JH (2011) Parametric optimization of individual and hybridized AOPs of Fe2+/H2O2 and UV/S2O82—for rapid dye destruction in aqueous media. Desalination 279:298–305. https://doi.org/10.1016/j.desal.2011.06.024

Saikat S, Kreis I, Davies B, Bridgman S, Kamanyire R (2013) The impact of PFOS on health in the general population: a review. Environ Sci Process Impacts 15:329–335. https://doi.org/10.1039/c2em30698k

Sasi S, Rayaroth MP, Devadasan D, Aravind UK, Aravindakumar CT (2015) Influence of inorganic ions and selected emerging contaminants on the degradation of Methylparaben: a sonochemical approach. J Hazard Mater 300:202–209. https://doi.org/10.1016/j.jhazmat.2015.06.072

Sauvé S, Desrosiers M (2014) A review of what is an emerging contaminant. Chem Cent J 8:1–7. https://doi.org/10.1186/1752-153X-8-15

Schweitzer L, Noblet J (2018) Water contamination and pollution. Green Chem An Incl Approach, pp 261–290. https://doi.org/10.1016/B978-0-12-809270-5.00011-X

Secondes MFN, Naddeo V, Belgiorno V, Ballesteros F (2014) Removal of emerging contaminants by simultaneous application of membrane ultrafiltration, activated carbon adsorption, and ultrasound irradiation. J Hazard Mater 264:342–349. https://doi.org/10.1016/j.jhazmat.2013.11.039

Shakil MR, El-Sawy AM, Tasnim H, Meguerdichian AG, Jin J, Dubrosky JP, Suib SL (2018) Single-doped and multidoped transition-metal (Mn, Fe Co, and Ni) ZnO and their electrocatalytic activities for oxygen reduction reaction. Inorg Chem 57:9977–9987. https://doi.org/10.1021/acs.inorgchem.8b01153

Shankar A, Xiao J, Ducatman A (2011) Perfluoroalkyl chemicals and chronic kidney disease in US adults. Am J Epidemiol 174:893–900. https://doi.org/10.1093/aje/kwr171

Sharma D, Jha R (2017) Transition metal (Co, Mn) co-doped ZnO nanoparticles: effect on structural and optical properties. J Alloys Compd 698:532–538. https://doi.org/10.1016/j.jallcom.2016.12.227

Sharma S, Ruparelia J, Patel M (2011) A general review on advanced oxidation processes for waste water treatment. International conference on current trends in technology. Nirma university, Ahmedabad, Institute of Technology, pp 8–10

Shayegan Z, Haghighat F, Lee CS (2019) Photocatalytic oxidation of volatile organic compounds for indoor environment applications: three different scaled setups. Chem Eng J 357:533–546. https://doi.org/10.1016/j.cej.2018.09.167

Shayegan Z, Lee CS, Haghighat F (2018) TiO2photocatalyst for removal of volatile organic compounds in gas phase—a review. Chem Eng J 334:2408–2439. https://doi.org/10.1016/j.cej.2017.09.153

Shen Y, Wang L, Wu Y, Li X, Zhao Q, Hou Y, Teng W (2015) Facile solvothermal synthesis of MnFe2SO4 hollow nanospheres and their photocatalytic degradation of benzene investigated by in situ FTIR. Catal Commun 68:11–14. https://doi.org/10.1016/j.catcom.2015.04.025

Singh KP, Malik A, Mohan D, Sinha S (2005) Persistent organochlorine pesticide residues in alluvial groundwater aquifers of gangetic plains, India. Bull Environ Contam Toxicol 74:162–169. https://doi.org/10.1007/s00128-004-0563-1

Sivagami K (2011) Studies on photocatalytic degradation of monocrotophos in an annular slurry reactor using factorial design of experiments. J Water Sustain 1:75–84

Söderström H, Lindberg RH, Fick J (2009) Strategies for monitoring the emerging polar organic contaminants in water with emphasis on integrative passive sampling. J Chromatogr A 1216:623–630. https://doi.org/10.1016/j.chroma.2008.08.030

Solak S, Vakondios N, Tzatzimaki I, Diamadopoulos E, Arda M, Kabay N, Yüksel M (2014) A comparative study of removal of endocrine disrupting compounds (EDCs) from treated wastewater using highly crosslinked polymeric adsorbents and activated carbon. J Chem Technol Biotechnol 89:819–824. https://doi.org/10.1002/jctb.4315

Staehelin J, Hoigne J (1985) Decomposition of ozone in water in the presence of organic solutes acting as promoters and inhibitors of radical chain reactions. Environ Sci Technol 19:1206–1213. https://doi.org/10.1021/es00142a012

Stasinakis AS, Kotsifa S, Gatidou G, Mamais D (2009) Diuron biodegradation in activated sludge batch reactors under aerobic and anoxic conditions. Water Res 43:1471–1479. https://doi.org/10.1016/j.watres.2008.12.040

Tan L, He M, Men B, Lin C (2009) Distribution and sources of organochlorine pesticides in water and sediments from Daliao River estuary of Liaodong Bay, Bohai Sea (China). Estuar Coast Shelf Sci 84:119–127. https://doi.org/10.1016/j.ecss.2009.06.013

Tang L, Huang H, Hao H, Zhao K (2013) Development of plasma pyrolysis/gasification systems for energy efficient and environmentally sound waste disposal. J Electrostat 71:839–847. https://doi.org/10.1016/j.elstat.2013.06.007

Tay KS, Madehi N (2015) Ozonation of ofloxacin in water: by-products, degradation pathway and ecotoxicity assessment. Sci Total Environ 520:23–31. https://doi.org/10.1016/j.scitotenv.2015.03.033

Ternes T, Joss A (2006) Human pharmaceuticals, hormones and fragrances: the micropollutant challenge for urban water management

Tichonovas M, Krugly E, Racys V, Hippler R, Kauneliene V, Stasiulaitiene I, Martuzevicius D (2013) Degradation of various textile dyes as wastewater pollutants under dielectric barrier discharge plasma treatment. Chem Eng J 229:9–19. https://doi.org/10.1016/j.cej.2013.05.095

Tonks L, Langmuir I (1929) A general theory of the plasma of an Arc. Phys Rev 34:876–922

Toro-Vélez AF, Madera-Parra CA, Peña-Varón MR, Lee WY, Bezares-Cruz JC, Walker WS, Cárdenas-Henao H, Quesada-Calderón S, García-Hernández H, Lens PNL (2016) BPA and NP removal from municipal wastewater by tropical horizontal subsurface constructed wetlands. Sci Total Environ 542:93–101. https://doi.org/10.1016/j.scitotenv.2015.09.154

Trapido M, Hirvonen A, Veressinina Y, Hentunen J, Munter R (1997) Ozonation, ozone/UV and UV/H2O2 degradation of chlorophenols. Ozone Sci Eng 19:75–96. https://doi.org/10.1080/01919519708547319

Trovó AG, Nogueira RFP (2011) Diclofenac abatement using modified solar photo-Fenton process with ammonium iron(III) citrate. J Braz Chem Soc 22:1033–1039. https://doi.org/10.1590/S0103-50532011000600005

United States Environmental Protection Agency (2012) Biosolids. https://www.epa.gov/biosolids. Accessed 28 May 2020

Vaiano V, Sacco O, Pisano D, Sannino D, Ciambelli P (2015) From the design to the development of a continuous fixed bed photoreactor for photocatalytic degradation of organic pollutants in wastewater. Chem Eng Sci 137:152–160. https://doi.org/10.1016/j.ces.2015.06.023

Vanraes P, Willems G, Nikiforov A, Surmont P, Lynen F, Vandamme J, Van Durme J, Verheust YP, Van Hulle SWH, Dumoulin A, Leys C (2015) Removal of atrazine in water by combination of activated carbon and dielectric barrier discharge. J Hazard Mater 299:647–655. https://doi.org/10.1016/j.jhazmat.2015.07.075

Vieira GB, Scaratti G, Rodembusch FS, De Amorim SM, Peterson M, Puma GL, Moreira RDFPM (2019) Tuning the photoactivity of TiO2 nanoarchitectures doped with cerium or neodymium and application to colour removal from wastewaters. Environ Technol (united Kingdom). https://doi.org/10.1080/09593330.2019.1651402

Vymazal J, Březinová T (2015) The use of constructed wetlands for removal of pesticides from agricultural runoff and drainage: a review. Environ Int 75:11–20. https://doi.org/10.1016/j.envint.2014.10.026

Wang F, Li F, Zhang L, Zeng H, Sun Y, Zhang S, Xu X (2017) S-TiO2 with enhanced visible-light photocatalytic activity derived from TiS2 in deionized water. Mater Res Bull 87:20–26. https://doi.org/10.1016/j.materresbull.2016.11.014

Wang Y, Nowack B (2018) Environmental risk assessment of engineered nano-SiO 2, nano iron oxides, nano-CeO 2, nano-Al 2 O 3, and quantum dots. Environ Toxicol Chem 37:1387–1395. https://doi.org/10.1002/etc.4080

WHO (2019) 1 in 3 people globally do not have access to safe drinking water—UNICEF, WHO New report on inequalities in access to. https://www.who.int/news-room/detail/18-06-2019-1-

in-3-people-globally-do-not-have-access-to-safe-drinking-water-unicef-who. Accessed 22 May 2020

Weavers LK, Ling FH, Hoffmann MR (1998) Aromatic compound degradation in water using a combination of sonolysis and ozonolysis. Environ Sci Technol 32:2727–2733. https://doi.org/10.1021/es970675a

Wilkinson J, Hooda PS, Barker J, Barton S, Swinden J (2017) Occurrence, fate and transformation of emerging contaminants in water: an overarching review of the field. Environ Pollut 231:954–970. https://doi.org/10.1016/j.envpol.2017.08.032

Xiao R, He Z, Diaz-Rivera D, Pee GY, Weavers LK (2014) Sonochemical degradation of ciprofloxacin and ibuprofen in the presence of matrix organic compounds. Ultrason Sonochem 21:428–435. https://doi.org/10.1016/j.ultsonch.2013.06.012

Yan JH, Du CM, Li XD, Sun XD, Ni MJ, Cen KF, Cheron B (2005) Plasma chemical degradation of phenol in solution by gas-liquid gliding arc discharge. Plasma Sources Sci Technol 14:637–644. https://doi.org/10.1088/0963-0252/14/4/001

Yang X, Flowers RC, Weinberg HS, Singer PC (2011) Occurrence and removal of pharmaceuticals and personal care products (PPCPs) in an advanced wastewater reclamation plant. Water Res 45:5218–5228. https://doi.org/10.1016/j.watres.2011.07.026

Yang Y, Ok YS, Kim KH, Kwon EE, Tsang YF (2017) Occurrences and removal of pharmaceuticals and personal care products (PPCPs) in drinking water and water/sewage treatment plants: a review. Sci Total Environ 596–597:303–320. https://doi.org/10.1016/j.scitotenv.2017.04.102

Yangali-Quintanilla V, Maeng SK, Fujioka T, Kennedy M, Li Z, Amy G (2011) Nanofiltration vs. reverse osmosis for the removal of emerging organic contaminants in water reuse. Desalin Water Treat 34:50–56. https://doi.org/10.5004/dwt.2011.2860

Zaki AH, Lee MJ (2019) Effects of K+, Mg2+, Ca2+, Zn2+, La3+, Cr3+, Ce3+, Ce4+, and Mo5+ doping on the adsorption performance and optical properties of sodium titanate nanotubes. ACS Omega 4:19623–19634. https://doi.org/10.1021/acsomega.9b02229

Zeng Z, Zou H, Li X, Arowo M, Sun B, Chen J, Chu G, Shao L (2013) Degradation of phenol by ozone in the presence of Fenton reagent in a rotating packed bed. Chem Eng J 229:404–411. https://doi.org/10.1016/j.cej.2013.06.018

Zepp RG, Faust BC, Jürg H (1992) Hydroxyl radical formation in aqueous reactions (pH 3–8) of iron(II) with hydrogen peroxide: The photo-Fenton reaction. Environ Sci Technol 26:313–319. https://doi.org/10.1021/es00026a011

Zhang W, Ding L, Luo J, Jaffrin MY, Tang B (2016) Membrane fouling in photocatalytic membrane reactors (PMRs) for water and wastewater treatment: a critical review. Chem Eng J 302:446–458. https://doi.org/10.1016/j.cej.2016.05.071

Zhao Y, Kuang J, Zhang S, Li X, Wang B, Huang J, Deng S, Wang Y, Yu G (2017) Ozonation of indomethacin: kinetics, mechanisms and toxicity. J Hazard Mater 323:460–470. https://doi.org/10.1016/j.jhazmat.2016.05.023

Zhu H, Li W (2013) Bisphenol A removal from synthetic municipal wastewater by a bioreactor coupled with either a forward osmotic membrane or a microfiltration membrane unit. Front Environ Sci Eng 7:294–300. https://doi.org/10.1007/s11783-013-0486-3

Zoeller RT, Brown TR, Doan LL, Gore AC, Skakkebaek NE, Soto AM, Woodruff TJ, Vom Saal FS (2012) Endocrine-disrupting chemicals and public health protection: a statement of principles from the Endocrine Society. Endocrinology 153:4097–4110. https://doi.org/10.1210/en.2012-1422

Zoschke K, Börnick H, Worch E (2014) Vacuum-UV radiation at 185nm in water treatment—a review. Water Res 52:131–145. https://doi.org/10.1016/j.watres.2013.12.034

Chapter 12
Surfaces and Modified Surfaces for Controlling the Pollution: Different Approaches

Shiju Abraham

Abstract Various pollutions such as air, water, plastic, noise, food waste, soil, and radioactive substance's pollutions are always a severe problem to the environment and to the humanity. To mitigate this issue, investigations were conducted by different researchers across the globe on several surfaces for their adsorption and desorption properties towards various physical, chemical and biological pollutants, such as particulate matter, heavy metals, oil spillage, micro-organisms, etc. Along with this, ample of surface modification approaches, both for nano structures as well as for bulk structures were tested over last few decades to strive these concerns. So, surfaces such as bulk, nano, patterned, hydrophobic, oleophobic, magnetic, membrane, and polymeric surfaces along with their various modifications are extensively discusses here. These surfaces and modified surfaces have been explored in detail focusing on their diverse environmental remedial applications like adsorption, photo catalytic degradation, oil–water separation and capture, self-cleaning, and anti-microbial/anti-bacterial or anti-foulants properties. Thus, the overall thrust of this chapter will be to provide a detailed outlook on the current status on the environmental remedial strategies especially by making use of various surfaces and modified surfaces and its probable scope in the years ahead.

Keywords Surfaces · Modified surfaces · Pollution control

12.1 Introduction

Pollution, always a severe global issue, is the introduction of various contaminants into the natural environment which result an adverse change to the same environment (Akimoto 2003; Li et al. 2019a; Schwarzenbach et al. 2010). The incredible increase of human population along with the rapid growth of industrialization in the last couple of decades have triggered serious environmental pollution across the

S. Abraham (✉)
Department of Physics, St. Pius X College Rajapuram (Kannur University), Kasaragod, Kerala 671532, India
e-mail: shiju@post.bgu.ac.il

S. P. Singh et al. (eds.), *Nanomaterials and Nanocomposites for Environmental Remediation*, Energy, Environment, and Sustainability,
https://doi.org/10.1007/978-981-16-3256-3_13

globe. Along with the rapid increase in the pollutant volume into the environment, the development of specialized and modern industries have added the diversity of pollutants flowing into the environment. There are different forms of pollution; but, air pollution, water pollution, light pollution, noise pollution, and soil contamination are the main forms of pollution (Akimoto 2003; Schwarzenbach et al. 2010; Zwolak et al. 2019; Stansfeld and Matheson 2003; Longcore and Rich 2004). Always, there were continuous effort from the individuals and from the organizations to mitigate the issues related to pollution. Scientists and technologists have developed several techniques based on various scientific concepts and are still continuing their efforts to discover new methods, or to improve the existing approaches.

Recent decades have observed widespread problems related to air and water pollution (Akimoto 2003; Schwarzenbach et al. 2010; Xue et al. 2017), the two major kinds of pollution among different forms of pollution present. The elevated concentrations of various pollutants; for instance, pesticides, dyes, particulate matter, phenols, toxic heavy metal ions, humic substances, inorganic anions, etc. has been largely reported to cause air and water pollution around the globe. Further, the discharge of these harmful pollutants into the water bodies has adverse effect on the ecological balance and can successively cause destructive effects on the flora and fauna. Adding to these, most of these contaminants are resistant against different chemical or biological degradation and have great environmental mobility and possess strong inclination towards bioaccumulation in the food chain. Further, the accumulation of unwanted material (may be living organisms or non-living inorganic and/or organic substances) on solid surfaces to harm the functions of such surfaces, or simply known as the 'fouling', also discussed in some parts of this chapter to show those surface's fouling control characteristics.

In this context, the current chapter focuses on the importance of different surfaces and modified surfaces which includes bulk, nano, patterned, hydrophobic, oleophobic, magnetic, membrane, and polymeric surfaces with an emphasis given to their function of working as a pollution control framework. The following Fig. 12.1 briefly describes an overview of the present chapter and the findings discussed in this chapter are mostly taken from the recent reports.

12.2 Surfaces

The term 'surface', can be generally used to define the outermost or uppermost layer of a physical object or space. Specifically, in physical science, it is the interface of two phases. These interfaces can be either solid–liquid, solid–vacuum, solid–gas, or liquid–gas interfaces. It is important to note that the term "surface" or 'modified surface' were used in an extensive manner to cover 2-dimensional planar substrates, surfaces of bulk materials, surface of nanostructures, scaffolds, and colloidal surfaces. The surfaces can be as simple like a silicon substrate to surfaces of quantum dots or complex modified surfaces such as polyvinylidene fluoride/ perfluorosulfonic

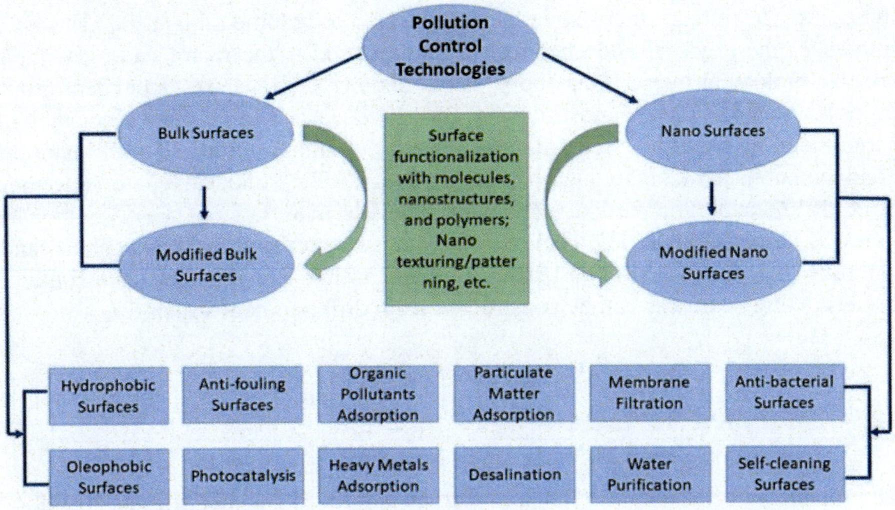

Fig. 12.1 A flow chart for the pollution control strategies reported in the present chapter

acid/ graphene oxides functionalized with polyethyleneimine (PVDF/PFSA/PEI-GO) based hybrid membrane (Abraham et al. 2018a; Kumar et al. 2014; Hao et al. 2020). Surfaces can be broadly classified into bulk or nano, depends on its size, structure and by their various properties. The main difference between them is that the nanostructure's sizes ranges from 1 to 100 nm at least in one dimension, and for bulk structures have sizes more than 100 nm in all dimensions.

12.2.1 Bulk Surfaces

In general, regardless of its size, bulk material has constant physical properties. Here, the grain size and its distribution, bulk density, moisture, and temperature are the main parameters that lead to the properties of such surfaces. The percentage of atoms present at the surface is infinitesimal comparative to the total number of atoms present in the material if it is bulk in nature and have the size larger than one micrometre.

12.2.2 Nano Surfaces

Nanostructured surfaces have long term potential to actively participate in the environmental curative research. For nanostructures, as the size shrinks, the surface area/volume ratio increase or in other words the fraction of atoms on the surface increases. This intense increase in the surface area to volume ratio at the nanoscale influences the physical and chemical properties, and is the major cause for their specific biological interactions and possible toxicity. Various size-dependent properties are found in these nanostructures, for instance, surface plasmon resonance in some metal nanoparticles (eg. Au, Ag, Pt, etc.) (Abraham et al. 2018b), quantum confinement in semiconducting quantum dots (eg. CdSe, PbSe, CdTe, etc.) (Holmes et al. 2012), and superparamagnetism in magnetic materials (eg. Fe_2O_3 nanoparticles) (Laurent et al. 2011). Nano surfaces can be prepared hard and robust, and can tune their functionality and sustainability. Further, dirt-repellent, anti-fouling, bacteria killing surfaces can be constructed from different nanostructures.

12.3 Modified Surfaces

The functionality of different surfaces can be improved by applying the functionalization protocols of different molecules/ materials on these surfaces. Generally, there are three kinds of functionalization approaches to modify the surfaces. It can be done either by different molecules, nano structures or with polymers. The three ways of functionalization approaches mentioned above along with an another surface modification method—surface patterning and the tools using in characterising those surfaces are described briefly in the following sub-sections.

12.3.1 Functionalization of Surfaces with Molecules

Various functional groups/ molecules such as amino groups, hydroxy groups, proteins, peptides, surfactants, DNA, RNA, etc. can be functionalised on various surfaces such as gold, silica, carbon nanomaterials, etc. to enhance the performance of the surfaces for numerous applications like adsorption, photocatalysis, desalination, sensing, etc. (Rathinam et al. 2019; Oliveira et al. 2015; Thamer et al. 2019; Zhu et al. 2020a; Srivastava et al. 2015). The approaches for the functionalization of surfaces may vary for different functional entities. For instance, the experimental protocol to make a surface functionalised with amino groups will be different than that of making a surface functionalised with proteins. Further, there will be different approaches to make the same functional group on a particular surface. Surface functionalization will obviously change the surface properties; for instance, the surface charge, hydrophilicity, etc.

12.3.2 Functionalization of Surfaces with Nanostructures

Different nanostructures like nanoparticles, nanorods, nanotubes, nanowires, nanodiscs, and nanosheets can be used on different surfaces either to make nanocomposites or heterostructures, which may have improved properties than that of the bare surfaces (Abraham et al. 2015; Zhang et al. 2010). For instance, gold nanoparticles grown on MoS_2 nanosheets (Su et al. 2014), gold nanoparticles on indium tin oxide (ITO) surface (Ma et al. 2009), and gold nanoparticles decorated on graphene surfaces (Abraham et al. 2015).

12.3.3 Functionalization of Surfaces with Polymers

Different polymers, as like small molecules can be used to modify the surfaces to make them more suitable for the desired function. For instance, membranes modified with polydopamine and graphene oxide (Li et al. 2020). Adding to this, well promising polymer brushes, like poly(acrylamidoxime-co-acrylic acid) and polyacid were fabricated and reported for their different applications such as heavy metals adsorption, protein binding, etc. (Sun et al. 2006; Dolatkhah and Wilson 2016; Chi et al. 2019).

12.3.4 Nanotextured Surfaces

Another way to modify the surface is by changing the surface profile by making specific pattern or arranging particles on the surface in a such manner that it may make a specific pattern (Zhao et al. 2011). Soft lithography, photo-lithography, and electron beam lithography techniques are extensively used to fabricate such surfaces.

12.3.5 Tools Used for the Surface Characterization

The tools employed to characterise the surfaces have much importance in surface science. Many characterization techniques such as x-ray photoelectron spectroscopy (XPS), x-ray diffraction (XRD), Fourier-transform infrared spectroscopy (FTIR), atomic force microscopy (AFM), scanning tunnelling microscopy (STM), transmission electron microscopy (TEM), scanning electron microscopy (SEM), quartz crystal microbalance (QCM), time-of-flight secondary ion mass spectrometry (TOF–SIMS), energy-dispersive x-ray spectroscopy (EDX), etc. have been extensively used in this context.

12.4 Surfaces and Modified Surfaces for Pollution Control

12.4.1 Bulk Surfaces and Modified Bulk Surfaces

Often, the surfaces get covered by various dust particles over time. Such dust particles attached on the surfaces may reduce the performance of the devices (eg, solar panels, optical devices, solar water heaters, and more) they cover (Zaihidee et al. 2016; Heckenthaler et al. 2019). To overcome this issues, there are many studies on the self-cleaning mechanism (also known as the 'lotus effect') to replicate such mechanism on engineered surfaces (Heckenthaler et al. 2019; Cheng and Rodak 2005). During the self-cleaning process, water droplets roll or slide along a dust covered surface, so the dust particles can adsorb to the water—air interface and thereby be removed from the surface. So, self-cleaning surfaces are in forefront of research owing to their wide range of applicability in solar cell panels, windows, food containers, paints, fabrics, oil tubing, kitchen equipment, etc.(Dalawai et al. 2020; Peng et al. 2019). Some surfaces have intrinsic cleansing property owing to their molecular structure, reduced particle adhesion, water repellence, and through their surface roughness. For instance, water repellent plants (eg. lotus leaf), animal skins, butterfly's wings, mussel inspired materials, etc. (Cheng and Rodak 2005; Hong et al. 2014; Cheng et al. 2006). Further, there are many attempts to modify the bulk surfaces by the introduction of different functional groups, surface charges, making the surfaces porous, and coating the surfaces with other materials to make the surfaces more resistant to the pollutants, and or to the foulants. Going briefly, Rathinam et al. (2019) recently reported the role of surface charge and surface functional group in controlling the silica scaling in brackish water desalination by employing AFM and QCM techniques. They have concluded that the surfaces with positively charged groups (eg. $-NH_2 \sim -N + (CH_3)_3$) induced fast silica precipitation than that of negatively charged or neutral groups (eg. $-NH_2/-COOH$, $-H_2PO_3$, $-OH$, $-COOH$, $-CH_3$) (shown in Fig. 12.2). So, such negatively charged or neutral surfaces are more appropriate for membrane-based water desalination.

On a different perspective, biochar and modified biochar possess promising potential to work as an environment friendly adsorbent for heavy metals from the aquatic environment. Recently, Zhu et al. (2020a) had studied two modified biochar made from porous biochar (PBC), namely, surface oxidized and surface aminated biochar (OPBC and APBC respectively) and were tested for its cadmium (Cd) adsorption capability. The APBC and OPBC showed an adsorption capacity of 23.54 and 19.04 mg g^{-1} respectively, and is about three times higher than that of bare PBC (7.02 mg g^{-1}). It was observed that the regions rich in π electrons, lone pair electrons as well as electron donor groups had showed more adsorption. Among APBC and OPBC, OPBC had plenty of electron donor groups, like $-COOH$ and $-OH$, and it was found that the main interaction for adsorption is in between these functional groups and with Cd(II). In case of APBC, the amino group's lone pair electrons dominated the complexation of APBC with that of Cd. Further, in an another biochar based work, thiol-modifed biochar (BMS-biochar) was synthesized by ball milling

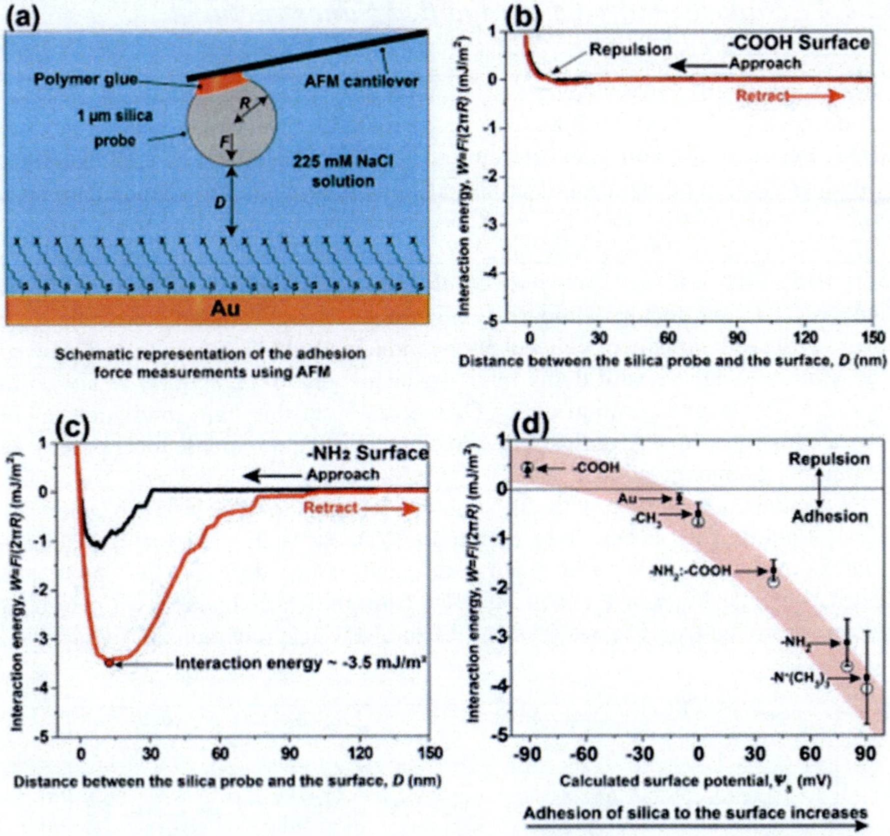

Fig. 12.2 **a** Schematic representation for measuring the adhesion force between silica colloid and different functional surfaces using AFM. **b** Interaction energy versus distance curve between the negatively charged -COOH surface and silica colloidal probe. **c** for the positively charged $-NH_2$ surface. **d** Adhesion/repulsion energy between silica probe and six different surfaces, plotted against the calculated surface potential. Here the white circles represent the calculated energy and black solid markers represent the measured energies. (Reprinted, with permission from Rathinam et al. (2019). Copyright (2019) American Chemical Society)

the biochar with 3-mercaptopropyltrimethoxysilane (3-MPTS) (Lyu et al. 2020). Ball milling enabled BMS-biochar to have large surface area, plentiful functional groups, and to possess more negatively charged surface, there by more electrostatic attraction. These properties resulted their higher removal efficiency of aqueous Hg^{2+} and CH_3Hg^+, which was 320.1 and 104.9 mg/g respectively. There are many more reports describing the usefulness of bulk surfaces and modified surfaces to mitigate the pollution (Cheng et al. 2006; Bhushan and Jung 2011; Blossey 2003; Lu et al. 2015; Genzer and Marmur 2008).

12.4.2 *Nanostructured and Modified Nanostructured Surfaces*

Recent couple of decades witnessed intense research on nanomaterials, functionalised nanomaterials, core–shell structures, 2-dimensional (2-D) layered materials, quantum dots, composite nanomaterials, etc. on pollution control applications, such as photocatalysis, heavy metals/ toxic substances capture, particulate matter adsorption, oil removal, water desalination, etc. (Umrao et al. 2014; Leng et al. 2012; Yu et al. 2018). Villa et al. (2019) recently developed a visible-light-driven bismuth vanadate ($BiVO_4$) micromotors without any functionalization. These micromotors were able to swim individually or collectively by inducing light illumination and showed high affinity to adhere with living microorganisms (eg. *S. cerevisiae*) as shown in Fig. 12.3. The photoactivation of $BiVO_4$ motors by visible light irradiation might generate highly oxidizing radicals, that work as disinfection motile tools to remove the microbial contamination in water.

Environmental pollution with different heavy metals such as lead (Pb), mercury (Hg), cadmium (Cd), arsenic (As), chromium (Cr), etc. is always a major challenge to tackle as their accumulation in our nature and its non-degradability. Adding to that, it is crucial to remove cobalt {Co(II)} from the water bodies as it may be a cause to many human diseases, besides it can also cause mutations in living cells.

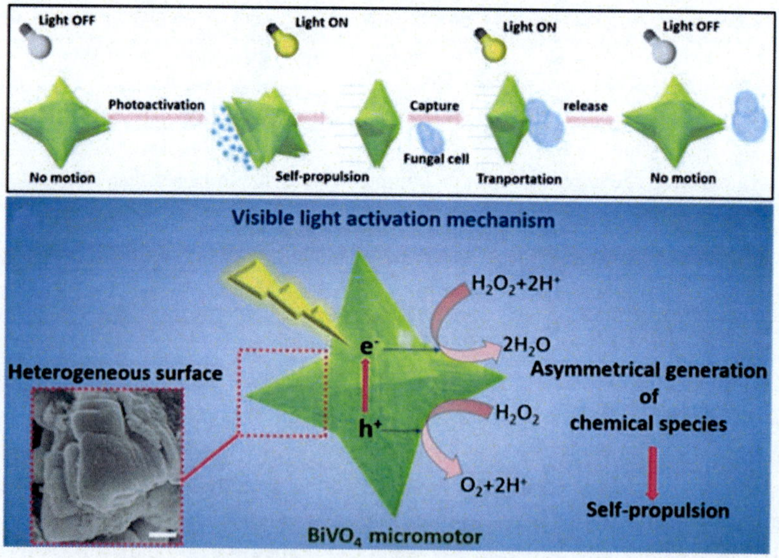

Fig. 12.3 **a** Schematic representation for the photoactivation of $BiVO_4$ micromotors under visible-light for controlled cargo transportation, **b** another schematic to represent the mechanism for the photocatalytic activation and their self-propulsion. On the inset showing a SEM image of one edge of a $BiVO_4$ motor with a scale bar of 500 nm (Reprinted, with permission from Villa et al. (2019). Copyright (2019) American Chemical Society)

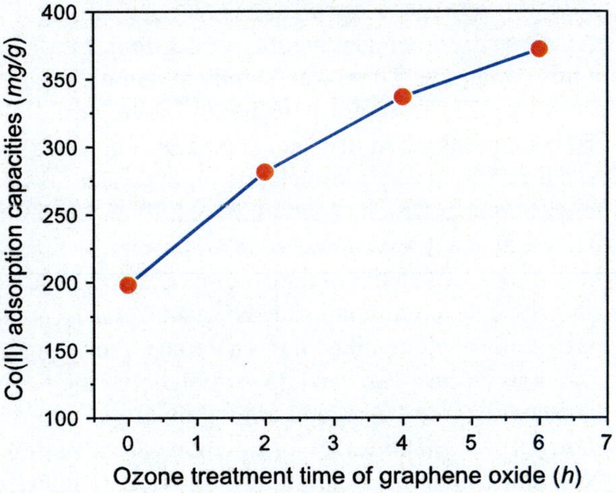

Fig. 12.4 Plot representing Co(II) adsorption capacities at various ozone treatment time of graphene oxide

Interestingly, promising graphene oxides (GOs) with controlled oxidation degrees by ozone treatment, showed Co(II) adsorption properties in aqueous solutions (Liu et al. 2016). So, the increased oxidation rises the amount of oxygen-containing functional groups present on the GOs surfaces and at the same time decreases the sizes of GOs sheets and this might facilitate their dispersion capabilities in aqueous solution. The adsorption processes observed here was endothermic and spontaneous in nature and followed Langmuir adsorption models and the pseudo-second-order kinetic models. Here, the cobalt can make complex with the oxygen-containing functional groups present on the GOs surfaces. For an increased ozone treatment time of GO showed improved Co(II) adsorption capacities as shown in Fig. 12.4, which may be the outcome of both the improved dispersion abilities and the increased oxygen species. In an another report, chemically synthesised graphene (reduced graphene oxide) was tested for its adsorption properties towards antimony {Sb(III)} from aqueous solution (Leng et al. 2012). The study was carried out under various conditions, like the contact time, solution temperature, pH, and initial concentration. Further, the data were modelled using Langmuir and Freundlich isotherms and the experimental data showed a well agreement with the pseudo-second-order kinetic model. The adsorption capacity calculated as 8.056 mg/g and the experimental data showed a well matched value 7.463 mg/g. Again, the system can also be re-used by taking the desorbing agent 0.1 mol/L of EDTA solution and could be used as a promising adsorbent in wastewater treatment. Yet in an another study, GO was evaluated for removing sulfamethoxazole (SMX) and ciprofloxacin (CIP), two antibiotics in the aquatic environment that have adverse effects to the microbial communities in the aqueous solution (Chen et al. 2015). GO system showed an adsorption capacity 379 and 240 mg g^{-1} for CIP and SMX respectively. Here, the electrostatic attraction is the main cause for CIP adsorption and the SMX sorption was mainly controlled by the –EDA attraction on the GO basal planes.

There were many more studies on bare nanostructured materials which focusing on their environmental remedial applications. Few of the reports are briefly discussed in this paragraph. There was a study to demonstrate the photocatalysis of zinc oxide nanowires (ZnO NWs) grown directly on a microfluidic reactor for the efficient chemical decontamination of water (Azzouz et al. 2018). As the photocatalytic medium used here was in very close vicinity of the water flow path, the interaction time for the purification performance was minimal. Even within <5 s and in one-pass, the degradation efficiency was noted as 95%. Further, the system was successfully tested for the removal of volatile organic compounds, which were usually difficult to remove using other decontamination techniques. Yet in an another report, different crystalline manganese oxide (MnO_2) nanomaterials, including α-, β-, γ-, δ-, and ε- were verified for its laccase-like catalytic activities and their capabilities in pollutant oxidation in wastewater treatment and among them γ-MnO_2 exhibited the best performance (Wang et al. 2017). Further, two carbon nanomaterials (CNMs), namely, nanocarbon (NC) and nanoporous carbon (NPC) were synthesized using the chemical vapour deposition (CVD) arrangement by varying the process parameters (Ruparelia et al. 2008). Among NC and NPC, NPC tested greater sorption ability for heavy metals like cadmium, lead, nickel and zinc compared to that of NC.

While looking on the various applications of modified nanostructures, the major contribution arrives from photocatalytic activity. The principle of photocatalytic activity lies on the absorption of photons of appropriate energy to overcome band gap threshold. Here, the absorption of light by the specific system yields an excited state electron and it is raised to the conduction band. This will facilitate the reduction of electron acceptors like H_2 and O_2. The electron holes which remain in the valence band of these system and are able to oxidize the substrates adsorbed to the surface. This oxidation of the surrounding environment assists single electron interactions and successively stimulate the sequential degradation reaction (Kanan et al. 2020). Titanium dioxide (TiO_2), Zinc Oxide (ZnO), Copper (Cu), based nanostructures were widely used in photocatalysis (Yoong et al. 2009; Lu et al. 2011). Titanium dioxide (TiO_2) is one of the most popular photo catalysts present owing to its non-toxicity, self-cleaning, de-polluting, low cost, high oxidizing capabilities, and is easy to immobilize on various surfaces (Kanan et al. 2020; Allen et al. 2008). TiO_2-based photocatalysts were extensively used for the degradation of organic pollutants and pesticides and for the evolution of clean energy since last several decades. TiO_2-graphene surfaces were successfully examined for their photocatalytic degradation ability either under UV or visible light through a number of reports (Zhang et al. 2010; Umrao et al. 2014; Wang et al. 2019). For instance, nanocomposites of TiO_2-graphene had been tested for its gas-phase degradation of a volatile aromatic pollutant- benzene present in air (Zhang et al. 2010). Akram et al. recently reported a cobalt hydroxide modified copper oxide {$Co(OH)_2/CuO$} nanocomposite for their successful organic pollutants (eg. Rhodamine B) degradation under visible light domain with a reaction rate constant of $k = 0.864\ \text{min}^{-1}$ and a degradation time of 8 min (Akram et al. 2020). The light absorption capacity semiconductor materials can be improved by tuning the bandgap of the materials and thereby can improve the photocatalytic performance. Siligenes (gersiloxenes) are honeycomb-like 2D Ge/Si alloy equivalents of germanane and silicane (Zhao et al. 2020). These freestanding material, terminated with − H/ − OH, have a general formula of $Ge_{1-x}Si_xH_{1-y}(OH)_y$, $x = 0.1 - 0.9$). The direct-gap gersiloxenes have a wide range of light absorption capabilities and are suitable for light driven H_2O reduction into H_2 as well as CO_2 reduction to CO. It showed a CO production rate of 6.91 mmol g^{-1} h^{-1}, along with an apparent quantum efficiency of 5.95% at 420 nm (Zhao et al. 2020).

Very recently, Sun et al. (2020a), reported about GO based porous sponges (GPF sponges) made up of GO, poly(vinyl alcohol) and formaldehyde solution for the dye based pollution control or for the wastewater purification. The sponges showed a maximum adsorption capacity of 476 mg g^{-1} for methylene blue along with a recycling efficiency of above 86% after the 10th cycle. A study were there on surface modified electrospun carbon nanofbers (ECNFs) for the removal of Pb^{2+} ions by studying its adsorption onto the surface (Thamer et al. 2019). In this regard melamine-functionalized ECNFs as well as poly(mphenylene diamine)-functionalized ECNFs were examined for their improved Pb^{2+} ions adsorption capacity along with adsorption rate. Yet another promising material, laser-induced graphene (LIG), which is a three-dimensional (3D) porous carbon nanostructure formed by direct laser writing on various polymers. There was a study where LIG was combined with 2-D GO to make a separation membrane with improved properties (Thakur et al. 2019). In the LIG-GO hybrid membrane fabricated, while increasing the quantity of crosslinked GO on the LIG surface had shown an increased rejection of bovine serum albumin (BSA) and an increased bacterial rejection from 20 to 99.9% for a mixed culture of bacteria (~10^6 CFU mL^{-1}).

Zhang et al. have reported a hybrid nanostructure prepared by encapsulating Zirconium Oxide (ZrO_2) nanoparticles into spherical polystyrene beads (MPS) (Zhang et al. 2013a). Further, the negatively charged sulfonate groups were immobilized onto this polymeric matrix, which would have enhanced the diffusion of positively charged ions (here Pb(II) ions) inside the adsorbent through electrostatic attraction between them. The Zr-MPS based resultant adsorbent showed more sorption toward Pb(II) than its simple equivalent mixture of MPS and ZrO_2. They had demonstrated further the role of surface functional groups by encapsulating ZrO_2 onto two other macroporous polystyrene with positively charged quaternary ammonium group ($-N(CH3)_3^+$) and neutral chloromethyl group ($-CH_2Cl$) and also with commercial granular activated carbon. These composite systems were denoted respectively as Zr–MPN, Zr–MPC, and Zr–GAC and the sequence of sorption capacities found as Zr–MPS > Zr–MPN > Zr-GAC > Zr–MPC. It could be concluded that the charged groups in the host resins increase the dispersion of embedded nanoparticles and further enhance its reactivity and capacity for metal ions adsorption, thereby cleansing of water bodies from toxic metals and other charged pollutants.

The crystalline material of metal—organic frameworks (MOFs), which possess high porosity, tunable pore size, and with different functionalities showed their promise towards pollutants capture (Zhang et al. 2016). Four unique MOF structures, namely ZIF-8, UiO-66-NH_2, MOF-199, Mg-MOF-74 were processed and embedded within three different polymers: polyacrylonitrile (PAN), polyvinylpyrrolidone (PVP), and polystyrene (PS) to form nanofibrous filters (MOFilter) with a MOF

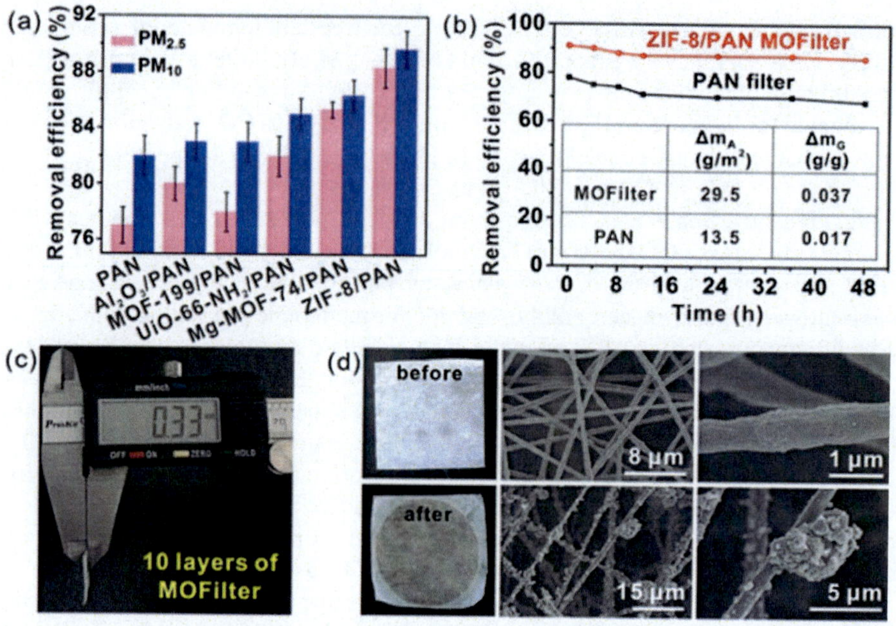

Fig. 12.5 **a** Describes Particulate matter removal efficiencies of different filters tested at T = 23.4 °C, RH = 58.6%, $PM_{2.5}$ = 350, PM_{10} = 720 μg/m^3). **b** Represents the long-term removal efficiencies of $PM_{2.5}$ for ZIF-8/PAN MOFilter in comparison with PAN filter. Inset table shows the gravimetric and areal mass changes of the above two filters before and after PM capture. **c** Measured thickness of ZIF-8/ PAN MOFilter for 10 layers is 330 μm. **d** Optical and SEM micrographs of the sample ZIF-8/PAN MOFilter before and after PM capture. (Reprinted, with permission from Zhang et al. (2016). Copyright (2016) American Chemical Society)

loading of up to 60 wt %. This MOFilters showed the particulate matter removal efficiency up to 88.33 ± 1.52% for $PM_{2.5}$ and 89.67 ± 1.33% for PM_{10} and this performance was even stable for over 48 h of continuous filtration ((shown in Fig. 12.5).

AFM is a powerful instrument to study the adsorption in low dimension (nano and sub-nanoscale) through its topographical imaging and force measurements and is able to provide more insights into the interfacial interactions between different species. Zhu et al. (2020b) studied the interaction between nanocellulose surface with water pollutants by exploring the above technique by modifying the AFM probes with two types of nanocellulose, {2,2,6,6-tetramethylpiperidine-1-oxylradical-mediated oxidized cellulose nanofibers (TOCNF) and cellulose nanocrystals (CNC)}. They took the model water pollutants Cu(II) ions and the Victoria blue B dye for their investigation and found TOCNF modified AFM tip showed higher adhesion force towards Cu(II) ions and to the dye molecules in comparison to CNC modified tip. Further to support this observation, they conducted classical reactive molecular dynamics simulations and showed the Cu(II) ions adsorbed onto the nanocelluloses surface through their co-operative chelating action of carboxyl and hydroxyl groups.

Adding to above mentioned studies on nanostructured and modified nanostructured surfaces in a broad angle of pollution and or fouling control, there were hundreds of reports further on the same spectrum and few more were discussed in brief here. A study on boron nitride nanosheets and nanotubes with different charge states showed the adsorption capabilities towards CO_2, CH_4, and H_2 (Sun et al. 2013). Airborne elemental mercury (Hg^0) adsorption was reported for 2D- MoS2 nanosheets owing to its structural defects from *S* vacancy in the MoS_2 (Zhao et al. 2019). ZnO/carbon quantum dots based nanocomposites showed excellent photocatalytic activity (degradation efficiency over 80%, 24 h) for the degradation of toxic gases like benzene and methanol under visible light irradiation (Yu et al. 2012). Core–shell structure, magnetic Fe_3O_4@poly(m-phenylenediamine) demonstrated for their Cr(VI) removal performance by utilising both the separation ability of magnetic nanoparticles along with the adsorption property of the polymers (Wang et al. 2015).

12.4.3 Membrane and Modified Membrane Surfaces

In general, a membrane performs like a barrier that prevents the passage of certain constituents, however allowing others to pass through it. Selectivity and flux rate are the key factors to determine the performance of a membrane. The removal mechanisms of membranes are commonly governed by their size exclusion or steric hindrance based mechanism by low pressure membranes, adsorptive capability towards specific contaminants by low pressure membranes, size exclusion/steric hindrance mechanism by thin film composite (TFC) membrane, and by Donnan exclusion/ charge-charge repulsion effect by TFC membrane (Abdullah et al. 2019) (Fig. 12.6).

Different membranes based separation and/or purification is an evolving technology and have enormous potentials for several applications (Li et al. 2020; Werber et al. 2016; Jiang et al. 2017). The applications ranges from wastewater treatment, water desalination to separation and purification of large spectra of pollutants, and above all shrink the global issues related to freshwater scarcity (Werber et al. 2016; Cao et al. 2020). So, efficient and cost effective water purification and separation methods are the requirement of the present world to mitigate the demand of a fast growing world population. In this regard, several researchers are working across the world for developing more productive membrane surfaces, mainly by modifying the existing polymer based membranes. Desalting techniques based on nanofiltration (NF) and electrodialysis (ED) and reverse osmosis (RO) technologies have been extensively used in seawater and brackish water desalination along with waste water treatment in last few decades (Werber et al. 2016; Walha et al. 2007).

Yet in an another study, halloysite nanotubes (HNTs) were directly anchored on commercial nanofiltration (NF) membrane surface by dopamine modification (Liu et al. 2020a). It was found that the hydrophilicity, surface roughness, anti-fouling property and stability of the membrane got enhanced owing to the introduction of HNTs to the NF membrane. The improved permeability was found to be 13.9

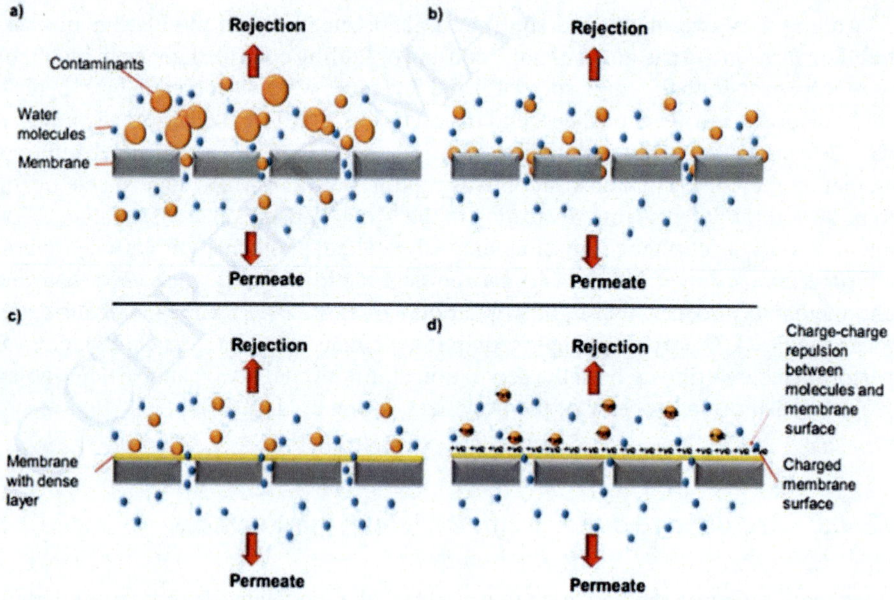

Fig. 12.6 Schematic to represent the solute removal mechanisms for different types of membranes. **a** size exclusion/steric hindrance mechanism, **b** adsorption mechanism, **c** size exclusion/steric hindrance mechanism by thin film composite (TFC) membrane and, **d** Donnan exclusion/charge-charge repulsion mechanism by TFC membrane. (Reprinted, with permission from Abdullah et al. (2019). Copyright (2019) Elsevier)

$L{\cdot}m^{-2}{\cdot}h^{-1}{\cdot}bar^{-1}$ and the Cu^{2+} rejection ratio obtained was 74.3% for this modified NF membrane. Here the enhancement in rejection ratio was a result of the synergy between the increase in dopamine concentration on the membrane surface and HNTs incorporation on the same surfaces. The active dopamine layer, that act as a separation layer, could efficiently prevent the penetration of heavy metal ions like Cu^{2+} and Ni^{2+}. Along with this, (–OH) and ($-NH_2$) groups existed in the dopamine provided effective adsorption sites to undertake chelation reactions with heavy metal ions. As HNTs have huge specific surface area, abundant hydroxyl groups and plenteous micropores, it adheres on the membrane surface by the adsorbent dopamine. This result the effective adsorption of heavy metal ions and higher rejection ratio.

Multi-level structured membranes were prepared by metal–organic framework, ZIF-8 and polyacrylonitrile (PAN), and then alternatingly stacking layers of PAN nanofibers, via electrospinning (Wang et al. 2020). Subsequent acid and base treatments on the membranes were allowed them to tailor the hydrophilicity of the fibre surfaces and induced multiscale surface roughness, which facilitated both moisture wicking and PM adsorption. This unique multilayer membrane composed of an outer layer of super-hydrophilic nature, made of PAN fibres, a hydrophilic intermediate layer consisting porous PAN-ZIF-8 microfibers along with PAN nanofibers, and a polystyrene fibrous inner layer which is hydrophobic in nature. The above resulting membrane exhibits excellent removal efficiencies of particulate matter (PM) of various sizes; 99.973% for 0.3–μm sized particles and ≧99.99% for particles of with other sizes, at a low air resistance (pressure drop of 80.1 Pa).

A poly(vinyl alcohol)/chitosan (PVA/Chi) based nanofbers (NF) membranes were fabricated using electro-spun technique and were tested for their selective heavy metals (Pb(II) and Cd(II) ions) adsorption from wastewater samples in the presence of divers metal ions (Karim et al. 2019). In this report, a maximum adsorption capacity of 266.0 mg/g was calculated for Pb(II) ions and 148.0 mg/g for Cd(II) ions. Here the pseudo-second-order model obeyed the adsorption kinetics for these ions and the equilibrium data were fitted well with the Langmuir adsorption isotherms model. Further, parameters such as pH of solution, adsorbent dosages, contact time, and initial concentrations were also having influence in the adsorption process.

A study by Efome et. al. showed the fabrication of electrospun nanofiber composite membranes that contain water-stable MOFs particles supported on polyacryloni-trile (PAN) nanofibers (Efome et al. 2018). The membrane performance towards heavy metal ion adsorption in batch filtration was estimated on the basis of Cd^{2+} and Zn^{2+} ions sequestration. The maximum adsorption capacities observed for Cd^{2+} was 225.05 mg g^{-1} and for Zn^{2+} was 287.06 mg g^{-1}. Here, the activation of porous water-stable material (hydractivation) produced an expanded MOF with improved adsorption potentials.

There were many more studies based on membranes and modified membrane surfaces for environmental remedial applications. For instance, positively charged NF membrane based on polyamide thin film with poly(amidoamine) dendrimer for the rejections for toxic elements (Cu^{2+}, Ni^{2+} and Pb^{2+}) (Li et al. 2017); silicified membrane for crude oil repulsion, improved protein interception (94.1%), and for good antifouling (73.2% enhancement) capacity (Yang et al. 2020); modified poly-acrylonitrile (PAN) membrane for heavy metal ions adsorption (Zhang et al. 2018); sodium ion modified carbon quantum dot based thin-film nanocomposite membranes for arsenic and selenium removal (He et al. 2018); silver nanoparticles (Ag NPs) immobilized on membranes for antibacterial properties (Qi et al. 2018); porous polyvinylidene fluoride–graphene oxide nanofibrous membranes for selective sepa-ration and filtration (Ghaffar et al. 2018); a composite membrane of zwitterionic polymers and silver nanoparticles for antimicrobial activity (Liu et al. 2017); and electrospun polyacrylonitrile nanofibers (EPNFs) for $PM_{2.5}$ capture air-filter (Kim et al. 2018) are few of them.

12.4.4 Polymer and Modified Polymer Surfaces

Polymers and modified polymers are extensively employed in pollution control strategies. The major contribution of pollution in our aquatic system arises from the increased industrial activities in its surroundings (Schwarzenbach et al. 2010; Xue et al. 2017). So, it is essential to eliminate these aromatic contaminants such as anilines, phenols, bisphenol, etc. from the wastewater. Many methods like oxidation, electrochemical degradation, photocatalytic degradation, adsorption-based separation and membrane filtration have been applied for tackling such contaminated waste water. Different polymer structure or with its different functionalised form are of special interest to deal such pollution. For instance, a phenyl-rich β-cyclodextrin porous cross-linked polymer solution system, namely, PCD-PCP was employed to eliminate the aromatic pollutants (here, 2,4,6-trichlorophenyl and bisphenol A was taken as model adsorbates) from water (Huang et al. 2020). The properties of the composite system were able to be tuned by adjusting the crosslinking degree, porous features, the nanostructure, and by the phenyl contents and thereby controlling the adsorption capacities. In wide pH along with ionic strength, PCD-PCP(H) system with high crosslinking density maintained a high adsorption capacity.

Porous cyclodextrin polymer (PBCD-B-D) was prepared and tested by Tu et al. (2020) for the speedy removal of organic micro pollutants. Water-soluble pollutants such as 3-phenylphenol, p-nitrophenol, 2-naphthol, and 4-chlorophenol were used for this adsorption study. The PBCD-B-D based system showed high adsorption rate with a removal efficiency above 95% in 1 min. Further, the PBCD-B-D system is amphiphilic as it can be used to adsorb both the hydrophilic and hydrophobic pollutants and have the potential to be used in water treatment. In an another study using reflux procedure, seven different benzimidazole polymers (BINP) were prepared and conjugated with porous silica through amine and aldehyde-based materials (Maruthapandi et al. 2020). These polymers were then tested by thermal analysis for its quantification towards the adsorption and desorption of CO_2. All those polymers were found good for adsorbing CO_2 from a flowing stream (80 mL/min at 25 °C) but the amorphous BINP showed better adsorption than the crystalline. For these polymers, the adsorption capacities dependent upon their nitrogen content, pore size and specific surface area.

There were reports on polymer stabilizers to mitigate issues of dust emission from the bauxite residue drying areas (RDA) by forming a crust on those surface of treated material (Ding et al. 2020). By using synthetic polymer polyacrylamide and natural polymers xanthan gum and guar gum as stabilizers, it was found that the water retention property of treated red sand was greatly improved. Yet in an another report, a renewable natural polymer—Konjac glucomannan (KGM) and graphene oxide (GO) were used to make a three-dimensional sponge network structure (KGM/GO) and adsorption experiment was conducted for malachite green (MG) and radionuclide uranium U(VI) (Chen et al. 2018a). The sponge network showed excellent selectivity for capturing U(VI) and adsorption capacities of 266.97 and 189.96 mg/g for U(VI) and MG respectively.

A promising, wearable and self-powered air filter was recently prepared and reported by exploring an ionic liquid–polymer (ILP) composite sponge network (Zhang et al. 2020). This filter was made up of 1-alkyl-3-methylimidazolium acetate along with few hydrophilic polymers {poly(acrylamide), poly(vinyl alcohol) and poly (vinylpyrrolidone)} onto a melamine–formaldehyde (MF) resin sponge network. This could be used to remove the particulate matter (PM) pollutants including nanoscale particles (NPs) as shown in Fig. 12.7. Owing to the exceptional electrochemical properties, the fabricated ILP@MF filter was able to remove $PM_{2.5}$

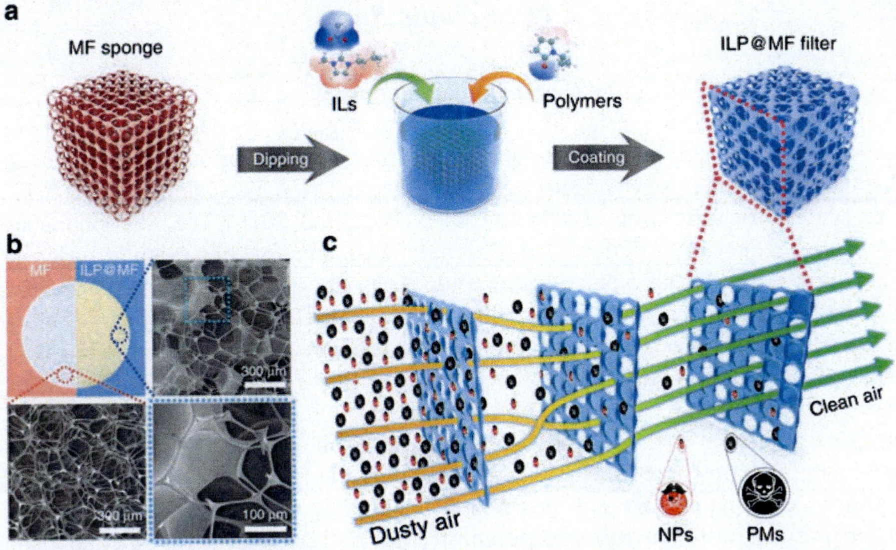

Fig. 12.7 **a** Schematic of the dip-coating process for the ILP@MF filter fabrication. **b** Optical and SEM micrographs of the MF sponge (red background) and the ILP@MF filter (blue background). **c** A schematic to illustrate the removal mechanism of multilayer ILP@MF filter. (Reprinted, with permission from Zhang et al. (2020). Copyright (2020) Springer Nature)

and PM_{10} with efficiencies as high as 99.59 and 99.75%, respectively, once applying a low voltage (3 V). Along with this, the charged ILP@MF filter also showed an improved removal efficiency (93.77%.) for NPs.

High nitrogenous fertilizer inputs increase the reactive nitrogen levels in soil, water and air, and this can change into a pollutant to the natural resources. By using corn starch, a biodegradable carbohydrate polymer, was processed into a coating material for reacting with urea and borate and showed better stability along with a larger release time than that of uncoated urea (Ibrahim et al. 2020). In an another study, graphene oxide (GO) is enwrapped on Polyethersulfone (PES) to make a porous structure and tested for its function to remove dyes and other pollutions from wastewater (Zhang et al. 2013b). The porous system showed an adsorption capacity up to 62.5 mg/g for the dye Methylene Blue (MB). Further, there are many reports which focuses on the development and fabrication of surfaces which have environmental remedial, antifouling and/or foul release properties while taking different polymers as examples (Lee et al. 2019; Zhan et al. 2011; Huang et al. 2015, 2014; Tungittiplakorn et al. 2004).

12.4.5 *Hydrophobic and Oleophobic Surfaces*

There are many plants, insects, and animals in our nature that can repel water and many other low surface tension liquids like oil and organic solvents. Lotus leaves are one of the best examples that exhibit superhydrophobicity. Various studies showed that the surface chemistry and surface texture are the two major factors to be considered while fabricating liquid repellent surfaces (Peng et al. 2019). The hydrophobicity of surfaces reduces the frictional force between water and that of particular surfaces. This allow the water droplets to roll or slide along the surfaces with more easiness than on hydrophilic surfaces. Hydrophobicity and superhydrophobicity (the contact angles of water droplet on the superhydrophobic surfaces exceed 150°) have received tremendous attention among researchers over the past few decades owing to their extensive potentials in enhancing self-cleaning, anti-icing, and in antibacterial coatings (Heckenthaler et al. 2019; Peng et al. 2019; Ruan et al. 2013). Many existing polymeric materials like PET, PTFE, PDMS, polyethylene, and polypropylene are hydrophobic in nature. So, it is possible to make large scale hydrophobic surfaces and that can be utilised in different pollution control strategies and for other applications. Generally, there are three methods to create a superhydrophobic surface and are listed below: (Yan et al. 2019).

(i) Roughening of an intrinsically hydrophobic substrate;
(ii) Roughening of an intrinsically hydrophilic substrate and successive conformal deposition of a thin hydrophobic coating.
(iii) Fabrication of re-entrant roughness geometry on the substrate

Above mentioned processes can be achieved by various techniques like chemical reaction, self-assembly, e-beam lithography, photolithography, electrospinning, and by electrochemical deposition (Yan et al. 2019; Park et al. 2010). Yan et al. recently made such a superhydrophobic surface having self-cleaning property (Yan et al. 2019). Initially, hierarchical copper oxide (CuO) of high surface energy was made from copper (Cu) by the laser processing as well as from the thermal oxidation. Successively, by a passive and repeatable atmospheric adsorption of different hydrophobic volatile organic compounds (VOCs); made up of alkanes, halogenated hydrocarbons, aromatic hydrocarbons, and oxygen-containing organic molecules, made the surface a stable superhydrophobic surface with an apparent advancing contact angle $\approx$160°. So, by exploiting the volatile 'pollutants'—thereby cleaning the environment; attaining a self-healing, and long-lasting superhydrophobic surface without the use of organosilane or polymeric coatings.

In an another report, a novel superhydrophobic sponge system was fabricated based on porous polyurethane (PU) sponge by coating it on with zinc oxide (ZnO), iron oxide particles (Fe_3O_4) and stearic acid (SA) (Tran and Lee 2017). A static water contact angle of 161° is observed for this PU@ZnO@Fe_3O_4@SA sponge confirming its superhydrophobicity. The above stable and re-usable sponge was capable of separating different oils from water with an efficiency in separation over 99%. It can additionally control by magnets to rapidly absorb oil floating on the water surface to

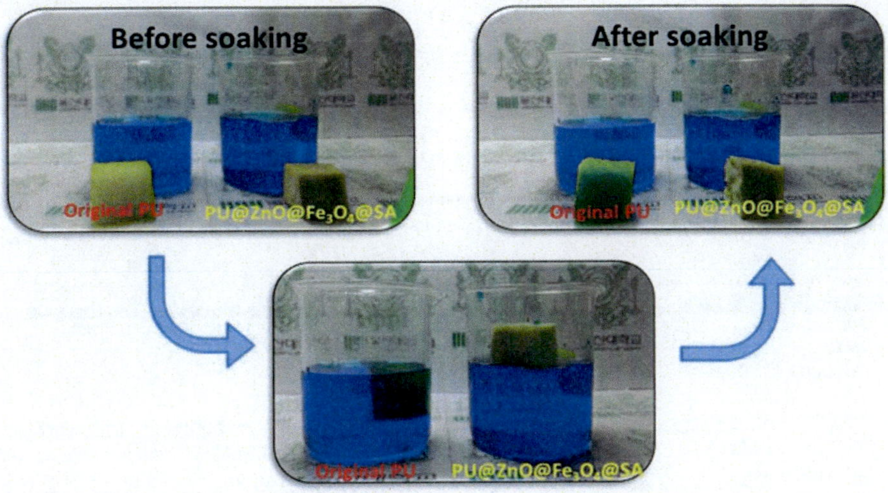

Fig. 12.8 Soaking experiment conducted on hydrophilic PU and superhydrophobic PU@ZnO@Fe3O4@SA sponges with methylene blue dye solution showed different wetting behaviours. (Reprinted, with permission from Tran and Lee (2017). Copyright (2017) Springer Nature)

have potential in marine oil spills and in other oil–water separation problems (shown in Fig. 12.8).

Yet in another study on hydrophobicity and its cleansing property, ZnO nanoparticles and SiO_2 nanoparticles were coated on a glass substrate using an epoxy resin and the nanoparticles were treated with 1H, 1H, 2H, 2H-perfluorodecyltriethoxydsilane (FAS-17) (Peng et al. 2019). This surface showed excellent superhydrophobicity along with high oleophobicity and the water and ethylene glycol contact angles could be as high as $172 \pm 2°$ and $157 \pm 2°$, respectively. The system is checked for its self-cleaning and antifouling behaviours for liquid and solid pollution.

Kobayashi et al. reported about the polymer brushes, with two types of functional groups; hydrophobic and hydrophilic and studied their surface wettability properties (Kobayashi et al. 2012). The polyelectrolyte brushes made by surface-initiated atom-transfer radical polymerization method on silicon wafer surfaces showed surface free energies of 70–74 mN/m which is close to the value for water. Here, the superhydrophilic polyelectrolyte brush showed the oil (for instance, silicone oil and hexadecane) detachment characteristics in water owing to its low adhesion force among the hydrate brush and the oil (shown in Fig. 12.9). These findings contributed to the antifouling and self-cleaning properties of these prepared polymer brushes towards strategic pollution control. Finally, and interestingly, this approach for making a superoleophobic water/solid interface relies on an ionic polymer based superhydrophilic surface, which is distinct from the usual methods of using superhydrophobic surfaces.

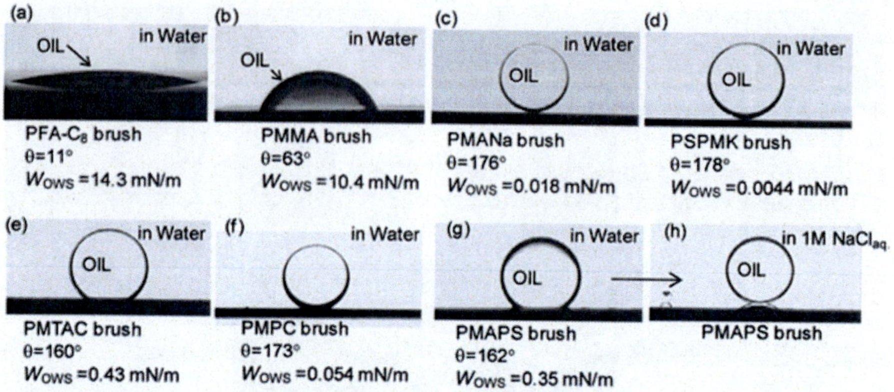

Fig. 12.9 Optical images (side view) of a silicone oil droplet on different polymer brush surfaces. **a** PFA-C8. **b** PMMA; **c** PMANa. **d** PSPMK. **e** PMTAC. **f** PMPC. **g** PMAPS brushes in water. and **h** PMAPS brush in 1.0 M NaCl aqueous solution at 298 K. Contact angles of the oil droplet in water and work of adhesion are mentioned under the images. (Reprinted, with permission from Kobayashi et al. (2012). Copyright (2012) American Chemical Society)

Along with superhydrophobic surfaces, superoleophobic surfaces also have broad potential applications in the area of pollution control. It can widely use in oil capture, transportation, anti-oil adhesion, and for oil/ water separation (Chen et al. 2018b). Chen et. al. recently reported such a surface made up of tannic acid (TA); a member of the polyphenols family, coated copper surfaces which have exceptional underwater super oil-repellent property (Chen et al. 2018b). Further, it was tested as a general strategy to make the underwater super oil-repellent surfaces by the validation of it for different oils and on various metal sheets as described in the Fig. 12.10.

In an another study using biomimetic TiO_2—titanium meshes (BTTMs), successively demonstrated about a stable smart surface which was capable of treating corrosive oily wastewater in a controllable manner (Gao et al. 2019). This BTTM surfaces possessed controlled wettability and could easily separate oil or water from an oil–water mixture. So, this surface can be tuned its wettability between underwater superoleophilicity to underwater superoleophobicity or vice versa by changing the surface hydroxyl content by UV irradiation or heating. Further, even for extremely corrosive oily wastewater, the BTTM surfaces upheld high separation efficiency (>92%). Along with this, BTTM surface have self-cleaning and healing ability as well as anti-oil-fouling properties, which is very much crucial for oil pollution control strategies.

Guix et al. developed a SAM-Au/Ni/PEDOT/Pt based micromachines for the successful interaction, capture, transport, besides the removal of oil droplets (Guix et al. 2012). Here, to modify such surface to interact strongly with oily liquids, self-assembled monolayers (SAMs) of alkanethiols was coated on the device's rough outer gold surface. This superhydrophobic layer make the machine to adsorb oil by means of its strong adhesion with the SAM. The oil collection efficiency, and the micromotor-oil interaction can be tuned by controlling the alkanethiol chain length,

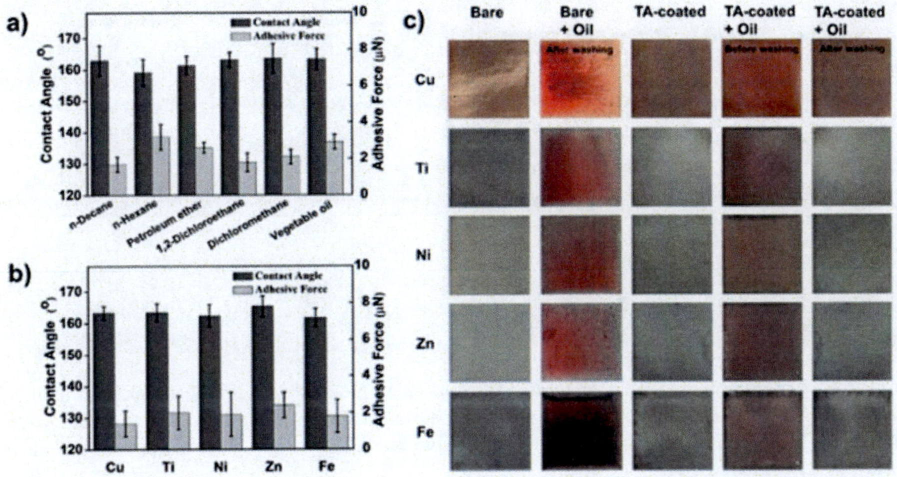

Fig. 12.10 Bar plots shows the underwater superoleophobicity and ultralow oil adhesion forces of **a** various oils, **b** different TA-coated metal sheets. **c** Optical images show the absence of residual oil (here, it was silicon oil dyed by oil red O with the viscosity of 14,000 cst) stained on different TA-coated sheets subsequently washed with water compared to that of bare sheets. (Reprinted, with permission from Chen et al. 2018b. Copyright (2018) American Chemical Society)

polarity, and the head functional groups present. So, such a system has huge potential in collection of oils in oil-contaminated water OR for other water–oil separation technologies.

Many other studies were there to explore the hydrophobicity/hydrophilicity/ oleophobicity of the surfaces to environmental remediation applications. For instance, a super-hydrophobic surfaces was made by the smart use of pyrolytic oils and gases which were the outcome of recycling technologies of waste printed circuit boards (Liu et al. 2020b); a hydrophobic and oleophilic methyltrichlorosilane (MTCS) coated chitin based sponge made by freeze-drying method as an effective remover of oil from water (Duan et al. 2014); and many more (Wu et al. 2018; Feng et al. 2004; Xue et al. 2011; Zhu et al. 2011).

12.4.6 Patterned/Textured Surfaces

Among the manmade contaminants and pollutants present in the world are mainly composed of organic materials. Such materials usually will stick on the surfaces which are superoleophilic in nature and are difficult to be removed from the surfaces. So, to tackle this issues of oils and greases from the surfaces, it is crucial to develope superoleophobic surfaces. Zhao et al. reported a method to make such surfaces on a silicon wafer by creating some textured surfaces through photolithography technique and successive chemical modification with a fluorosilane (Zhao et al. 2011). The

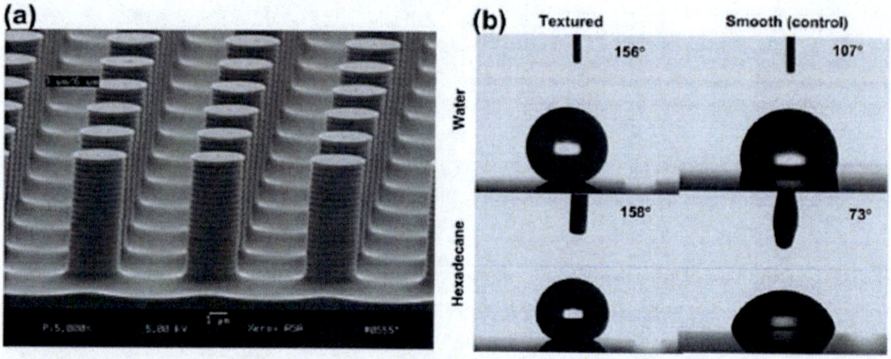

Fig. 12.11 **a** SEM image of the textured FOTS surface on Si substrate. **b** Static contact angles measured for water and hexadecane on the same surface, keeping a control measurement with smooth FOTS/Si surface. (Reprinted, with permission from Zhao et al. (2011). Copyright (2011) American Chemical Society)

patterns were made in such a manner that arrays of pillars with ~3 μm diameter, ~7 μm height with a 6 μm center-to-center spacing (shown in Fig. 12.11). Further, the side walls of each pillar were not smooth; but had some wavy structure of a ~300 nm from top to bottom. The molecular vapor deposition technique was employed for fluorosilane (FOTS) coating to make the surface oleophobic. Such surface exhibited excellent repellency toward water and oil (eg. hexadecane) with contact angles at 156 and 158, respectively. The control experiments with the bare silicon surfaces (both smooth and textured) for their contact angles showed that the superoleophobicity was an outcome of both surface texturing and fluorination. Further, they had compared two types of textured surfaces, one with smooth straight side wall pillars and other with straight side wall pillars having a 500 nm re-entrant structure (made of SiO_2) and made the surfaces oleophobic with FOTS. In this case, the textured surfaces with the re-entrant structure were both superoleophobic as well as superhydrophobic. So, the study concludes that the wavy structure at the top of every pillar is the core geometrical contributor to the surface's superoleophobic property.

The intrinsic nanotexturing of hydrophobic surfaces improves the self-cleaning efficiency as it further reduces the frictional force between water and the surface. Further, the forces that act on the attached particles is an another important matter to be take into account. The two prominent forces need to be considered are the particle–surface adhesion force and the force of the water-particle-air line tension put on to the particles. Very recently, Heckenthaler et al. (2019) demonstrated the importance of textured surfaces in self-cleaning mechanism by conducting various experiments on smooth/textured surfaces. They found that the superhydrophobic Si-based nanotextured surfaces they made were much superior to hydrophilic smooth Si-based surfaces in particle removal experiments. Here, the particle removal was ~41% for hydrophilic smooth Si surface and 98% for superhydrophobic nanotextured surface (shown in Fig. 12.12). So, the particle removal efficiency increases with

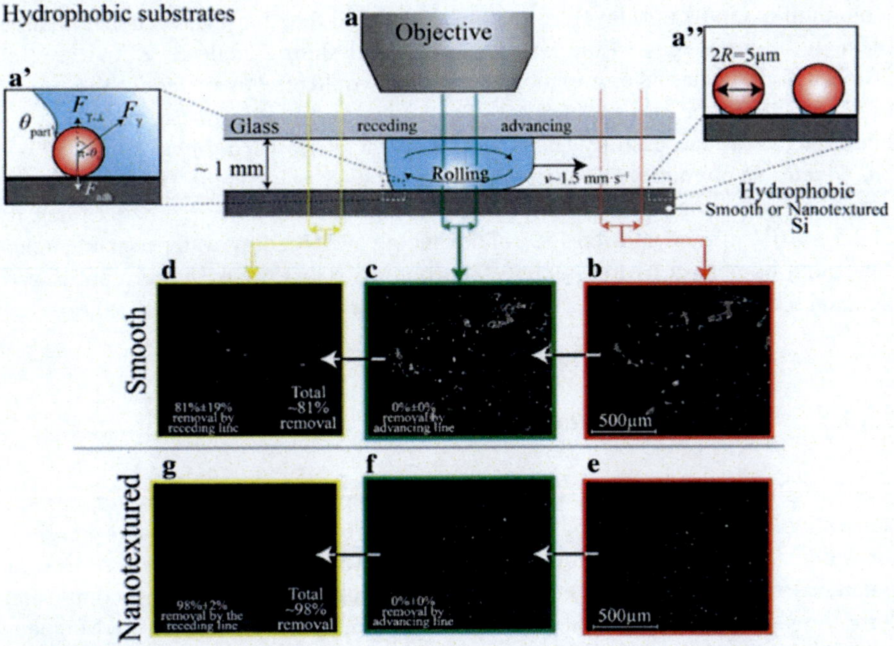

Fig. 12.12 **a** and **a″** represent the schematic of the drop of water moving along the hydrophobic smooth or hydrophobic nanotextured substrate through the microfluidic device, **b–d** represent the fluorescent images of the smooth substrate, and **e–g** for the nanotextured substrates. Here, (**b** and **e**) denote the particles present before the self-cleaning mechanism. The removal percentage is based on four measurements in both smooth hydrophobic substrates and for the nanotextured hydrophobic substrate. (Reprinted, with permission from Heckenthaler et al. (2019). Copyright (2019) American Chemical Society)

hydrophobicity of the surface and the hydrophobicity enhances with nanotexturing on the surfaces.

In an another report about an ordered polystyrene microsphere-patterned surface (MSPS), fabricated by the spin-coating technique, was used for the adhesion and friction studies (Zhang et al. 2012). Further, 1-octadecanethiol (ODT) self-assembled monolayer (SAM) was deposited on this MSPS system to monitor the effect of this hydrophobic coating on the patterned surfaces. AFM technique using a colloidal probe was used to check the adhesion and friction on the MSPS and the results showed that the MSPS reduced the adhesion and friction compared to that of flat surface. It is mainly due to the reduction in the contact area between the contacting interfaces. Additionally, the introduction of a hydrophobic film onto this MSPS can further reduce the forces of adhesion and friction owing to the meniscus-mediated adhesion and friction reduction between interacting bodies.

Yet in an another study on aluminum substrate for the prevention of bacterial attachment, oil-impregnated nanoporous teflon-coated hydrophobic anodic aluminum oxide (AAO) layers were syntesized and they have showed an enhancement in the repellency to the water with a small sliding angle of 2° (Villa et al. 2019). This indicates its omniphobicity and that would effectively reduce the contact of bacteria onto the surface of AAO. The antibacterial performance on this modified urfaces was successfully tested with model bacteria *Escherichia coli* K-12 and *Salmonella* Typhimurium. Gyu Park et al. showed polystyrene (PS) nanoneedle arrays of protruding tips having a tip diameter of around 20 nm, a pattern density of $2.5 \times 10^8/cm^2$ and showed 180° of advancing and receding water contact angles, confiming its perfect hydrophobicity. (Park et al. 2010). Many more reports were there on the same filed (Ma et al. 2017; Ding et al. 2019; Kharraz et al. 2020).

12.4.7 Magnetic Surfaces

A unique physical property; magnetism, which independently helps in water purification technologies by influencing the physical properties of the contaminants present in the water. Adsorption is a well-practised and effective technology to remove several pollutants from the environment, especially pathogenic and polluting composites from the water (Maksoud et al. 2020; Yu et al. 2020; Chen et al. 2009). Magnetic materials/ magnetic nanoparticles have shown intense research interest in recent years owing to its function to work as an efficient adsorbent (Maksoud et al. 2020; Latour 1973; Kumar et al. 2020; Ambashta and Sillanpää 2010). By applying a suitable magnetic field, it is easy to remove pollutants from the waste water or from any other contaminated system. Adding to the magnetism, the extraordinary surface charge along with the redox activity characteristics contributes their qualification to compete with other materials as an adsorbent.

There were a number of oil spill incidents and chemical leakages across the world and it causes major threats to the coastal environment and oceanic ecosystem. Therefore, to mitigate this severe environmental damage, it is the need of the time to remove those large amounts of oil and other organic pollutants from the water surfaces. Yu et al. (2020) has recently developed a three dimensional (3D) magnetic, superhydrophobic/superoleophilic porous composite material for the efficient adsorption of organic pollutants from water. The composite structure was made up of polyurethane sponge along with high-density polyethylene (HDPE) containing magnetic (Fe_3O_4) nanoparticles. Here, the magnetic-controlled elimination of underwater oil is supported by Fe_3O_4 particles. The superhydrophobic nature of the functionalized polyurethane sponge network allowed the separation of oil/water mixtures along with the demulsification of toluene/water emulsions. The present system showed an oil absorption capacity of 15–52 gg^{-1} subject to the organic solvent used and was stable for even 10 cycles. The presence of HDPE coating contributed its exceptional stability of this composite system with acid, base, salts, seawater or even with a temperature range, −20 to 105 °C. In an another study to remove the oil and other organic pollutants from the water surfaces, magnetic graphene foam loaded with magnetite (Fe_3O_4) nanoparticles were made to validate its applicability in the adsorption study (Yang et al. 2014). Advantages from the porous structure along with the magnetic properties, the composite system showed exceptional oil adsorption capacity in addition to its stability for different cycle, high restoration for absorbates and excellent recyclability. The adsorption was calculated according to the Eq. 12.1 mentioned below:

$$Q = \frac{W_{\text{After}} - W_{\text{Before}}}{W_{\text{Before}}} \tag{12.1}$$

where W_{Before} and W_{After} were respectively the weight of magnetic graphene foam before and after dipping in target liquids.

Yet, in another study, a composite system based on polyurethane foams functionalized with superparamagnetic iron oxide nanoparticles and sub-micrometre particles of polytetrafluoroethylene was investigated for separating oil from water (Calcagnile et al. 2012). Here, the functionalization of the polyurethane foams with the polytetrafluoroethylene made the system water-repellent and oil-absorbing (shown in Fig. 12.13). Adding to this, the functionalization with iron oxide nanoparticles provided its magnetic responsivity. Further, as the composite foams have light weight, they can float easily on water. So, moving these foams around the oil polluted waters using a magnet, the system can absorb the floating oil from the polluted areas; results, cleansing the water underneath.

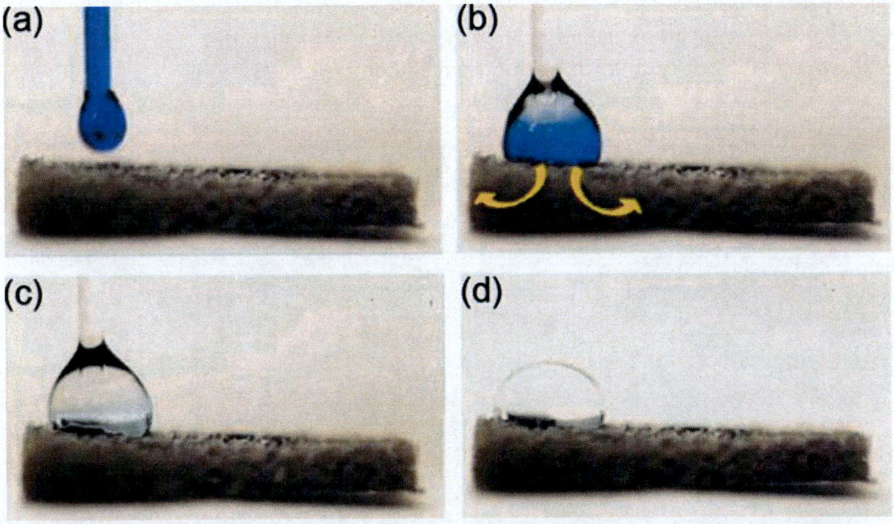

Fig. 12.13 **a–d** Represents the optical images of mixed oil (blue colour) as well as water drop, which are phase separated. Here, oil was absorbed immediately {shown by arrows in **b**} while water remained on the surface. The time interval between (**b** and **c**) was less than 1 s (Reprinted, with permission from Calcagnile et al. (2012). Copyright (2012) American Chemical Society).

A magnetic bifunctional β-cyclodextrin nanocomposite (Fe_3O_4@β-CD-CDI) was tested against persistent organic pollutants (POPs) by Liu et al. (2020c) for its adsorption and degradation properties in aqueous solution. As it has a magnetic core, the nanocomposite system can be separated and recycled from the solution by applying an external magnetic field. Further, the same system is tested to function as a heterogeneous catalyst to remove bisphenol A (BPA) in presence of hydrogen peroxide (H_2O_2). Yet another magnetic adsorbent based on $CuFe_2O_4$/activated carbon has the synergic effect of adsorption features from the activated carbon and the magnetic and the catalytic properties arises from the powdered $CuFe_2O_4$ (Zhang et al. 2007). This system is used to adsorb acid orange II in water and can be easily separated from the medium by the simple magnetic technique.

Self-cleaning and antifouling capabilities of membranes are always a challenge to achieve in the fabrication process of membranes. So, a novel method was implemented by arranging magnetic TiO_2@Ni particles onto a polyether sulfone (PES) membrane surface by applying an external magnetic field (Sun et al. 2020b) (shown in Fig. 12.14). A flux of 871.2 ± 2.9 L.m^{-2}.h^{-1}.bar^{-1} was achieved by this composite membrane and had five times as large as the flux of the pristine PES membrane. Further, it had an advantageous rejection (95.85%) of bovine serum albumin (BSA) than that of the pristine PES membrane. Along with this, the composite membrane system possessed a self-cleaning property under UV light and sunlight. A flux recovery ratio of 75.4, 99.56, 92.11 and 98.26% for BSA, yeast extract fermentation, ammonium alginate and humic acid solutions respectively were found after a self-cleaning process.

The coal combustions usually cause the emission of mercury (Hg) and it is one of the main sources of anthropogenic mercury emission. These Hg species mainly

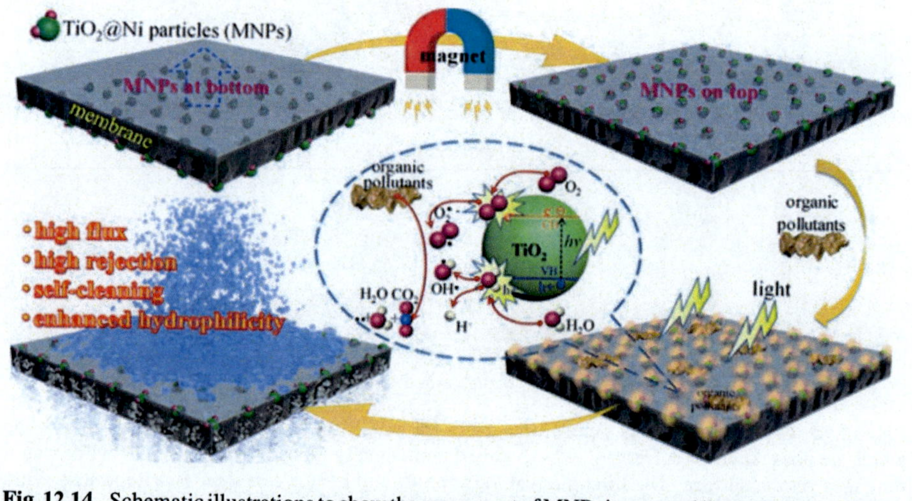

Fig. 12.14 Schematic illustrations to show the movement of MNPs in a casted liquid PES membrane and the photodegradation pathway. (Reprinted, with permission from Sun et al. (2020b). Copyright (2020) Elsevier Inc.)

present as elemental mercury (Hg^0), particulate-bound mercury (Hg^p) and oxidized mercury (Hg^{2+}). Flue gas desulfurization (FGD) and electrostatic precipitator (ESP) can be employed to remove Hg^p and Hg^{2+} but cannot remove Hg^0. So, among different Hg species, gaseous Hg^0 is the main species emitted by coal-fired plants to the environment. So a H_2S-modified Fe–Ti spinel was developed by Zou et al. (2017) as a recyclable magnetic sorbent to recover the Hg^0 species from flue gas as an additional advantage from the wet electrostatic precipitators. At, 60° C, the system showed an approximate adsorption rate of 1.92 $\mu g\ g^{-1}\ min^{-1}$ along with a capacity of 0.69 $mg\ g^{-1}$ owing to the presence of $S_2{}^{2-}$ on its surface. Further, the system is almost stable for the five cycles tested for Hg^0 capture, Hg^0 recovery, as well as the sorbent regeneration. The chemical adsorption reaction of Hg^0 onto the H_2S-modified Fe–Ti spinel is approximately described as follows:

$$Hg^0_{(g)} \rightarrow Hg^0_{(ad)} \tag{12.2}$$

$$Hg^0_{(\mathrm{ad})} + S_2^{2-} \rightarrow HgS + S^{2-} \tag{12.3}$$

Equation 12.2 represents the physical adsorption of gaseous Hg^0 on the surface, while the formed Hg species is Hg*S* as in Eq. 12.3.

In an another study, adsorption and catalytic reduction of environmental pollutants were investigated by employing a magnetic carbon nanotube-reduced graphene oxide-silver nanocomposite based system (Islam et al. 2020). The catalytic activity as well as the fast and efficient adsorption of the nanocomposite system was confirmed at variable pH. The model toxic dye, methylene blue (MB), as well as the aromatic nitro compound, 4-nitrophenol (4-NP) were tested with this composite structure and the system found to be highly recyclable (more than 15 cycles). Further, Hu et al. (2020) prepared a magnetic β-cyclodextrin polymer (MNP-CMCDP) with several macro pores and ultramicropores in aqueous phase. Using that polymer system, several pollutants such as BPA, MB, RhB, Cr(III), Pb(II), and Cu(II)) were removed efficiently from the aqueous solution. Yet, in an another study based on magnet-responsive oil-degrading bacteria, *Brevibacillus parabrev*, was conducted successively to remove various oil pollutants from the water under an external magnetic field (Cheng et al. 2020). In this regard, the oil-degrading bacteria were coated with magnetic Fe_3O_4 nanoparticles shell using polycations and those bacteria showed excellent pickering emulsification for oil.

Many more other reports were there in utilising the magnetic property of the materials to adsorb pollutants from the water and air medium. Magnetic MCM-41 was prepared by embedding iron oxide nanoparticles in MCM-41 and is tested for its selective adsorption of chromium(VI) as well as arsenic(V) (Chen et al. 2009). A shape tuned Ag core and Fe_3O_4 shell nanohybrids synthesised which is water-soluble in nature and possessed substantial superparamagnetic properties, resulting efficient magnetic separation (Zhai et al. 2011). In an another study, magnetic Fe–Zr binary oxide was prepared and explored as an adsorbent to remove phosphate from aqueous solution (Long et al. 2011). Fe_3O_4/SiO_2-Thiotet-Cu (II) used as a nanocatalyst in the reduction of Congo Red, Methylene Blue, hexavalent chromium, 4-nitrophenol in aqueous solution (Nasrollahzadeh et al. 2020). Further, more studies were reported on the same area (Ambashta and Sillanpää 2010; Chen et al. 2011; Mohammed et al. 2017; Li et al. 2019b; Özdemir et al. 2019).

12.5 Summary, Conclusions and Future Research

The present chapter discussed in detail with recent reports, the various environmental remedial strategies focusing on some major forms of surfaces along with their modifications to improve the performance. In this regard, various surfaces have been extensively described in this chapter based on their size, structure and physical–chemical properties. So the surfaces are classified into bulk, nano, polymeric, membrane, hydrophobic, oleophobic, patterned, and magnetic surfaces in this chapter. So, several materials and modified materials like biochar, $BiVO_4$, TA-coated metal sheets, GO, TiO_2, GO@PDA-M, GO-PEI-PES, etc. have been explored in detail focusing on their different environmental remedial applications like adsorption, oil–water separation and capture, photo catalytic degradation, anti-bacterial properties, etc. In general, the modifications of the surfaces with different molecules, nanoparticles, and polymers found to have improved the environmental remedial capacity of the surfaces as it increases the active sites and that would favour higher uptake of contaminants. On the same time, it is still premature to predict the wide scale applications of these different surfaces, especially the nano structured/patterned and hydrophobic/oleophobic surfaces in environmental remedial needs owing to the cost-effectiveness, mass production difficulties, technical challenges, and fundamental questions of these material's influence in long term to the ecosystem and to the human beings. Nevertheless, these surfaces and modified surfaces could offer great potentials and we can expect more revolutionary applications in the area of environmental healing in accordance with the advances in the fundamental physics, chemistry and engineering techniques and may expand the horizons and open up new windows to save our planet.

References

Abdullah N, Yusof N, Lau WJ, Jaafar J, Ismail AF (2019) Recent trends of heavy metal removal from water/wastewater by membrane technologies. J Ind Eng Chem 76:17–38

Abraham S, Nirala NR, Panday SA, Srivastava M, Srivastava SK, Walkenfort B, Srivastava A (2015) Functional graphene-gold nanoparticles hybrid system for enhanced electrochemical biosensing of free cholesterol. Anal Methods. https://doi.org/10.1039/C5AY00050E

Abraham S, Koenig M, Srivastava SK, Kumar V, Walkenfort B, Srivastava A (2018b) Carbon nanostructure (0–3 dimensional) supported isolated gold nanoparticles as an effective SERS substrate. Sens Actuators B Chem 273:455–465